T0360757

Four Open Questions for the N-Body Problem

The N-body problem has been investigated since it was posed by Isaac Newton in 1687. However, vast tracts of the problem remain open. Showcasing the vibrancy of the problem, this book describes four open questions and explores the progress made over the last 20 years. After a comprehensive introduction, each chapter focuses on a different open question, highlighting how the stance taken and tools used vary greatly depending on the question.

Progress on question one, "Are the central configurations finite?", uses tools from algebraic geometry. Question two, "Are there any stable periodic orbits?", is dynamical and requires some understanding of the KAM theorem. The third question, "Is every braid realised?", requires topology and variational methods. The final question, "Does a scattered beam have a dense image?", is quite new, and formulating it precisely takes some effort.

This book is an excellent resource for students and researchers of mathematics, astronomy, and physics interested in exploring state-of-the-art techniques and perspectives on this classical problem.

RICHARD MONTGOMERY is Professor Emeritus of Mathematics at UC Santa Cruz. He is a co-rediscoverer of the surprisingly stable figure eight orbit for the classical three-body problem. His work on the N-body problem uses variational, topological, and differential geometric methods to say new things about this old problem.

Four Open Questions for the N-Body Problem

RICHARD MONTGOMERY

University of California, Santa Cruz

CAMBRIDGE
UNIVERSITY PRESS

Shaftesbury Road, Cambridge CB2 8EA, United Kingdom

One Liberty Plaza, 20th Floor, New York, NY 10006, USA

477 Williamstown Road, Port Melbourne, VIC 3207, Australia

314–321, 3rd Floor, Plot 3, Splendor Forum, Jasola District Centre,
New Delhi – 110025, India

103 Penang Road, #05–06/07, Visioncrest Commercial, Singapore 238467

Cambridge University Press is part of Cambridge University Press & Assessment,
a department of the University of Cambridge.

We share the University's mission to contribute to society through the pursuit of
education, learning and research at the highest international levels of excellence.

www.cambridge.org
Information on this title: www.cambridge.org/9781009200585

DOI: 10.1017/9781009200608

First published 2025

A catalogue record for this publication is available from the British Library

Library of Congress Cataloging-in-Publication Data
Names: Montgomery, Richard, 1956– author.
Title: Four open questions for the n-body problem / Richard Montgomery, University of
California, Santa Cruz.
Description: Cambridge, United Kingdom ; New York, NY : Cambridge University Press,
2025. | Includes bibliographical references and index.
Identifiers: LCCN 2024018152 (print) | LCCN 2024018153 (ebook) |
ISBN 9781009200585 (hardback) | ISBN 9781009200608 (ebook)
Subjects: LCSH: Many-body problem.
Classification: LCC QB362.M3 M66 2025 (print) | LCC QB362.M3 (ebook) |
DDC 521–dc23/eng20240802
LC record available at https://lccn.loc.gov/2024018152
LC ebook record available at https://lccn.loc.gov/2024018153

ISBN 978-1-009-20058-5 Hardback

I dedicate this book to the memory of Jerry Marsden, who introduced me
to geometric mechanics, and to Bill Burke, who listened to my first seminar
on the N-body problem, to Chris Golé, who was also there at that seminar
and who insisted that I introduce myself to Alain Albouy and Alain
Chenciner, and to Albouy and Chenciner and their institution, the IMCCE,
for support, friendship, a sense of history, surprising ideas, and their patience
with my French. Finally, I have to dedicate it to my family as well and,
in particular, my grandson Jude for his unending fascination and joy at
being an intermediary in the interactions of elastic spherical objects
with gravity and pavement.

Contents

Preface

The N-body problem, despite having been formulated more than 350 years ago, is alive and well. My goal is to convince you of this by exploring four open questions in the field. Each question gets a chapter, and these make up Chapters 1, 2, 3, and 4. These chapters begin by stating their question. They go on to explain why that question is important, to explore partial answers, and to dive into some of the methods used to arrive at these partial answers.

This book is an outgrowth of a colloquium I gave (online) at Vanderbilt in the Fall of 2020 during the midst of the COVID-19 pandemic. I tried to write the book to be accessible by colleagues in any field of mathematics as well as for strong undergraduates. I expect readers to have proficiency in linear algebra, and to understand what a differential equation is. I hope that the more traditional clientele of experts in celestial mechanics, dynamical systems, and mathematical physics will get something out of the book. I do not expect readers to need a working knowledge of introductory physics. For this reason I've included Chapter 0 and Appendix A, which go into some of the physics, posing the N-body problem as a differential equation, and discussing the assumptions underlying the posing of the problem, as well as its symmetries, conservation laws, and other structures implicit to the equations. Preceding Chapter 0 is Chapter -1, which is a pictorial tour of some known solutions.

Three of these four open questions are stated in the succinct collection of open problems due to Albouy, Cabral, and Santos [6]. The question making up Chapter 1 is their problem 9. The question of Chapter 2 is their problem 7. The question of Chapter 3, for $N = 3$, is (essentially) their problem 4. One will find many other open problems in this collection.

◇

Humanity has been working on the N-body problem since Newton posed it in 1687 [163]. You might think we'd be done with it by now. We're not. The problem remains very much alive, and this book was written to convince my readers of this fact. Substantial results have been achieved in the last decade, but new open questions continue to arise.

Perturbation theory has dominated the analytical work on the problem. One expands solutions in some series – Taylor, Fourier – about known solutions to limiting cases of the problem, successively working out the terms in the series by various ingenious schemes. Taylor series are expanded with respect to small parameters – typically mass or distance ratios – that arise naturally in the problem. The limiting cases are often some version or other of the two-body problem. Laplace, Lagrange, Poisson, Gauss, Legendre, Delanay, Hill, Poincaré, Moser, Arnol'd – the names go on and on – have all contributed to perturbation theory with great success. I have avoided perturbation theory in this book as well as in my own career. I cannot compete with the old masters and would likely end up with nothing new to say. An exception to my "avoid perturbation theory" rule is the discussion of KAM theory in Chapter 2.

My favorite case of the N-body problem is the equal mass zero angular momentum planar three-body problem. Studying it forces one to replace perturbation theory with other tools. All four open questions except the first one are open for this case. The third open question has been solved for the equal mass planar three-body problem when the angular momentum is small but nonzero.

◇

Newton [163] posited that any two masses in the universe attract each other by a gravitational force. He took this force to be proportional to $1/r^2$ where r is the distance between the masses. Supposing the universe to be populated by a finite number N of point masses subject only to their mutual gravitational attractions and his laws of mechanics, he formulated a set of differential equations (see Equation (0.4) in Chapter 0) that describe the motions of these N point masses. **The classical N-body problem is the study of this dynamical system.**

Newton solved his two-body problem. Supposing these two masses to be the Sun and a planet, he derived Kepler's three laws of planetary motion and thereby gained himself a preeminent role in the history of science. See Section −1.1.

More than two hundred years later, Poincaré proved that a limiting case of the three-body problem is unsolvable in a certain technical sense: it admits "homoclinic tangles" and therefore is not "integrable by analytic functions." The effect of his proving unsolvability was analogous to that of Galois' proving the unsolvability of the general quintic. Rather than killing their subjects, they developed methods to establish their impossibility results that opened up vast and previously unimagined vistas of research. For Galois these vistas included group theory, algebra, and number theory. For Poincaré the vistas were nonlinear qualitative dynamical systems (popularly referred to as "chaos theory") and their interaction with topology and analysis.

PART ONE

Tour, Problem, and Structures

−1

A Tour of Solutions

We start off with a tour of some of the solutions to the classical N-body problem, using the number N of bodies as an organizing parameter for our tour. The problem is posed in terms of a system of ODEs whose solutions consist of N synchronized curves in the plane or in space, the curves representing the motion of stars or planets interacting solely through classical gravitational attraction. Solutions can be drawn by indicating the trace of the curves. The reader may wish to consult some animations on the web for a more visceral feel.[1]

−1.1 The One- and Two-Body Problems

Newton posed and solved the gravitational two-body problem. Figure −1.1 shows a solution to this problem when the two masses are equal. The general solutions are as follows. The center of mass of the two bodies moves in a straight line at constant speed. Each body moves around that center along a conic section according to an ODE for a single body known nowadays as Kepler's problem. In this way, the two-body reduces to a one-body problem.

The Kepler problem, our "one-body problem," is the limiting ODE we get from the two-body problem when one of the two masses tends to infinity while the other remains finite. Then the infinite mass, the "Sun," does not budge while the finite mass, the "planet," moves about the Sun according to Kepler's three laws. The first of Kepler's laws says that the planet moves in a conic section having the Sun as one focus. We discuss these laws and derive Kepler's problem from the two-body problem in Chapter 0.

---

[1] We often get a better sense of solutions through animations instead of stills, but this here is a book. Please see https://peopleweb.prd.web.aws.ucsc.edu/~rmont/Nbdy/NbdyC1.html? for animations.

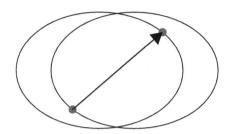

Figure −1.1 A solution to the equal mass two-body problem.

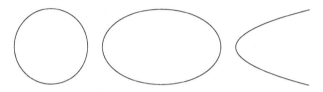

Figure −1.2 Some Kepler conics.

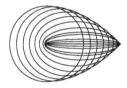

Figure −1.3 A family of constant negative energy Keplerian conics.

Some solutions to Kepler's problem are shown in Figure −1.2. The Kepler problem admits a conserved energy and an angular momentum. Its bounded solutions all have negative energy, the absolute value of which is proportional to the reciprocal of the ellipse's semi-major axis. Figure −1.3 shows a family of Keplerian ellipses having fixed negative energy. As the angular momentum tends to zero in the family, the ellipses degenerate to line segments colliding with the Sun, as indicated in the figure. These line segments are the collision solutions and play a central role in the N-body problem, most notably in Chapter 3. The periods of all these orbits are the same.

Returning to the two-body problem, the difference vector joining the two bodies satisfies Kepler's problem, as does the difference vector joining the center of mass to either body. Our standard Newtonian view of the solar system is as N planets moving around the Sun in bounded Keplerian orbits that are nearly circular. This model ignores the gravitational attraction between planets, which one treats later as a perturbation.

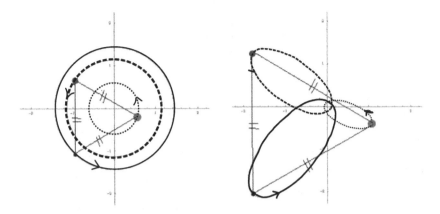

Figure −1.4 Lagrange's solution. Courtesy of Richard Moeckel. See www
.scholarpedia.org/article/3-body_problem.

Stars are not infinitely massive. Our own Sun moves in a small Keplerian
orbit about the center of mass of our solar system, a point buried inside the Sun.
The discovery of the first exoplanets was based on detecting small periodic
oscillations in the velocities of distant stars as they moved about the center of
mass of their own solar systems. These velocity changes – known as "radial
velocity dispersion" – can be seen in the Doppler shifts of the light frequencies
detected.

−1.2 The Three-Body Problem

−1.2.1 The Central Solutions of Euler and Lagrange

We have closed-form analytic expressions for five families of solutions to the
three-body problem collectively known as the central configuration solutions.
Euler [54] found three of these families in 1767. Lagrange found the remaining
two families in 1772 [99, 100]. In their solutions, the three bodies travel on
three similar Kepler conics, their motion synchronized so that the shape of the
triangle they form does not change. The three conics making up such a solution
are scaled, rotated versions of each other, as shown in Figure −1.4.

Lagrange's Solution

In Lagrange's solution the bodies form an equilateral triangle at each instant.
If we take the Kepler conic to be a circle, then this equilateral triangle rotates
rigidly about the center of mass of the triangle. If we take a general Keplerian

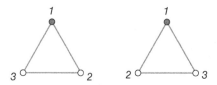

Figure −1.5 The two orientations of a labeled planar triangle.

conic, then each body moves in a Keplerian conic about the center of mass, with the motion of the conics synchronized so that at each instant they form an equilateral triangle. Included in the family is the degenerate zero-angular momentum solution where the bodies are dropped from rest in an equilateral configuration, in which case each body moves on a line segment ending in simultaneous triple collision at the center of mass. The Lagrange families exists for all mass ratios. We count the Lagrange solutions as two families of solutions since we distinguish between "right-handed" and "left-handed" planar labeled triangles. See Figure −1.5. The circular Lagrange solutions are linearly stable only when one of the three masses is much greater than the other two.

Euler's Solutions

In Euler's solutions the bodies are collinear at every instant, and the line they lie on is typically spinning. They comprise three families labeled according to which of the three bodies sits between the other two during the motion. Throughout the motion the ratio between the distances of the bodies remains constant. (That ratio is determined as the positive real solution to a fifth-order polynomial whose coefficients depend on the masses.) When the masses are all equal this middle mass must lie at at the midpoint of the segment formed by the other two. If we choose the circular solution to Kepler's equations, then the ends of the segment rotate as the ends of a diameter of a circle about this center. For an Euler solution when the masses are all distinct see Figure −1.6. The Euler solutions are linearly unstable for all mass distributions. (See Chapter 2 regarding linear stability and instability.)

−1.2.2 The Circular Restricted Limit

The Euler and Lagrange solutions persist if we let one of the three masses tend to zero while the other two are fixed. If we take the circular Kepler element of the central configuration family, then these limiting solutions become special solutions to the *circular restricted planar three-body problem,*

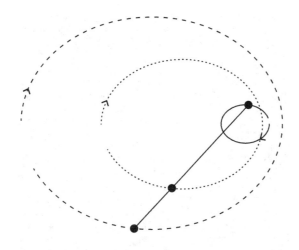

Figure −1.6 Euler's solution.

which is probably the most studied of all three-body problems. In this limiting version, the two big masses, or "primaries," rotate in a circular Keplerian orbit. Newton's equations, written in coordinates rotating at the angular frequency of the primaries, reduces to a second-order autonomous ODE in the rotating plane for the remaining infinitesimal mass. The corresponding ODE in the rotating frame has five fixed points corresponding to these five limiting central configurations. These fixed points are named L1–L5 and are indicated in Figure −1.7. (Somehow Euler's name disappeared in the labeling.)

−1.2.3 The Eight

At the other extreme of having one of the three masses close to zero, take all three masses to be equal. In what is called the "figure eight," or simply "eight," solution, three *equal* masses chase each other around a fixed figure of eight curve drawn (here laid on its side) on the plane. See Figure −1.8. You can generate the figure eight yourself by using a numerical integrator to integrate the three-body equations (0.1)–(0.2). Take $m_1 = m_2 = m_3 = G = 1$ in these equations. For inital positions take $q_1 = -q_2, q_3 = 0$, and for initial velocities take $\dot{q}_1 = \dot{q}_2, \dot{q}_3 = -2\dot{q}_1$ where

$$q_1 = (-0.97000436, 0.24308753), \dot{q}_3 = (0.93240737, 0.86473146).$$

I'm indebted to Carlés Simó for zeroing in on these initial conditions. See [192].

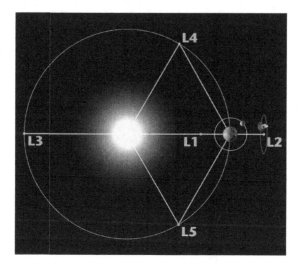

Figure −1.7 The central configurations marked within the circular restricted three-body problem. In this context, the central configurations are usually labeled L1, L2, L3, L4, and L5. Image credit: NASA/WMAP Science Team.

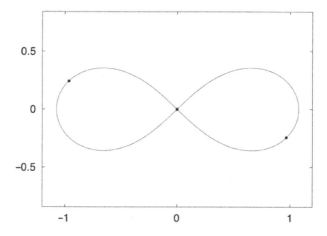

Figure −1.8 The figure eight, a periodic zero-angular momentum KAM stable solution to the three-body problem in which three equal masses chase each other around a figure eight–shaped curve in the plane. From Chenciner et al. [35].

The figure eight makes guest appearances in Chapter 2 on stability and Chapter 3 on braids. Unexpectedly, the eight is about as stable as one can hope for a periodic solution in celestial mechanics. Figure −1.9 depicts a perturbed eight, a solution resulting from near-eight initial conditions. We can imagine this orbit as being "nestled between" two KAM 2-tori near the eight.

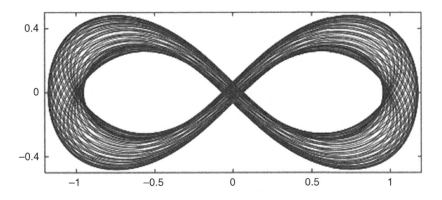

Figure −1.9 A near-eight. From Simó [192].

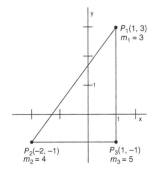

Figure −1.10 Initial conditions for the Pythagorean three-body problem. Reprinted from Szebehely and Peters [204] ©AAS. Reproduced with permission.

−1.2.4 Escape and Scattering

Drop three bodies. In other words, let them go from rest. What can happen? A special case of this problem was investigated for nearly a century in a long and convoluted series of papers for a specific case known as the Pythagorean three-body problem, and which became a kind of benchmark problem for N-body integrators. For initial conditions place three masses at the vertices of a triangle whose edge lengths are in the ratio 3:4:5. Choose the masses to have the same ratio, and place mass 5 opposite the edge of length 5, and so on. See Figure −1.10. Now drop the bodies: Take all velocities to be zero, and numerically integrate. What happens? See Figure −1.11.

Now throw a binary system at a distant isolated third body. Hut and collaborators [85] did this numerically in order to understand questions

(a)

(b)

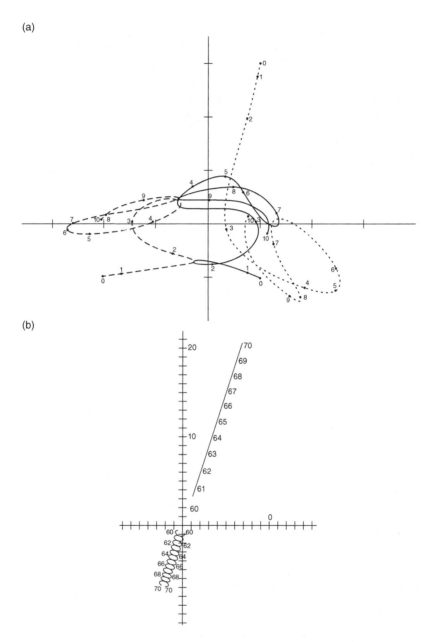

Figure −1.11 The Pythagorean three-body problem. (a) The solution from being dropped at $t = 0$ to time $t = 10$ with one body's curve depicted as a solid line, one as a dashed curve, and the other as a dotted curve. (b) The motion from $t = 50$ to 70 is depicted. We see that two of the masses form a tight binary and escape to infinity. Reprinted from Szebehely and Peters [204] ©AAS. Reproduced with permission.

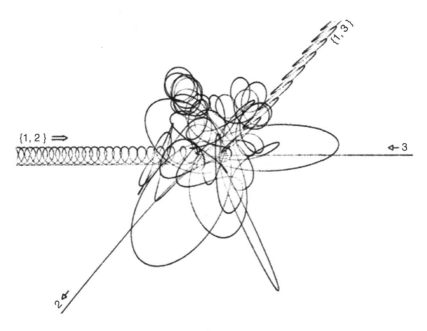

Figure −1.12 Chaotic interchange. Reprinted from Hut & Bahcall [85] ©AAS.
Reproduced with permission.

regarding the prevalence of binary stars in galaxies and their effects on galactic
evolution. A sample solution is depicted in Figure −1.12.

−1.2.5 Schubart and Broucke–Henon

In 1956, Schubart [188] discovered the periodic solution to the equal mass
collinear three-body problem sketched in Figure −1.13. The middle body
shuttles back and forth between the two on the extremes, colliding with each
once per period. These collisions are regularized according to Levi-Civita, a
transformation of space–time that analytically extends the planar N-body flow
through isolated binary collisions. See Appendix G. Schubart's orbit is stable
within the context of the collinear problem and plays a central role in the phase
portrait of the negative energy equal mass collinear three-body problem. See
[205, 206].

In 1974, Broucke [22] found many new orbits for the three-body problem.
See the 3rd and 4th orbits in Figure −1.14. Henon [80] connected the Schubart
and Broucke orbits through an analytic family of relative periodic orbits by
using the angular momentum as a bifurcation parameter. In the process, Henon
discovered orbits now called the Broucke–Henon orbits.

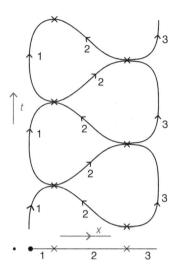

Figure −1.13 Space–time diagram of Schubart's collinear equal mass three-body solution. The x's mark collisions.

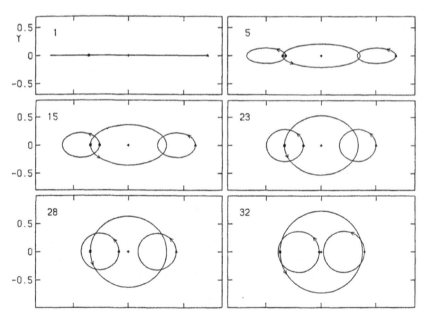

Figure −1.14 Broucke and Henon's continuation of the Schubart orbit (top left). The orbits are shown in a rotating frame with respect to which the orbit is periodic. For the Schubart collinear orbit the frame does not rotate. From [80].

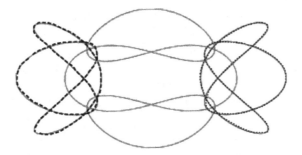

Figure −1.15 Another equal mass zero angular momentum solution, one of hundreds found by Danya Rose.

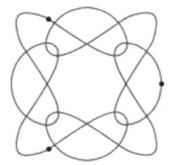

Figure −1.16 The three-body "Celtic Knot" choreography denoted $D(3,4)$ by Montaldi and Steckles [143].

−1.2.6 A Bestiary

In 2015, Danya Rose [177] combined simultaneous regularization of all binary collisions, shape space thinking, and careful numerical integrations to catalogue hundreds of equal-mass zero angular momentum solutions to the three-body problem. Figure −1.15 is an unstable orbit that Rose christened F1.2.7.

In 2013, James Montaldi and Katrina Steckles [143] compiled an artistic bestiary of a different nature, based on equivariant homotopy ideas combined with the symmetry ideas that led to the figure eight. Figure −1.16 shows a representative orbit of theirs whose existence has not yet been rigorously established.

−1.3 The Restricted Three-Body Problem

When the value of one of the masses tends to zero in the three-body equations (see Equations (0.1) at the beginning of Chapter 0) the limiting equations decouple. The two nonzero masses – called "the primaries" – move according to the two-body problem. So the primaries move in Keplerian orbits about their common center of mass, unaffected by the third infinitesimal mass. The infinitesimal mass feels the changing gravitational pull of the primaries. How does it move? This is the restricted three-body problem.

The restricted three-body problem is a singular limit of the full three-body problem. It is also, historically speaking, the most studied three-body problem. Satellite and space mission planning problems are treated as restricted three-body problems, with perturbations added as needed. The Earth–Moon–Sun system is also typically treated as a restricted three-body problem,[2] the ratios of masses Sun : Earth : Moon being of the order of $1 : 10^{-6} : 10^{-8}$. Much of Poincaré's work on the three-body problem and, in particular, his three-volume "Methodes Nouvelles de Mecanique Celeste" [169] is aimed at the restricted three-body problem.

The restricted three-body problem is really a family of problems parameterized by the choice of conic section for the Keplerian orbit of the primaries, the mass ratios of the primaries, the residual energy of the third body (called the Jacobi constant), and whether or not the tiny body moves in the plane of the primaries or is allowed to wander in space. The most studied problems within this family of problems are those where the orbits of the primaries are circles. This sub-problem is called the circular restricted planar three-body problem By going to a rotating frame, rotating at the rotation rate of the primaries, this problem becomes a nonautonomous two-degree of freedom system, while the full planar three-body problem, after reduction by symmetries, is three-degree of freedom. We saw solutions to this problem arising from central solutions earlier. See Figure −1.7. In the rotating frame these central configuration solutions of Euler and Lagrange became equilibria. Another solution to the circular restricted planar three-body problem is represented in Figure −1.17.

In the early 1930s, Strömgren organized a group of human computers to survey some periodic solutions for the planar circular restricted three-body problem in the case where two masses are equal [200]. The remarkable variety of orbits he found are summarized in Figure −1.18. The variety, complicated character, and implications of the orbits he depicted have not been explored.

The Sitnikov problem (Figure −1.19) concerns a perturbed circular restricted *spatial* three-body problem, again with the two primaries having equal mass. The infinitesimal mass moves on the line orthogonal to the plane of motion of the primaries, passing through their center of mass. By symmetry,

---

[2] The Earth–Moon–Sun system turns out to be especially challenging since, due to the distances involved, the magnitudes of the force exerted on the Moon by the Sun and by the Earth are of roughly the same order.

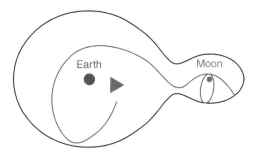

Figure −1.17 Transit orbit. The primaries, Earth and Moon, move in circular orbits and are viewed in a frame rotating with them so they become stationary. The infinitesimal mass, which we may think of as a moon mission, transits back and forth. Courtesy of Richard Moeckel www.scholarpedia.org/article/3-body_problem [137].

the tiny mass stays on this line, and this remains true if the conics on which the primaries move are not circular.

A good chunk of the book by Moser [159] is dedicated to the Sitnikov problem. If the primaries move in circular orbits, then the force on the infinitesimal is not time-dependent, and the problem becomes a one-degree-of-freedom autonomous Hamiltonian system, and hence is integrable. Let them move in eccentric orbits and the problem becomes a nonautonomous time-periodic one-degree-of-freedom system. Moser uses the eccentricity of the ellipse as a perturbation parameter away from the integrable case. He uses analysis at infinity and a study of Smale horseshoes to prove the following. Let time be counted by full revolutions of the primary once around their center of mass. For negative energies (Jacobi constant) and for most solutions, the infinitesimal mass crosses the plane of motion of the primaries over and over. Every time it crosses, mark down the integer number of years passed since the last crossing. Then there is a (large) integer N_0 such that every infinite sequence $n_{-1}n_0n_1 \cdots$ with $n_i \geq N_0$ is realized: There is some solution that ticks off this sequence in its travels.

It was within the context of the circular restricted three-body problem that Poincaré established his famous non-integrability results: that the three-body problem is not an integrable system, as that term is meant nowadays in Hamiltonian dynamics. He did so by establishing what would later be called "homoclinic tangles," which imply the presence of "Hamiltonian chaos." Often, Poincaré took as the perturbation parameter the ratio of masses between the primaries, with the case where one is infinitely massive relative to the other limiting the problem to an integrable problem: two uncoupled Kepler problems.

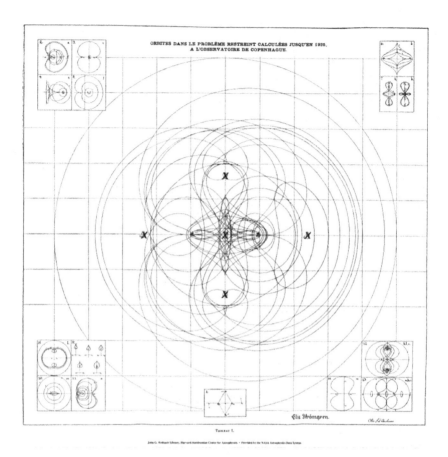

Figure −1.18 A few dozen periodic orbits for the planar restricted circular three-body problem in the case where the two primaries have equal mass, viewed in the rotating frame. The filled-in gray circles are centered on the primaries, the X's indicate the central configurations of Euler and Lagrange. Strömgren [200] found these orbits in 1933 through numerical integration before digital computers were available.

−1.4 The Four-Body Problem

−1.4.1 Central Configurations

Solutions such as the Euler and Lagrange three-body solutions in which the N bodies maintain their shape throughout their motion are called central configuration solutions. The shapes themselves are referred to as "central configurations." For the equal-mass planar four-body problem there are precisely four possible shapes. See Figure −1.20(a). When the vertices of the

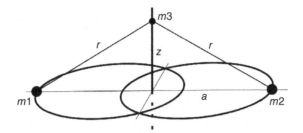

Figure −1.19 The Sitnikov problem.

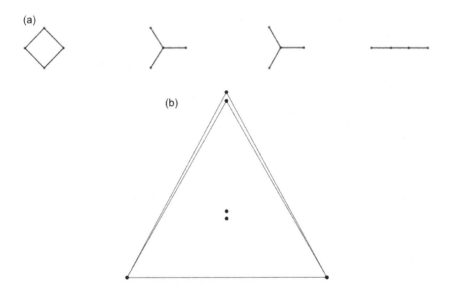

Figure −1.20 Four-body central configurations for equal masses. (a) Two of the four shapes look the same. One of these two is an equilateral triangle with the fourth body placed at the center of mass. The other is isosceles but not equilateral. (b) We have enlarged the two identical-looking configurations and super-imposed them after a rotation to show that they are actually different. Reprinted from Hampton [74].

4-gons shapes are labeled they yield 50 distinct labeled shapes modulo scaling and rotation, and hence 50 families of solutions, the four-body analogues of the Euler and Lagrange three-body solutions. Counting the number of central configurations for $N \geq 4$ is the subject of Question 1 of the book (see Chapter 1). A simple four-body central configuration solution to visualize is a rotating square for the equal-mass four-body problem (Figure −1.20(a)).

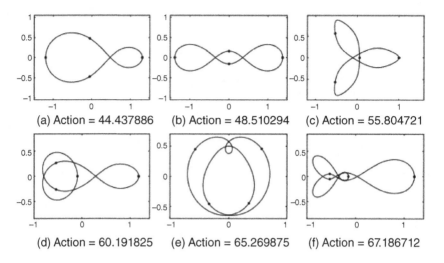

Figure −1.21 Four-body choreographies. Panel b) is known as the Gerver super eight and was discovered by Joseph Gerver at a conference during the last week of the twentieth century. From Chenciner et al. [35].

−1.4.2 Dancing Quadrilaterals

A vast array of equal-mass periodic orbits with high symmetry have been discovered in the last 25 years by combining symmetry and variational methods. Figure −1.21 illustrates some four-body choreographies. A *choreography* is an N-body solution for which all N masses travel the same curve in d-space. Figure −1.22 depicts the *hip-hop* solution, which enjoys the symmetry group $\mathbb{Z}_2 \times \mathbb{Z}_4$.

−1.4.3 A Non-collision Singularity

A solution has a singularity if there is a finite time t_* beyond which it cannot be continued as a solution. Along a singular solution the lim inf of at least one pair r_{ij} of relative distances must tend to zero. The singularity is called a collision singularity provided all the bodies have limiting positions as t_* is approached, in which case at least two of the bodies have collided. Otherwise, the solution is called a non-collision singularity. In 2022, Joseph Gerver, Guan Huang, and Jinxin Xue [67] described solutions with non-collision singularities for the four-body problem. We give a cartoon of their solutions in Figure −1.23.

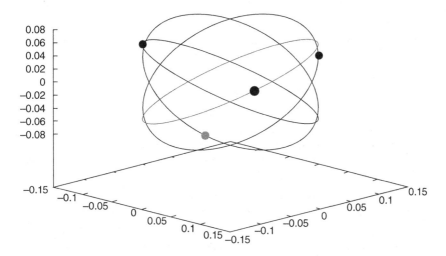

Figure −1.22 The hip-hop solution discovered by Chenciner and Venturelli [34] has four equal masses oscillating between the square and tetrahedral configurations and enjoys the symmetry group $\mathbb{Z}_2 \times \mathbb{Z}_4$. Courtesy of Davide Ferrario.

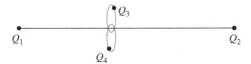

Figure −1.23 Cartoon of non-collision singularity solutions with four bodies found by [67]. None of the bodies have finite limiting positions as $t \to t_*$, the singularity time. $Q_3(t)$ and $Q_4(t)$ accumulate onto the entire x axis with the distance between them going to zero, while $Q_1(t) \to (-\infty, 0)$ and $Q_2(t) \to (+\infty, 0)$ as $t \to t_*$. If $r_{ij}(t) = |Q_i(t) - Q_j(t)|$ is the distance between bodies i and j at time t then $\limsup_{t \to t_*} r_{ij}(t) = \infty$ and $\liminf_{t \to t_*} r_{ij}(t) = 0$ provided $i = 1$ or 2 and $j = 3$ or 4. Think $\frac{1}{|t|}|\sin(1/t)|$ with $t_* = 0$. Reprinted from [67].

−1.5 The Many-Body Problem

−1.5.1 Central Configurations and Dark Matter

The ansatz leading the central configurations (CCs) of Euler and Lagrange works for any N. See Chapter 1. The number of distinct similarity classes of CCs grows at least as fast as $(3/4)N!$ for planar CCs. Is this number always finite? That's our first open question.

Our galaxy contains about 10^{11} stars. Saari [186] argues that CCs involving this number of bodies might organize the structure of many galaxies and thereby dispense with the need for dark matter. Imagine a galaxy roughly as a spinning disc full of stars. For many galaxies, as we go out radially away from the galactic center, after a certain distance, the velocities v of the stars have been observed to increase with radius r in a nearly linear manner with distance, as in $v = \omega r$ – as if the stars were glued to a rotary saw rotating at angular velocity ω. This graph of v versus r is called the rotation curve. In contrast, if we take a continuum approach, supposing the stars move under the gravitational potential induced by a spherically or cylindrically symmetric potential whose mass density is estimated by counting luminous stars in the galaxy, we get a very different rotation curve. The observed linear behavior of the rotation curve in a certain range of r, and its discrepancy from the one derived by continuum models is one of the arguments for the existence of dark matter. Plunk enough dark matter into the density distribution and you can adjust the rotation curve to fit the data and also estimate how much dark matter there is. Saari doesn't buy this. He says, in essence, "drop this Keplerian – spherically symmetric, star soup – thinking and use central configurations to explain the rotational curve data." I don't buy Saari's counterargument among other reasons because equal mass CCs all seem to be dynamically unstable (see [89] for a detailed critique). But it is fun to think about.

−1.5.2 Large Dances

For any positive integer N, there are choreographic solutions for the equal mass N-body problem. The existence of infinite families of choreographies has been established. Figure −1.24 is a sample. One such family consists of an odd number N of bodies traveling a chain made of $N − 1$ loops, the eight being the case $N = 3$. For any fixed finite postive integer N there may be an infinite number of choreographies with N bodies. We don't know.

The variational and symmetry methods that established the existence of the figure eight solution have allowed researchers to discover and establish the existence of solutions having the symmetries of each of the Platonic solids. Figure −1.25 depicts a 60-body solution enjoying dodecahedral symmetry, 60

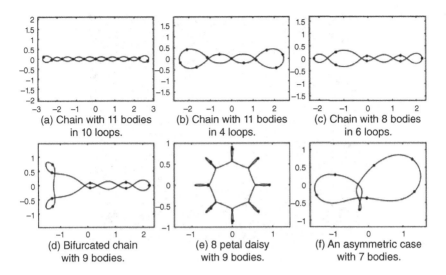

Figure −1.24 A sampling of N-body choreographies. From Chenciner et al. [35].

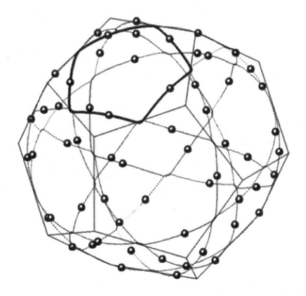

Figure −1.25 A 60-body solution enjoying dodecahedral symmetry. Associated to each of 12 faces are 5 bodies moving in a choreography. Reprinted from [65].

being the order of the subgroup of symmetries of the dodecahedron that are orientation preserving.

Figure −1.26 Galaxy NGC 1414 as viewed from the Hubble space telescope. Image credit: The Hubble Heritage Team (AURA/STScI/NASA).

−1.5.3 Globular Clusters and Galaxies

For astronomers, N is often quite large. See Figure −1.26. Our own Milky Way contains around 10^{11} stars. Galaxies contain up to 10^{12} stars. Globular clusters are spherical collections of around 10^6 stars that are found orbiting galaxies.

There is an enormous body of work on galactic dynamics. Researchers tend to use continuum models and statistical mechanical thinking. See the classic book by Binney and Tremaine [18] for a sense of this field. Closer to the spirit of the book in your hands, large N-body simulators have been an important tool. See the books "The Gravitational Million-Body Problem" by Heggie and Hut [79] and "Gravitational N-Body Simulations" by Aarseth [1] for an overview and surveys of this kind of work.

0

The Problem and Its Structure

0.1 The Problem, Its Symmetries, and Conservation Laws

0.1.1 Three Point Masses

Imagine three point masses alone in the universe. Each exerts an attractive force on the other two (see Figure 0.1). How do they move? Upon following Newton's slogan "force = mass times acceleration" we arrive at the coupled system of nonlinear ordinary differential equations (ODEs),

$$m_1 \ddot{q}_1 = F_{21} + F_{31},$$
$$m_2 \ddot{q}_2 = F_{12} + F_{32}, \tag{0.1}$$
$$m_3 \ddot{q}_3 = F_{23} + F_{13},$$

governing their motion. We take the bodies to be points – the point masses – and label the points by the index a with $a = 1, 2, 3$. The ath body has mass $m_a > 0$ and position $q_a = q_a(t)$, and exerts the force F_{ab} on body b. The positions depend on time $q_a = q_a(t)$, where $t \in \mathbb{R}$ is Newtonian "universal" time. The double dots over q_a denote acceleration: $\ddot{q}_a = d^2 q_a / dt^2$.

In order to turn Equations (0.1) into a self-contained ODE we need to know the forces. Newton, taking the suggestions of Galileo, Hooke, and Halley, settled on

$$F_{ba} = -\frac{G m_a m_b}{r_{ab}^2} \hat{q}_{ab} \quad \text{with} \quad \hat{q}_{ab} = \frac{q_a - q_b}{r_{ab}}, \tag{0.2}$$

where

$$r_{ab} = |q_a - q_b| \tag{0.3}$$

is the distance between bodies a and b. The vector $\hat{q}_{ab}$ is the unit vector pointing *from body b to body a*, since $q_a = q_b + (q_a - q_b)$. This choice of forces says that the force between two bodies is attractive, directed along the

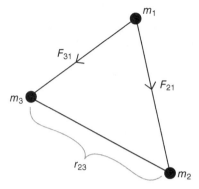

Figure 0.1 The three-body problem. The force of one body on another is directed along the corresponding edge of the triangle formed by the masses. The force on a body is the sum of the forces exerted by the other two.

line connecting them, with magnitude inversely proportional to the square of the distance between them and directly proportional to the product of their masses. The parameter G, the gravitational constant, is a physical constant needed if the physical units on both sides of Equations (0.1) are to agree. Being mathematicians, somewhat unconcerned with units, we can and generally will set $G = 1$.

The standard choice for the universe in which our point masses move is Euclidean three-space $\mathbb{R}^3$, so that $q_a \in \mathbb{R}^3$. In this book we will generally take the universe to be d-dimensional Euclidean space $\mathbb{R}^d$ so that $q_a \in \mathbb{R}^d$. Our primary concern is the case $d = 2$ of the planar three-body problem. The planar problem embeds as a special subproblem of the spatial problem (see Lemma 0.4 and Exercise 0.8) so is physically relevant. At crucial points in Chapters 2 and 3 we take $d = 4$. Taking the universe to be a Riemannian manifold X instead of Euclidean space is not without merit. See the notes in Section 0.6.

We depicted a number of solutions to the planar three-body problem in Chapter -1. The last three of the four open questions posed in this book are open for the planar three-body problem. *The reader may want to turn to any one of these questions now.*

0.1.2 Any Number, Any Dimension

The reader can now write down the equations of motion for N point masses moving in Euclidean d-space:

$$m_a \ddot{q}_a = F_a, \quad a = 1, \ldots, N, \quad F_a = \Sigma_{b \neq a} F_{ba}, \quad q_a \in \mathbb{R}^d. \tag{0.4}$$

Definition 0.1 The standard N-body problem in d-space is the system of ODEs (0.4), with Equations (0.2) and (0.3) being understood, and with $m_a > 0$ fixed parameters called the masses. A solution to the N-body problem is a solution $q(t) = (q_1(t), \dots, q_N(t)) \in (\mathbb{R}^d)^N$ to this system of ODEs.

The standard N-body problem in d-space is not a single problem, but rather a family of problems parameterized by the choice of masses m_a and the values of the conserved energy and angular momentum (see Section 0.1.7). At times, notably in Chapter 3, it is useful to think of the force law itself as a parameter, usually in the form of the exponent of the power law. See Equation (0.15) in Section 0.1.5.

0.1.3 Configuration Space, Phase Space, and Flows

Set

$$\mathbb{E} = \mathbb{E}(d, N) = (\mathbb{R}^d)^N = \mathbb{R}^d \times \mathbb{R}^d \times \cdots \times \mathbb{R}^d.$$

We call this space *configuration space* A solution to the N-body problem is a curve in configuration space satisfying the ODEs (0.2) and (0.4). The letter $\mathbb{E}$ stands for "Euclidean" since $\mathbb{E}$ comes with a particular inner product that we call the *mass inner product* and which is central to our perspective. See Equation (0.36).

A collision configuration is a point of $\mathbb{E}$ at which two masses collide: $q_a = q_b$ for some pair $a \neq b$. Equivalently, a collision occurs when $r_{ab} = 0$. Forces blow up at collisions, and consequently solutions typically fail to exist past collison times.[1]

Definition 0.2 The collision locus Δ is the subset of configuration space

$$\Delta = \{q = (q_1, \dots, q_N) \in \mathbb{E} : q_a = q_b \text{ for some } a \neq b\} \subset \mathbb{E}, \qquad (0.5)$$

where two or more bodies collide.

Set

$$\hat{\mathbb{E}} = \mathbb{E} \setminus \Delta, \qquad (0.6)$$

and call it collision-free configuration space.

---

[1] Binary collisions can be regularized so that solutions sensibly extend beyond them. Collisions involving three or more bodies play the role of an essential singularity and one cannot sensibly extend solutions beyound such collisions. See Appendix G.

Definition 0.3 The phase space for the N-body problem in d dimensions is the space

$$\mathcal{P} = \mathcal{P}(d, N) = \hat{\mathbb{E}} \times \mathbb{E}. \tag{0.7}$$

We can rephrase the N-body problem in terms of a vector field on phase space. Introduce the N velocities $v_a = \dot{q}_a \in \mathbb{R}^d$ as additional variables, and combine them together into $v = (v_1, \ldots, v_N) \in \mathbb{E}$. Then the equations defining the N-body problem take the first-order form

$$\dot{q} = v, \tag{0.8}$$

$$\mathbb{M}\dot{v} = F(q). \tag{0.9}$$

Here, $F \colon \hat{\mathbb{E}} \to \mathbb{E}$ is the vector of forces, $F(q) = (F_1(q), \ldots, F_N(q))$ where F_a is the force on the ath body as in Equations (0.2) and (0.4). The linear operator $\mathbb{M} \colon \mathbb{E} \to \mathbb{E}$ is a diagonal matrix in the standard basis with entries corresponding to the masses

$$\mathbb{M}(v_1, \ldots, v_N) = (m_1 v_1, \ldots, m_N v_N). \tag{0.10}$$

We call $\mathbb{M}$ the mass matrix. The mass matrix $\mathbb{M}$ is invertible since all the masses are positive, so we can rewrite these last equations as $\frac{d}{dt}(q, v) = X(q, v)$ where $X(q, v) = (v, \mathbb{M}^{-1} F(q))$, thus defining a vector field X on $\mathcal{P}$ whose corresponding ODE is equivalent to Newton's equations.

We write the flow of the Newtonian vector field X as

$$\Phi_t \colon \mathcal{P} \dashrightarrow \mathcal{P}, \quad \Phi_t(q(0), v(0)) = (q(t), v(t)), \tag{0.11}$$

so that $q(t)$ is the solution to the N-body problem with initial conditions $(q(0), v(0)) \in P$ and $v(t) = \dot{q}(t)$. The broken arrow notation for the flow indicates that, for fixed time $t \neq 0$, the domain of the time t flow Φ_t is not all of $\mathcal{P}$. Some solutions fail to exist all the way up to time t due to collisions. In other words, the N-body flow is incomplete.[2] By general theory, Φ_t is a one-parameter family of analytic diffeomorphisms; Φ_t also depends analytically on t and satisfies the semigroup property $\Phi_t \circ \Phi_s = \Phi_{t+s}$ on points $\zeta \in \mathcal{P}$ for which $t, s, t + s$ all lie on the interval of existence of the solution with initial condition ζ. It is worth restating this business about initial conditions in mundane terms. Initial conditions consist of N initial positions $q_a(0) \in \mathbb{R}^d$ and N initial velocities $v_a(0) = \dot{q}_a(0) \in \mathbb{R}^d$ at time $t = 0$. We put them all together to get the point $(q(0), v(0))$ in phase space.

---

[2] The set of inital conditions yielding incomplete solutions is non-empty. It is believed to be of measure zero. For more see the notes at the end of this chapter (Section 0.6).

0.1.4 The Planar Problem Sits in the Spatial One

The spatial N-body problem contains the planar N-body problem.

Lemma 0.4 *Consider* $\mathbb{R}^2 \subset \mathbb{R}^3$. *If initial conditions* $(q(0), v(0)) \in \mathcal{P}(3, N)$ *satisfy* $q_a(0), v_a(0) \in \mathbb{R}^2$, *then the corresponding solution* $\Phi_t(q(0), v(0))$ *of the spatial N-body problem is the curve in* $\mathcal{P}(2, N) \subset \mathcal{P}(3, N)$ *that solves the planar N-body problem with these same initial conditions.*

Lemma 0.4 follows from the fact that, under the assumptions of the lemma and Equation (0.2), the forces $F_a(q) = \Sigma_b F_{ba}$, being linear combinations of the q_a, also lie in the plane $\mathbb{R}^2$. For an alternative proof of Lemma 0.4 based on reflectional symmetry, see Exercise 0.8.

A similar lemma holds for the line $\mathbb{R} \subset \mathbb{R}^2$ so that the planar N-body problem in turn contains within it the N-body problem on the line $\mathbb{R}$. More generally, this lemma holds for any d and any $m < d$ by taking $\mathbb{R}^m \subset \mathbb{R}^d$. The m-dimensionsal N-body problem embeds into the d-dimensional N-body problem as a subproblem.

0.1.5 Other Forces

The Newtonian two-body forces can be written $F_{ba} = \nabla_a \frac{Gm_a m_b}{r_{ab}}$ where ∇_a means the gradient with respect to the vector variable $q_a \in \mathbb{R}^d$. Thus, if we put orthonormal coordinates on $\mathbb{R}^d$ so that q_a has components $q_{ai}, i = 1, \ldots, d$, and if $f : (\mathbb{R}^d)^N \to \mathbb{R}$ is a differentiable function, then $(\nabla_a f)_i = \frac{\partial f}{\partial q_{ai}}$.

Definition 0.5 A general N-body problem is one for which the two-body forces F_{ab} of Equation (0.4) are given by

$$F_{ba} = \nabla_a f_{ab}(r_{ab})$$
$$= f'(r_{ab}) \hat{q}_{ab} \tag{0.12}$$

for some collection of $\binom{N}{2}$ smooth functions $f_{ab} : (0, \infty) \to \mathbb{R}$, subject to $f_{ba} = f_{ab}, a, b = 1, 2, \ldots, N$. The *negatives* of the functions f_{ab} are called the *pair potentials*.

The standard N-body problem corresponds to $f_{ab}(r) = m_a m_b / r$. For a general N-body problem, define

$$U(q) = \Sigma f_{ab}(r_{ab}).$$

This function plays a basic role and its negative is known as the total potential energy. For the standard N-body problem, we have

$$U = \Sigma \frac{m_a m_b}{r_{ab}} \quad \text{(standard gravity)}. \tag{0.13}$$

The standard N-body problem is one member of a one-parameter family of general N-body problems that we call power law N-body problems. The α member of the family, $\alpha > 0$, has pair potential

$$f_{ab}(r) = \frac{m_a m_b}{r^\alpha}. \tag{0.14}$$

Thus the negative of the total potential is

$$U := U_\alpha = \Sigma \frac{m_a m_b}{(r_{ab})^\alpha}. \tag{0.15}$$

Definition 0.6 The α-power law N-body problem is the general N-body problem whose pair potential has the form of Equation (0.14). The power law N-body problems with $\alpha \geq 2$ will be called *strong forces* and their potentials *strong force potentials*.

All four open questions in this book make sense for the general N-body problems in the planar case of $d = 2$. The third only makes sense when $d = 2$ and has been solved in the strong force case.

Caveat "The gravitational N-body problem in d dimensions" often refers to the power law N-body problem in $\mathbb{R}^d$ with power and dimension linked according to

$$\alpha = d - 2.$$

The link arises because the fundamental solution to the Laplacian in $\mathbb{R}^d$ is a constant times $1/r^{d-2}$. See Section 0.5.1. In this book we focus on the standard N-body problem ($\alpha = 1$) even when $d \neq 3$.

0.1.6 Symmetries

Galileo argued, about a half a century before Newton, that the laws of mechanics must be invariant under a transformation group acting on space-time $\mathbb{R}^d \times \mathbb{R}$. Newton designed his equations to be invariant under Galileo's transformation group. This group is generated by the isometries of space, the isometries of the timeline, and a special class of transformations called Galilean boosts which mix time with space.[3]

The Galilean invariance of Newton's N-body equations implies that if $q(t) = (q_1(t), \ldots, q_N(t))$ solves the equations, then, upon applying any

---

[3] The boosts of special relativity also mix space with time.

Galilean transformation simultaneously to each of the component curves $q_a(t)$ in $\mathbb{R}^d$, the result is again a solution. The Galilean transformations are

$$q_a(t) \mapsto q_a(t) + c \text{ (space translation)}, \tag{0.16}$$
$$q_a(t) \mapsto R(q_a(t)) \text{ (space rotation or reflection)}, \tag{0.17}$$
$$q_a(t) \mapsto q_a(t - t_0) \text{ (time translation)}, \tag{0.18}$$
$$q_a(t) \mapsto q_a(-t) \text{ (time reflection)}, \tag{0.19}$$
$$q_a(t) \mapsto q_a(t) + tv \text{ (boost)}. \tag{0.20}$$

In the formulae, $t \in \mathbb{R}$, $c, v \in \mathbb{R}^d$ while $R \in O(d)$, the group of orthogonal transformations of $\mathbb{R}^d$ with its standard inner product. (See Appendix D for the definition and information regarding the orthogonal group.) Any general N-body problem also has the Galilean group as a symmetry group.

Exercise 0.7 Check Galilean invariance of the standard N-body problem (Definition 0.1) by direct computation

Exercise 0.8 Use the symmetry $R \in O(3)$ of reflection about the plane $\mathbb{R}^2 \subset \mathbb{R}^3$ to prove Lemma 0.4.

Following standard Lie theory terminology, the isometry group of space will be denoted $E(d)$ (E for Euclidean), and is generated by the space translations, rotations, and reflections. If we exclude reflections from $E(d)$ we get the subgroup of rigid motions of space, denoted $SE(d)$.

0.1.7 Conservation Laws

Definition 0.9 A conservation law is a smooth function on phase space that is constant along any solution $(q(t), v(t))$ to the N-body problem.

The energy, linear momentum, and angular momentum are the basic conservation laws for the N-body problem. They can be obtained by applying an algorithm made explicit by Noether that associates a conservation law to each continuous symmetry of a *Lagrangian system*. See Section A.5

The N-body problem can be expressed as a Lagrangian system. The isometries of space and time form a continuous symmetry group – a Lie subgroup of the Galilean group. We further break up this subgroup into time translations, space translations, and space rotations. The associated conservation laws that follow from Noether are

$$E(q, \dot{q}) = K(\dot{q}) - U(q)$$

$$= \frac{1}{2}\Sigma m_a |\dot{q}_a|^2 - \Sigma_{a<b} f_{ab}(r_{ab}) \tag{0.21}$$

$$= \text{ energy (for time translations),}$$

$$P = \Sigma m_a \dot{q}_a = \text{ linear momentum (for space translations),} \tag{0.22}$$

$$J = \Sigma m_a q_a \wedge \dot{q}_a = \text{ angular momentum (for space rotations).} \tag{0.23}$$

In the last equations $(P, J) \in \mathbb{R}^d \oplus \wedge^2\mathbb{R}^d$ are understood as the Lie algebra of group of translations and rotations. See Appendix D. In the formula for angular momentum, the $\wedge$ is the wedge product on $\mathbb{R}^d$, which can be identified with the cross-product on $\mathbb{R}^3$ when $d = 3$. In Equation (0.21) for the energy E, we have split the energy into kinetic (K) and potential $(-U)$ energies. Thus

$$U = \Sigma \frac{m_a m_b}{r_{ab}} = -V \quad \text{(standard gravity).} \tag{0.24}$$

Remark 0.10 We use U in place of the standard potential energy $V = -U$ because U is positive for the standard N-body problem. Experience shows it is easier to keep track of positive functions versus negative ones when deriving and interpreting estimates.

These conservation laws were known centuries before Noether stated her theorem and can be derived by hand, starting from the equations defining the general N-body problem (Definition 0.5).

Exercise 0.11 Show that energy, angular momentum, and linear momentum are conservation laws for the general N-body problem (Definition 0.5) by using the equations that define this problem to compute the time derivatives of these functions along solutions to Newton's equations.

One can have forces more general than the general N-body problem and still get conservation of linear and angular momentum.

Exercise 0.12 Show that "equal and opposite reactions" implies conservation of linear momentum. In other words, prove that for any forces that satisfy $F_{ab} = -F_{ba}$, we have $dP/dt = 0$ along its solutions.

Show that "force directed along lines connecting bodies" implies conservation of angular momentum. In other words, prove that for any forces that satisfy $F_{ab} = \lambda_{ab}(q_a - q_b)$ for some symmetric matrix-valued function $\lambda_{ab} = \lambda_{ba}$ of q and $\dot{q}$ we have that $dJ/dt = 0$ along solutions. (Symmetry of λ_{ab} implies that $F_{ab} = -F_{ba}$.)

In Section 0.3.1 we will verify conservation of energy by putting Newton's equations in a particularly simple form.

Planar Angular Momentum

Planar angular momentum deserves special attention. It can be viewed as a scalar. Let e_1, e_2 be the standard orthonormal basis for $\mathbb{R}^2$. Then $(xe_1 + ye_2) \wedge (\dot{x}e_1 + \dot{y}e_2) = (x\dot{y} - y\dot{x})e_1 \wedge e_2$. The bivector $e_1 \wedge e_2$ is the standard basis for the one-dimensional space $\wedge^2\mathbb{R}^2 \cong \mathbb{R}$ so we can "divide out" by it and simply set

$$(x, y) \wedge (\dot{x}, \dot{y}) = x\dot{y} - y\dot{x}. \tag{0.25}$$

In polar coordinates we find that

$$(x, y) \wedge (\dot{x}, \dot{y}) = r^2\dot{\theta}. \tag{0.26}$$

The expression $(x, y) \wedge (\dot{x}, \dot{y})$ is often written in physics texts as $e_3 \cdot (xe_1 + ye_2) \times (\dot{x}e_1 + \dot{y}e_2)$. If we use complex variables and identify $\mathbb{R}^2$ with $\mathbb{C}$ by sending $(x, y) \mapsto z = x + iy$ then $(x, y) \wedge (\dot{x}, \dot{y}) = Im(\bar{z}\dot{z})$.

Continuing with the identification of $\mathbb{R}^2$ and $\mathbb{C}$, we get

$$\mathbb{E}(2, N) = \mathbb{C}^N, \tag{0.27}$$

$$J(q, \dot{q}) = \Sigma m_a Im(\bar{q}_a \dot{q}_a), \tag{0.28}$$

and the action of planar rotations on configuration space is given by Equation (0.20)

$$q_a \mapsto e^{i\theta}q_a, q_a \in \mathbb{C}, \tag{0.29}$$

where we have identified the planar rotation group $SO(2)$ with the unit complex numbers $\mathbb{S}^1 \subset \mathbb{C}$.

0.1.8 Scaling Symmetry

The N-body problems with α-power law potential admit an additional space–time scaling symmetry:

$$q(t) \mapsto \lambda q(\lambda^{-\beta}t); \quad \beta = \frac{\alpha}{2} + 1. \tag{0.30}$$

That this is a symmetry follows directly from the homogeneity of U. The scaling symmetry transforms velocities according to $v \mapsto \lambda^{-\alpha/2}v$ and energy and angular momentum according to $E \mapsto \lambda^{-\alpha}E, J \mapsto \lambda^{1-\alpha/2}J$.

In the Newtonian case of $\alpha = 1$, we have $\beta = 3/2$. Write $q^\lambda(t) = \lambda q(\lambda^{-3/2}t)$ for the transformed solution of a solution $q(t)$. If $q(t)$ is periodic with period T and typical size a then $q^\lambda(t)$ is periodic with period $\lambda^{3/2}T$ and size λa. The quantity $a^3 T^{-2}$ remains invariant under scaling. This invariance yields Kepler's 3rd law for the case $N = 2$. See K3 and the associated Equation (0.34) in Section 0.2.

Energy and angular momentum for the Newtonian case scale according to $E \mapsto \frac{1}{\lambda} E, J \mapsto \lambda^{1/2} J$ so that the quantity

$$\text{Dziobek constant} := E|J|^2 \tag{0.31}$$

is invariant under scaling. This quantity, referred to as the Dziobek constant, is the basic bifurcation parameter for the planar N-body problem once the mass distribution is fixed.

0.2 Kepler and the Two-Body Problem

Newton's two-body equations read

$$m_1 \ddot{q}_1 = F_{21},$$

$$m_2 \ddot{q}_2 = F_{12}.$$

Add the two equations and use $F_{21} + F_{12} = 0$ to obtain $\ddot{Q}_{cm} = 0$, where

$$Q_{cm} = \frac{1}{m_1 + m_2}(m_1 q_1 + m_2 q_2)$$

is the center of mass of the two bodies. Subtract the two equations after dividing each by its mass to obtain the evolution equation

$$\ddot{\vec{r}} = -\mu \frac{\vec{r}}{r^3}, r = |\vec{r}|, \mu = G(m_1 + m_2) \tag{0.32}$$

for the difference vector

$$\vec{r} = q_1 - q_2.$$

Here $M = m_1 + m_2$.

The ODE (0.32) is called Kepler's problem nowadays. Its solutions, $\vec{r}(t)$, lie in a fixed plane within $\mathbb{R}^d$, namely the two-plane spanned by $\vec{r}(0)$ and $\dot{\vec{r}}(0)$. The constancy of this plane during the motion can be established by verifying that the angular momentum bivector

$$J := \vec{r} \wedge \dot{\vec{r}}$$

is constant along solutions to Kepler's problem, where J is the usual angular momentum of Section 0.1.7, up to a mass-dependent constant, provided $Q_{cm} = 0$; J is a bivector so defines a two-plane when it is not zero. This two-plane is called the *invariable plane*. For the spatial Kepler problem, $J \in \mathbb{R}^3 (\cong \wedge^2 \mathbb{R}^3)$ and the invariable plane is the two-plane $J^\perp$.

Assuming that $\vec{r}(t)$ solves the Kepler problem we form the general solution to the gravitational two-body problem by adding constant multiples of $\vec{r}(t)$ to the motion of the center of mass $Q_{cm}(t)$. Specifically, writing

$$M = m_1 + m_2$$

for the total mass, we have that the general solution is given by $q_1(t) = Q_{cm}(t) + \frac{m_2}{M}\vec{r}(t); q_2(t) = Q_{cm}(t) - \frac{m_1}{M}\vec{r}(t)$, where $Q_{cm}(t) = tV + Q_0$ with V, Q_0 constant vectors in $\mathbb{R}^d$ and where $\vec{r}(t)$ is any solution to Kepler's problem. By this center of mass trick, we have reduced the study of the two-body problem to that of Kepler's problem. Conversely, Kepler's problem is equivalent to the two-body problem upon imposing $Q_{cm} = 0$.

Assume now that $Q_{cm} = 0$. Let e_1, e_2 be an orthonormal basis for the invariable plane. Introduce polar coordinates r, θ on the invariable plane by writing $\vec{r} = xe_1 + ye_2$ then $x = r\cos(\theta), y = r\sin(\theta)$. We find that $J = r^2\dot{\theta}e_1 \wedge e_2$, which we henceforth replace with the scalar

$$J = r^2\dot{\theta}$$

(see Equation (0.26)) and that the energy E is given by

$$\begin{aligned}
\frac{E}{m_{red}} &= \frac{1}{2}|\dot{\vec{r}}|^2 - \frac{\mu}{r} \\
&= \frac{1}{2}(\dot{r}^2 + r^2\dot{\theta}^2) - \frac{\mu}{r} \\
&= \frac{1}{2}\left(\dot{r}^2 + \frac{J^2}{r^2}\right) - \frac{\mu}{r},
\end{aligned} \tag{0.33}$$

where

$$m_{red} = \frac{m_1 m_2}{m_1 + m_2} = \frac{m_1 m_2}{M}$$

is called the *reduced mass*. (This expression for the Kepler energy can be derived either directly from Kepler's problem viewed as a Hamiltonian system or by re-expressing the total two-body energy E of Equation (0.21) in terms of Q_{cm} and $\vec{r}$ and then setting $\dot{Q}_{cm} = 0$.)

Kepler's three laws summarize the most important features of the solutions to Kepler's problem. They are:

K1. *Solutions form branches of conic sections in the invariable plane. The origin is one focus of these conics.*

The conics are ellipses (or circles) if the energy E is negative, parabolas if $E = 0$, and hyperbolas if $E > 0$. In the negative energy case, the energy H and size a of the orbit are related by

$$H = \frac{-\mu}{2a},$$

where "size" means semi-major axis of the ellipse (radius in case of a circle). The conics degenerate to rays ($E \geq 0$) or line segments ($E < 0$) having one endpoint collision with the Sun ($r = 0$) if and only if $J = 0$.

K2. *Solutions sweep out equal areas in equal times.*

The moving segment $[0, \vec{r}(t)]$ sweeps out an area $A(t)$ in the time interval $0 \leq t' \leq t$. K2 asserts that $A(t_2) - A(t_1) = c(t_2 - t_1)$ where c is a constant and $t_2 > t_1$. The law is a direct consequence of conservation of angular momentum $J = r^2 \dot{\theta}$. Rewrite this conservation law in the form $r^2 d\theta = J dt$ and notice that $d(\frac{1}{2} \int r^2 d\theta) = r dr \wedge d\theta = dx \wedge dy$ is the element of area on the invariable plane. Apply Stokes' theorem to arrive at $A(t) = \frac{1}{2} J t$, $t > 0$, so that $c = J/2$.

K3. *Periodic solutions satisfy the power-law relation*

$$T^2 = ca^3, \tag{0.34}$$

relating their period T and size a.

The constant $c = 4\pi^2/GM$ where $M = m_1 + m_2$ is the total mass of the two-body system. The size a is as per K1. This law follows from the scaling symmetry described in Section 0.1.8.

A remarkable thing about Kepler's problem is that for fixed negative energy all its orbits are periodic with the same period T. This fact plays a central role in Chapter 3. Indeed, the Kepler problem appears in various ways when facing any of the four open questions on our list.

The limiting case of angular momentum zero is worth discussing. We have $J = 0$ if and only if the space spanned by the vectors $\vec{r}(0)$ and $\dot{\vec{r}}(0)$ is one-dimensional, which in turn is true if and only if the solution is a solution to the one-dimensional Kepler problem, and thus travels along the ray joining $\vec{r}(0)$ to the origin $\vec{r} = 0$. Let us fix the two-plane, say $e_1 \wedge e_2$, and the energy at some negative value and let the angular momentum go to zero by setting $J = Le_1 \wedge e_2$ and letting $L \to 0$. In this way we get a family of ellipses crunching down to line segments.

Exercise 0.13   Consider the general two-body problem as per Section 0.1.5 where the negative of the pair potential is $-f(r)$ with $f_{21}(r) = -f_{12}(r) = f(r)$. Show that the relative motion vector $\vec{r}$ evolves according to $\ddot{\vec{r}} = f'(r)\frac{\vec{r}}{r}$. This ODE is refered to as the *central force problem*. Its potential function is $V(\vec{r}) = -f(r)$.

See Appendix H for more regarding the central force problem and Landau [101] for a detailed discussion.

0.3 Mass Inner Product

0.3.1 Mass Metric

The kinetic energy $K = \frac{1}{2}\Sigma m_a |\dot{q}_a|^2$ defines an inner product $\langle \cdot, \cdot \rangle$ on configuration space $\mathbb{E} = (\mathbb{R}^d)^N$

$$K(\dot{q}) = \frac{1}{2}\langle \dot{q}, \dot{q} \rangle, \tag{0.35}$$

which we call the mass metric. Explicitly,

$$\langle q, v \rangle = \Sigma m_a q_a \cdot v_a, \tag{0.36}$$

where $q_a \cdot v_a$ denotes the standard inner product of $q_a, v_a \in \mathbb{R}^d$.

Definition 0.14   The inner product (0.36) is called the mass inner product.

The mass inner product can also be written as

$$\langle q, v \rangle = \langle q, \mathbb{M}v \rangle_1, \tag{0.37}$$

where $\mathbb{M}: \mathbb{E} \to \mathbb{E}$ is the mass matrix (Equation (0.10)) and $\langle q, v \rangle_1 = \Sigma q_a \cdot v_a$ is the standard inner product, which is to say what the mass inner product would be if all the masses were 1.

Lemma 0.15 *The general N-body equations can be rewritten as*

$$\ddot{q} = \nabla U(q), \tag{0.38}$$

where ∇ is the gradient of $U = \Sigma f_{ab}(r_{ab})$ with respect to the mass inner product.

Proof   Equations (0.8), (0.9), and (0.37) show us that we can write the N-body problem as

$$\mathbb{M}\ddot{q} = F(q), \tag{0.39}$$

where $\mathbb{M}$ is the mass matrix and $F: \hat{\mathbb{E}} \to \mathbb{E}$ is the vector of forces. Thus $\ddot{q} = \mathbb{M}^{-1}F(q)$. Now $F(q)_a = F_a(q) = \nabla_a U(q)$ where ∇_a is the usual gradient with respect to $q_a \in \mathbb{R}^d$. It follows that $F(q) = \nabla^{(1)}U$ where $\nabla^{(1)}$ is the gradient with respect to the standard inner product. Recall, in general, the relation between the differential and the gradient of a function on an inner product space: $dU(q)(h) = \langle \nabla U(q), h \rangle$. The intertwining relation, Equation (0.37), implies that our two gradient operators, $\nabla^{(1)}$ and ∇, on $\mathbb{E}$ are related by the mass matrix $\nabla^{(1)} = \mathbb{M}\nabla$ or $\nabla = \mathbb{M}^{-1}\nabla^{(1)}$, so that $\mathbb{M}^{-1}F(q) = \nabla U(q)$. QED

Using the mass metric we have that the energy is given by

$$E = \frac{1}{2}\langle \dot{q}, \dot{q} \rangle - U(q).$$

Using the condensed form of the general N-body (Equation (0.38)) and the mass metric we can quickly verify the fact that energy is conserved; that is, it is constant along solutions:

$$\begin{aligned}\dot{E} &= \langle \dot{q}, \ddot{q} \rangle - \langle \dot{q}, \nabla U(q) \rangle \\ &= \langle \dot{q}, \ddot{q} - \nabla U(q) \rangle \\ &= 0,\end{aligned}$$

where in arriving at the last line we used Equation (0.38).

0.3.2 Moment of Inertia

Write

$$I = \langle q, q \rangle = \Sigma m_a |q_a|^2 \tag{0.40}$$

for the norm squared of our configuration and

$$M = \Sigma m_a \tag{0.41}$$

for the total mass. We will also call I the moment of inertia for reasons to be clarified below; I satisfies the identities

$$\ddot{I} = 4E + 2U, \tag{0.42}$$

$$MI = \Sigma_{a<b} m_a m_b r_{ab}^2 \quad \text{if } \Sigma m_a q_a = 0, \tag{0.43}$$

and

$$IK \geq \frac{1}{2}\|J\|^2. \tag{0.44}$$

The first identity, Equation (0.42), is called the Lagrange–Jacobi identity. In physics and astronomy texts it is often called the *virial identity*. In the identity, $\ddot{I}$ denotes the second derivative of $I(q(t))$ along a solution curve $q(t)$ having energy E, and $U = U(q(t))$ is the negative of the potential along $q(t)$.

Before proving the identities we use the Lagrange–Jacobi identity to obtain our first global qualitative result.

Corollary 0.16 *If a solution $q(t)$ to the gravitational N-body problem has positive energy and is defined for all time, then its motion is unbounded in both forward and negative time, and it size $|q(t)| = \sqrt{I(q(t))}$ achieves a minimum value at a unique time.*

Proof $U > 0$ everywhere. If the energy E is positive then the Lagrange–Jacobi identity gives strict convexity $\ddot{I} > 4E > 0$ for I, proving that $t \mapsto I(q(t))$ is unbounded in both time directions with a unique minimum. QED

The corollary continues to hold for energy zero, but the proof requires a bit more work. We postpone its proof to the end of this section.

We verify these identities now. For the Lagrange–Jacobi identity differentiate I once to get $\dot{I} = 2\langle q, \dot{q} \rangle$. Differentiate again to obtain

$$\begin{aligned} \ddot{I} &= 2\langle \dot{q}, \dot{q} \rangle + 2\langle q, \ddot{q} \rangle \\ &= 4K(\dot{q}) + 2\langle q, \nabla U(q) \rangle. \end{aligned} \tag{0.45}$$

The function U is homogeneous of degree -1 so by one of Euler's many identities we have $\langle q, \nabla U(q) \rangle = -U(q)$. It follows that

$$\begin{aligned} \ddot{I} &= 4K - 2U \\ &= 4E + 2U, \end{aligned} \tag{0.46}$$

where in arriving at the last equation we used $E = K - U$.

In order to prove the second identity, Equation (0.43), write Σ' for the sum over all pairs $a < b$ and Σ for the sum over *all pairs* a, b coming from the mass index set $\{1, \ldots, N\}$. The sum in Equation (0.43) is a Σ'. By symmetry and the fact that $r_{aa} = 0$ we have, on the one hand, $\Sigma m_a m_b r_{ab}^2 = 2\Sigma' m_a m_b r_{ab}^2$. On the other hand,

$$\Sigma m_a m_b |q_a - q_b|^2 = \Sigma m_a m_b |q_a|^2 - 2\Sigma m_a m_b q_a \cdot q_b + \Sigma m_a m_b |q_b|^2.$$

The first and last summands on the right-hand side of this equation equal MI where, as before $I = \Sigma m_a |q_a|^2 = \langle q, q \rangle$. The second, cross, term, $\Sigma m_a q_a \cdot \Sigma m_b q_b$, is zero since $\Sigma m_a q_a = 0$. Thus $2\Sigma' m_a m_b r_{ab}^2 = 2MI$.

For the final identity, Equation (0.44), we use that $\|x \wedge v\|_{o(d)} \leq |x||v|$ for $x, v \in \mathbb{R}^d$. Then

$$\begin{aligned} \|J\| &= \|\Sigma m_a q_a \wedge v_a\| \\ &\leq \Sigma m_a \|q_a \wedge v_a\| \\ &\leq \Sigma m_a |q_a||v_a| \\ &\leq \sqrt{\Sigma m_a |q_a|^2}\sqrt{\Sigma m_a |v_a|^2} \\ &= \sqrt{I}\sqrt{2K}, \end{aligned}$$

which yields the final identity.

Identity (0.43) enjoys a version free of the constraint $\Sigma m_a q_a = 0$. The center of mass of a configuration, denoted Q_{cm}, satisfies $MQ_{cm} = \Sigma m_a q_a$, where $M = \Sigma m_a$ is the total mass. (See also Equation (0.50).) The desired identity is

$$\langle q - Q_{cm}, q - Q_{cm} \rangle = \frac{1}{M} \Sigma_{a<b} m_a m_b r_{ab}^2, q \in \mathbb{E}_0. \qquad (0.47)$$

When $Q_{cm} = 0$ this identity is the already established identity (0.43). To establish Equation (0.47) in the general case, observe that both sides of the identity are invariant under translations.

We extend the fact that $E > 0 \implies$ unbounded to the case of energy zero.

Lemma 0.17 *If a solution $q(t)$ to the gravitational N-body problem has energy $E = 0$ and is defined for all time, then its motion is unbounded in both forward and negative time and its size $|q(t)| = \sqrt{I(q(t))}$ achieves a minimum value at a unique time.*

Proof $U = 0$ everywhere so that we still have that I is a strictly convex non-negative function of time, but its second derivative is no longer bounded away from zero since we might have that $U \to 0$. Such a function $I(t)$ is either unbounded in both directions with a unique minimum or it limits monotonically to a finite value I_* at one end of the time line or the other, like the convex functions e^t and e^{-t}. We must eliminate this latter possibility. QED

Definition 0.18 The normalized potential $\tilde{U}$ is the function

$$\tilde{U} = \sqrt{I} U. \qquad (0.48)$$

Identify the sphere $\mathbb{S} \subset \mathbb{E}$ as the set of rays through the origin of $\mathbb{E}$ so that $\tilde{U}$ becomes a function on the sphere with poles at the collision locus. It is positive and continuous as a function on $\mathbb{S}$ with values in $[0, \infty] = [0, \infty) \cup \{\infty\}$. Since the sphere is compact there is a positive constant c such that $\tilde{U} > c$ everywhere. Let us suppose now that $I \to I_*$ as $t \to \infty$. Since $U = \tilde{U}/\sqrt{I}$, it follows that $U > c/\sqrt{I_*} > 0$ in some neighborhood of $t = +\infty$, so that by the Lagrange–Jacobi identity $\ddot{I} \geq 2c/\sqrt{I_*}$ near $t = +\infty$. This inequality is inconsistent with $I(t) \to I_*$, a contradiction implying that the finite limit I_* cannot exist. A similar argument holds as $t \to -\infty$. QED

Mechanical Significance of Moment of Inertia

In classical mechanics "moment of inertia" denotes a rotational analogue of mass. The kinetic energy of a particle traveling at constant velocity v is $\frac{1}{2}m|v|^2$ where m is the particle's mass. Similarly, if a planar rigid body rotates at constant angular velocity ω about some center, then its kinetic energy is $\frac{1}{2}I\omega^2$

where I is called the moment of inertia about that center. We can think of a planar N-body configuration q as a rigid body simply by freezing all the distances between its points to make it rigid. In order to rotate our planar N-gon q at constant angular velocity ω about the origin of $\mathbb{R}^d$ form the curve $exp(i\omega t)q$, where we use the complex numbers notation as in Equation (0.28) so that $q \in \mathbb{E}(2, N) = \mathbb{C}^N$. The time derivative of our curve is $\dot{q} = i\omega q$ so that

$$K(\dot{q}) = \frac{1}{2}\Sigma m_a \omega^2 |q_a|^2 = \frac{1}{2}I\omega^2,$$

justifying our use of I for the moment of inertia for the planar case.

Exercise 0.19 For power law potentials with exponent α, derive the variant of the Lagrange–Jacobi identity

$$\ddot{I} = 4E + (4 - 2\alpha)U. \tag{0.49}$$

Exercise 0.20 Use the Lagrange–Jacobi identity to show that if $q(t)$ is a periodic solution for the standard N-body problem, then its energy E satisfies

$$E = -\frac{1}{2}\langle U \rangle,$$

where $\langle U \rangle := \frac{1}{T}\int_0^T U(q(t))dt$ is the average of the negative of the potential energy over one period T of the orbit.

Exercise 0.21 In the case of the power law potential in the limit $\alpha = 0$ appropriate to "real" planar gravity, we have $U = -\Sigma m_a m_b \log(r_{ab})$. Show that regardless of energy, every solution is bounded over its entire interval of existence.

0.3.3 Center of Mass

When we solved the two-body problem we split its motion up into that of its center of mass, which was uniform, and the relative motion of the two bodies with respect to that center of mass, which was governed by Kepler's problem. This same trick works for any number of bodies. Write

$$Q_{cm} = \frac{1}{M}\Sigma m_a q_a \in \mathbb{R}^d \tag{0.50}$$

for the center of mass of a configuration $q \in \mathbb{E}(d, N)$. Here M is the total mass (Equation (0.41)). Set

$$\mathbb{E}_0 = \{q \in \mathbb{E} : Q_{cm} = 0\}. \tag{0.51}$$

Call $\mathbb{E}_0$ the *centered configuration space* and its elements *centered configurations*. The space $\mathbb{E}_0$ plays the role in the N-body problem that the relative position vector $q_1 - q_2$ played in the two-body problem.

The moving center of mass Q_{cm} of a solution $q(t)$ to the N-body problem satisfies $M\dot{Q}_{cm} = P$, which is a constant vector, the linear momentum. It follows that Q_{cm} travels uniformly: $Q_{cm} = Q_0 + tV$ with $V = P/M$. As we did for $N = 2$, we can now Galilean boost by $-V$ and translate by $-Q_0$ to bring this moving center of mass to the origin, reducing the general N-body problem to the problem of solving the problem within $\mathbb{E}_0$. (Note that if $q(t) \in \mathbb{E}_0$ then $P(\dot{q}) = 0$, which is to say, motions tangent to $\mathbb{E}_0$ remain in $\mathbb{E}_0$.)

The center of mass has a close relationship to the action of the translation group on $\mathbb{E}$. (See the first transformation in the series, Equation (0.20).) Write the action of translation of a configuration q by $c \in \mathbb{R}^d$ as $q \mapsto q + c\vec{1}$, where we use $c\vec{1}$ to denote the vector $(c, c, \ldots, c) \in \mathbb{E}$. The orbit of the origin $0 \in \mathbb{E}$ under the translation group is the d-dimensional linear subspace $\{q = c\vec{1} : c \in \mathbb{R}^d\}$. We call this subspace the *translation subspace* and sometimes denote it by $\mathbb{R}^d\vec{1}$. A vector $q \in \mathbb{E}$ is orthogonal to the translation subspace if and only if its center of mass is zero. Indeed,

$$\langle c\vec{1}, q \rangle = (c \cdot \Sigma m_a q_a) = c \cdot MQ_{cm}, \tag{0.52}$$

and $\mathbb{E}_0$, being the orthogonal subspace to the translation subspace, represents the quotient space of $\mathbb{E}$ by translations.

Set

$$\hat{\mathbb{E}}_0 = \hat{\mathbb{E}}_0(d, N) := \mathbb{E}_0 \setminus \Delta$$

to be the collision-free part of the center-of-mass zero configuration space and

$$\mathcal{P}_0 = \mathcal{P}_0(d, N) := \hat{\mathbb{E}}_0 \times \mathbb{E}_0 = \{(q, v) \in \mathbb{E} \times \mathbb{E} : Q_{cm} = 0, P(v) = 0, q \notin \Delta\}, \tag{0.53}$$

where $\mathcal{P}_0 \subset \mathcal{P}$.

We have explained how solving the general N-body problem is equivalent to solving it within the submanifold $\mathcal{P}_0 \subset \mathcal{P}$. We refer to $\mathcal{P}_0$ as either "center-of-mass phase space" or "centered phase space." We work in $\mathcal{P}_0$ rather than $\mathcal{P}$, or $\mathbb{E}_0$ rather than $\mathbb{E}$, whenever it is more convenient.

Exercise 0.22 Take a centered solution $q(t) \in \hat{\mathbb{E}}_0$ and supply a Galilean boost in the velocity direction $V \in \mathbb{R}^d$, $V \neq 0$. Show that along the resulting boosted solution we have $I(t) \to \infty$. Conclude that any solution whose linear momentum is not zero has $I(t) \to \infty$. Argue that this is true despite the fact that such solutions might have negative energy E.

0.3.4 Qualitative Behaviors

We have the following two-sided implications for the solutions to the standard two-body problem viewed in the center-of-mass frame:

periodic $\Longleftrightarrow$ rel. periodic $\Longleftrightarrow$ quasi-periodic

$\Longleftrightarrow$ bounded $\Longleftrightarrow$ negative energy.

They follow from the explicit conic section solutions to Kepler's problem described in Section 0.2. (In order for the implications to hold for all solutions we must Levi-Civita regularize the collision solutions, in which case the negative energy collision solutions become periodic.) For the N-body problem with $N > 2$ we only get the one-sided implications:

periodic $\Longrightarrow$ rel. periodic $\Longrightarrow$ quasi-periodic

$\Longrightarrow$ bounded $\Longrightarrow$ negative energy.

We have counterexamples to all the reversed implications. A proof, or reminder, of the last implication "bounded $\Longrightarrow$ negative energy" is in order. Lemmas 0.16 and 0.17 combine to yield the implication "non-negative energy $\Longrightarrow$ bounded," the contrapositive of "bounded $\Longrightarrow$ negative energy." Formal definitions for the words just used are perhaps in order.

Definition 0.23 Let $q : \mathbb{R} \to \mathbb{E}_0$ be a curve in the centered N-body configuration space $\mathbb{E}_0 \cong (\mathbb{R}^d)^{N-1}$. Then we say that q is

- *periodic* (of period T) if there exists a $T > 0$ such that $q(T + t) = q(t)$.
- *relative periodic* if there exists a $T > 0$ and a $g \in O(d)$ such that $q(T + t) = gq(t)$. We call g the *monodromy of the orbit*.
- *quasiperiodic* if there exists an embedding $\varphi : \mathbb{T}^k \to \mathcal{P}$ of k-dimensional torus $\mathbb{T}^k$ into phase space $\mathcal{P}$ such that when viewed on the universal cover $\mathbb{R}^n$ of the torus the phase curve $\zeta(t) = (q(t), \dot{q}(t))$ becomes a straight line. In other words, if $\theta^i, i = 1, \ldots, k$, are standard coordinates on $\mathbb{T}^k = (\mathbb{R}/2\pi\mathbb{Z})^k$, then $\varphi^*\zeta(t) = (\theta^1(t), \ldots, \theta^k(t))$ with $\dot{\theta}^i = \omega^i = const$.
- *bounded* if there exists a positive constant C such that $\|q(t)\| \leq C$ for all time.
- *negative energy* if its energy E is negative.

In representing orbits, we have imagined $\mathbb{R}^d$ as a Euclidean space. But as far as we know, God did not paint Cartesian axes onto the fabric of the universe. Philosophically speaking, the positions q_a are not observable. What we can more or less directly observe is the relative positions and velocities and, hence, the interparticle distances $r_{ab} = |q_a - q_b|$. Notice that the mutual distances

r_{ab} are Galilean invariant. All the previous definitions except for periodicity can be reworded in terms of properties of the mutual distances $r_{ab}(t)$ along a solution $q(t)$.

Proposition 0.24 *Consider a curve in $\mathbb{E}$ and let $q : \mathbb{R} \to \mathbb{E}_0$ be this same curve viewed in the center-of-mass frame. Then q is*

- *relative periodic if there exists a $T > 0$ such that the $r_{ab}(t)$ are T-periodic functions.*
- *quasiperiodic if all the $r_{ab}(t)$ are almost-periodic functions whose span within the space $C(\mathbb{R}, \mathbb{R})$ of continuous functions lies on a k-dimensional torus $\mathbb{T}^k$, $\mathbb{R} \to \mathbb{R}$.*
- *bounded if the $r_{ab}(t)$ are all bounded.*

0.3.5 Total Collision and Zero Angular Momentum

We just used the Lagrange–Jacobi identity to relate $E \geq 0$ to $I \to \infty$. At the other extreme $I \to 0$, which means total collision: all $r_{ab} \to 0$. See Equation (0.47). Sundman related $J = 0$ to $I \to 0$ [201, 202].

Proposition 0.25 *If a solution suffers total collision then its angular momentum is zero in the center-of-mass frame.*

The bulk of our proof of this proposition of Sundman is taken from [195, p. 26–27].

Proof of Proposition 0.25 Let t_c be the collision time so that $I(t) \to 0$ as $t \to t_c$. Then we also have $U(t) \to +\infty$ as $t \to t_c$. Use the Lagrange–Jacobi identity $\ddot{I} = 4H + 2U$ to observe that $\ddot{I} > 0$ and $\dot{I} < 0$ in a one-sided neighborhood $[t_2, t_c)$ of t_c. Indeed, $U \to \infty$ at collision, so from $\ddot{I} = 4E + 2U$ and since E is constant we see that $\ddot{I} \to +\infty$. In particular, $\ddot{I} > 0$ in a one-sided neighborhood (t_2, t_c) of t_c. From $I(t) > 0$ and $I(t_c) = 0$ in this same neighborhood we get that $\dot{I} < 0$ in the neighborhood.

Use $H = K - U$ to rewrite the Lagrange–Jacobi identity as

$$\ddot{I} = 2K + 2H$$

and multiply it by $-2\dot{I}$ to get $-2\dot{I}\ddot{I} = -4\dot{I}K - 4H\dot{I}$. In Equation (0.44) we established that $K \geq \frac{1}{2}\frac{\|J\|^2}{I}$, from which it follows that

$$-4\dot{I}K \geq 2|J|^2 \frac{-\dot{I}}{I}, \quad (\dot{I} < 0),$$

so that, on our neighborhood $[t_2, t_c)$ where $\dot{I} < 0$, we get

$$-2\dot{I}\ddot{I} \geq |J|^2 \frac{-\dot{I}}{I} - 4H\dot{I}.$$

Observe that the left-hand side is the time derivative of $-\dot{I}^2$ while the right-hand side is the time derivative of $-2|J|^2 \log I - 4HI$. Integrating the inequality from t_2 to t with $t_2 < t < t_c$ yields

$$\dot{I}(t_2)^2 - \dot{I}(t)^2 \geq 2|J|^2 \log \frac{I(t_2)}{I(t)} - 4H(I(t_2) - I(t)).$$

Since $I(t_2) > I(t)$ and $\dot{I}(t)^2 > 0$, this inequality yields

$$\dot{I}(t_2)^2 > 2|J|^2 \log \frac{I(t_2)}{I(t)} - 4|H|I(t_2).$$

Fix t_2. The left-hand side of this inequality becomes a positive constant, but when $J \neq 0$, the right-hand side goes to $+\infty$ as $t \to t_c$ since $I(t) \to 0$, and hence $\log \frac{I(t_2)}{I(t)} \to +\infty$ as $t \to t_c$. This contradiction shows us that we must have $J = 0$ in order to arrive at total collision. QED

0.3.6 Metric Meaning of Conservation Laws

The mass metric lets us give a geometric meaning to linear and angular momentum. Recall that the isometry group $E(d)$ of $\mathbb{R}^d$ is the subgroup of the Galilean group, which does not involve time – and hence acts directly on configuration space $\mathbb{E}$.

Proposition 0.26 *The linear and angular momentum vanish at $(q, v) \in \mathbb{E} \times \mathbb{E}$ if and only if the velocity v is mass-metric orthogonal to the orbit of the isometry group $E(d)$ through q.*

Whenever a Lie group G acts smoothly on a manifold it induces an infinitesimal action of its Lie algebra $\mathfrak{g}$, which is to say, a family of linear maps $\sigma_q : \mathfrak{g} \to T_q Q$ obtained by differentiating $(g, q) \mapsto gq$ at the identity element of G. The vector field $q \mapsto \sigma_q(\xi)$ is called the *infinitesimal generator* of the action in the direction $\xi \in \mathfrak{g}$.

Our group $G = E(d)$ acts by affine transformations of $\mathbb{E}$ and so the infinitesimal generator map yields a collection of affine vector fields. Upon splitting the group into translations and rotations, this space of affine vector fields splits into constant vector fields and linear vector fields. The corresponding split of the Lie algebra is $Lie(d) = \mathbb{R}^d \oplus o(d)$. We have already seen the translational part. Its image is the translation subspace described in the third paragraph of Section 0.3.3:

$$\sigma_{trans} : \mathbb{R}^d \to \mathbb{E}; \sigma_{tr}(c) = (c, c, \ldots, c) := c\vec{1}.$$

The rotational part of the infinitesimal action is

$$\sigma_{rot,q} : o(d) \to \mathbb{E}; \sigma_q(\omega) = (\omega(q_1), \omega(q_2), \ldots, \omega(q_N)),$$

where $\omega \in o(d)$ is a skew-symmetric linear operator on $\mathbb{R}^d$. (When $d = 3$ one usually expresses $\omega(q_a)$ as $\vec{\omega} \times \vec{q}_a$ where $\vec{\omega} \in \mathbb{R}^3$ is the angular velocity vector.)

Proposition 0.27 *Linear and angular momentum are related to the infinitesimal generators of translation and rotation by the transpose. Explicitly, $P(v) = \sigma_{trans}^T(v)$ and $J(q, v) = \sigma_{rot,q}^T(v)$. (Compare with Equation (A.7).)*

Proof A computation identical to that of Equation (0.52) yields

$$\langle c\vec{1}, v \rangle = c \cdot P(v), \qquad (0.54)$$

which shows that $P(v) = \sigma_{trans}^T(v)$.

For the rotational equality we use the unique inner product $\langle \cdot, \cdot \rangle_{o(d)}$ on $o(d)$ for which $\langle \omega, e \wedge f \rangle_{o(d)} = \omega(e) \cdot f$ for all $\omega \in o(d)$ and all $e, f \in \mathbb{R}^d$, where $e \wedge f \in o(d)$ denotes the skew-symmetric operator $x \mapsto (e \cdot x)f - (f \cdot x)e$. (See Appendix D for more information.) Then

$$\langle v, \sigma_{rot,q}(\omega) \rangle_{\mathbb{E}} = \Sigma m_a v_a \cdot \omega(q_a)$$
$$= \Sigma m_a \langle \omega, q_a \wedge v_a \rangle_{o(d)}$$
$$= \langle \omega, \Sigma m_a q_a \wedge v_a \rangle_{o(d)}$$
$$= \langle \omega, J(q, v) \rangle_{o(d)},$$

showing that $\sigma_{rot,q}(v) = J(q, v)$. QED

Proof of proposition 0.26   The tangent space at q to the G-orbit through q is the image of the infinitesimal generator maps $(\sigma_{trans}, \sigma_q): e(d) = \mathbb{R}^d \oplus o(d) \to \mathbb{E}$. A vector $v \in \mathbb{E}$ is then orthogonal to the image if and only if it is in the kernel of the transpose: $\sigma_{transl}^T(v) = 0$ and $\sigma_q^T(v) = 0$. But we just saw that this means that $P(v) = 0$ and $J(q, v) = 0$. QED

Terminology   The image of σ_{trans} is called the translation subspace:

$$\text{translational subspace} = \{(c, c, \ldots, c) : c \in \mathbb{R}^d\}.$$

The image of $\sigma_{rot,q}$ is called the rotation subspace:

$$\text{rotational subspace at } q = \{(\omega(q_1), \ldots, \omega(q_N)) : \omega \in o(d)\}.$$

Exercise 0.28   Show that the translational subspace is orthogonal to the rotational subspace if and only if the center of mass of q is zero, that is, if and only if $\Sigma m_a q_a = 0$.

0.3.7 Saari Decomposition

At an infinitesimal level, we may split a deformation of a given configuration $q \in \mathbb{E}$ into translational, rotational, scaling, and what remains: the pure shape deformations. Identifying $\mathbb{E}$ with $T_q\mathbb{E}$, we get

$$\mathbb{E} = \text{translation} \oplus \text{rotation} \oplus \text{scaling} \oplus \text{pure shape} \quad (q - \text{dependent}).$$

Saari formalized this decomposition in [184]; hence his name became attached to it. The translation and rotation subspaces were defined in Section 0.3.6. The scaling subspace at q is $\mathbb{R}q$, the subspace generated by the scaling action $(\lambda, q) \mapsto \lambda q$, $\lambda > 0$. The *pure shape* subspace is, by definition, what's left: the orthogonal complement of the sum of the translation, rotation, and scaling subspaces. All subspaces in this decomposition except the translation subspace depend on q.

If $\Sigma m_a q_a = 0$, then this splitting is an orthogonal direct sum decomposition. The point of Exercise 0.28 was to show that the translation and rotation subspaces are orthogonal. To see the orthogonality of the translation and scaling spaces, use that a general element of the translation subspace looks like $c\vec{1} := (c, \ldots, c)$ for arbitrary $c \in \mathbb{R}^d$, and a general element of the dilation subspace looks like $\lambda q = \lambda(q_1, \ldots, q_N)$ for $\lambda \in \mathbb{R}$. We have $\langle c\vec{1}, \lambda q \rangle = \lambda \Sigma m_a c \cdot q_a = \lambda c \cdot \Sigma m_a q_a$, which is zero if the center of mass is zero. Similarly, a general element of the rotation subspace has the form $\sigma_{rot,q}(\omega) = (\omega(q_1), \ldots, \omega(q_N))$ and $\langle c\vec{1}, \sigma_{rot,q}(\omega) \rangle_{\mathbb{E}} = \Sigma m_a c \cdot \omega(q_a) = \Sigma \langle \omega, \Sigma m_a q_a \wedge c \rangle_{o(d)} = \langle \omega, (\Sigma m_a q_a) \wedge c \rangle_{o(d)} = 0$, provided $\Sigma m_a q_a = 0$.

Corresponding to our orthogonal direct sum decomposition of $\mathbb{E}$, the kinetic energy splits (assume $\Sigma m_a q_a = 0$) as:

$$K = \frac{1}{2} \frac{\|P\|^2}{M} + \frac{1}{2} \langle J, \mathbb{I}(q)^{-1} J \rangle + \frac{1}{2}\dot{r}^2 + K_{p.s.}$$
$$= \text{translation} + \text{rotation} + \text{scaling} + \text{pure shape.} \quad (0.55)$$

In the planar case this splitting simplifies to

$$K = \frac{1}{2} \frac{\|P\|^2}{M} + \frac{1}{2} \frac{1}{r^2} J^2 + \frac{1}{2}\dot{r}^2 + K_{p.s.}. \quad (0.56)$$

In these expressions

$$r^2 = I(q) := \langle q, q \rangle$$

and $\dot{r}$ is the time derivative of r. Differentiating I yields $2r\dot{r} = 2\langle \dot{q}, q \rangle$ so that

$$\dot{r} = \left\langle \dot{q}, \frac{q}{r} \right\rangle = \frac{1}{r} \langle v, q \rangle$$

represents the projection of the velocity $v = \dot{q}$ onto the one-dimensional scaling subspace.

The operator $\mathbb{I}(q)$ whose inverse occurs in the 2nd term of Equation (0.55) is a symmetric operator on $\wedge^2 \mathbb{R}^d = o(d)$ called the *moment of inertia tensor* and which measures the kinetic energy of rotations. Specifically, this tensor is defined by

$$\langle \omega, \mathbb{I}(q)\omega \rangle_{so(d)} = \langle \sigma_{rot,q}(\omega), \sigma_{rot,q}(\omega) \rangle_{\mathbb{E}}, \qquad (0.57)$$

so that $\mathbb{I}(q) = \sigma_{rot,q}^T \sigma_{rot,q}$. Here, the transpose of $\sigma_q : o(d) \to \mathbb{E}$ is computed relative to the bi-invariant inner product on $o(d)$ (see Appendix D) and the mass inner product on $\mathbb{E}$.

In order to get from the general version of the Saari decomposition (0.55) to the planar version given in Equation (0.56) use the fact, shown at the end of Section 0.3.2, that

$$\mathbb{I}(q) = I(q) = r^2 \quad \text{when } d = 2. \qquad (0.58)$$

Remark 0.29   When $d = 2$ we have $o(2) \cong \mathbb{R}$ so that $\mathbb{I}(q)$ is a scalar. When $d = 2$ the infinitesimal rotation operator $\sigma_{rot,q} : \mathbb{R} \to \mathbb{E} = \mathbb{C}^N$ is given by $\sigma_q(\omega) = i\omega q$.

The shape space term   The last two terms of the Saari decomposition (0.55) induce a Riemannian metric on the smooth points of shape space, which is the space of oriented congruence classes of labeled N-gons in d-space. See Definition 0.31. The $\dot{r}$ term has to do with size changes. The final pure shape space kinetic energy term $K_{p.s.}$ corresponds to a Riemannian metric on the space of oriented *similarity* classes of labeled N-gons in d-space, smooth points of normalized shape space. This metric has a simple concrete form for the planar N-body problem, in which case $K_{p.s.}$ corresponds to the standard Fubini–Study metric on $\mathbb{CP}^{N-2}$. See Equation (0.65) and Section 0.4.6.

Verification of Saari decomposition, equation (0.55)   We have already verified that the rotational, translational, and scaling subspaces are mutually orthogonal if the center of mass of the configuration is zero. It then remains to verify that the kinetic energy agrees with the given quadratic form from the decomposition on each subspace.

First, we take v in the translational subspace. Then we can write $v = (c, \ldots, c)$ in which case $K(v) = \frac{1}{2} M \|c\|^2$ while $P(v) = Mc$, from which it follows that $K = \frac{1}{2M}|P(v)|^2$. This yields the first term of the Saari decomposition.

Next, if v is in the rotational subspace, then $v = (\omega(q_1), \dots, \omega(q_N))$ for some $\omega \in o(d)$. It then follows that

$$K(v) = \frac{1}{2}\Sigma m_a \|\omega(q_a)\|^2$$
$$= \langle \omega, \mathbb{I}(q)\omega \rangle_{o(d)}. \tag{0.59}$$

But $J(q,v) = \sigma_{rot,q}^T v = \sigma_{rot,q}^T \sigma_{rot,q}\omega = \mathbb{I}(q)\omega$ so that $\omega = \mathbb{I}(q)^{-1}J(q,v)$. Substituting this relation for ω into Equation (0.59) yields $K(v) = \frac{1}{2}\langle \mathbb{I}(q)^{-1}J, J\rangle_{o(d)}$.

Finally, if v is in the dilational subspace we have that $v = \dot{r}(q/r)$ from which it follows that $K(v) = \frac{1}{2}\dot{r}^2$. QED

Remark 0.30 The relation

$$\mathbb{I}(q) \leq I(q)$$

holds between $I = r^2$ and $\mathbb{I}$, to be understood as an inequality between real quadratic forms, namely $\langle \omega, \mathbb{I}(q)\omega \rangle_{so(d)} \leq I\|\omega\|^2_{so(d)}$ for all $\omega \in o(d)$. To verify this relation, compute

$$\langle \omega, \mathbb{I}(q)\omega \rangle_{so(d)} = \langle \sigma_{rot,q}(\omega), \sigma_{rot,q}(\omega)\rangle$$
$$= \Sigma m_a \|\omega(q_a)\|^2$$
$$\leq \Sigma m_a \|\omega\|^2_{op}\|q_a\|^2$$
$$= \|\omega\|^2_{op}\Sigma m_a|q_a|^2$$
$$= \|\omega\|^2_{op}I$$
$$\leq \|\omega\|^2_{so(d)}I.$$

In the third line we used the definition of the operator norm for an operator $\omega: \mathbb{R}^d \to \mathbb{R}^d$. In obtaining the last line we used the inequality $\|\omega\|^2_{op} \leq \|\omega\|^2_{so(d)}$ derived in Appendix D.

By the spectral theory, the assertion $\mathbb{I} \leq I$ is equivalent to the assertion that all eigenvalues of $\mathbb{I}(q)$ are less than or equal to I. It follows that $\mathbb{I}(q)^{-1} \geq \frac{1}{I}$, again in the sense of quadratic forms. Hence $K \geq \frac{1}{2}\frac{\|P\|^2}{M} + \frac{1}{2}\frac{\|J\|^2}{I} + \frac{1}{2}\dot{r}^2 + K_{p.s.}$. In particular, $IK \geq \frac{1}{2}\|J\|^2$, providing another proof of inequality (0.44).

0.3.8 Jacobi Vectors

Any two real inner product spaces of the same dimension are linearly isometric. It follows by a dimension count that we have a linear isometry

$$\mathbb{E}_0 \cong (\mathbb{R}^d)^{N-1},$$

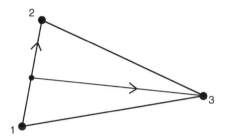

Figure 0.2 Jacobi vectors.

where we use the standard squared inner product on $(\mathbb{R}^d)^{N-1}$. Such a map sends $q \in \mathbb{E}_0$ to $(Q_1, \ldots, Q_{N-1})$ in such a way that $I(q) = \Sigma |Q_a|^2$. The $(N-1)$-tuples of $Q_a \in \mathbb{R}^d$ are called *normalized Jacobi vectors*. Jacobi vectors are not unique. Any linear isometry $R \in O(d(N-1))$ of $(\mathbb{R}^d)^{N-1}$ will take one system of Jacobi vectors to another.

Jacobi wrote down a standard method for computing his vectors when $N = 3$ (see [2] and references therein). It begins by splitting off a pair, say $\{1, 2\}$, from our mass label set $\{1, 2, 3\}$. For $N = 3$ and $q = (q_1, q_2, q_3) \in \mathbb{E}$, the two (un-normalized) Jacobi vectors are

$$Q_1 = q_2 - q_1, Q_2 = q_3 - (m_1 q_1 + m_2 q_2)/(m_1 + m_2). \qquad (0.60)$$

See Figure 0.2. Here, Q_2 connects the center of mass of 12 to mass 3. Let $Q_{cm} = \frac{1}{M}(m_1 q_1 + m_2 q_2 + m_3 q_3)$ be the center of mass. One computes

$$I(q) = M|Q_{cm}|^2 + \mu_1 |Q_1|^2 + \mu_2 |Q_2|^2, \qquad (0.61)$$

where

$$\frac{1}{\mu_1} = \frac{1}{m_1} + \frac{1}{m_2}, \quad \frac{1}{\mu_2} = \frac{1}{m_1 + m_2} + \frac{1}{m_3}. \qquad (0.62)$$

It follows that if $Q_{cm} = 0$ so that $q \in \mathbb{E}_0$, then $I(q) = \mu_1 |Q_1|^2 + \mu_2 |Q_2|^2$. The normalized Jacobi vectors are

$$Z_1 = \sqrt{\mu_1} Q_1, \quad Z_2 = \sqrt{\mu_2} Q_2. \qquad (0.63)$$

An iterative binary tree algorithm for computing Jacobi vectors for any N is explained nicely on Wikipedia.[4] This algorithm yields the standard $N = 3$ Jacobi vectors just given. Alternatives to this binary tree algorithm are achieved by realizing that, as an inner product space, $\mathbb{E} = \mathbb{R}^d \otimes \mathbb{R}^N$ provided we give $\mathbb{R}^d$ its standard inner product while giving $\mathbb{R}^N$ the mass-induced inner product.

---

[4] See https://en.wikipedia.org/wiki/Jacobi_coordinates. Accessed November 2021.

Write $\epsilon = (1, \ldots, 1) \in \mathbb{R}^N$ and observe that the generator of the translation space is $\mathbb{R}^d \otimes \epsilon$ so that $\mathbb{E}_0 = \mathbb{R}^d \otimes \epsilon^\perp$. Find an orthonormal basis $\epsilon_1, \epsilon_2, \ldots, \epsilon_{N-1}$ for $\epsilon^\perp \subset \mathbb{R}^N$ by any method you like. You could, for example, fill out ϵ to a basis $(\epsilon, f_1, f_2, \ldots, f_N)$ for $\mathbb{R}^N$ by any method and then apply Gram–Schmidt to this basis. Then any $q \in \mathbb{E}_0$ can be expanded uniquely as $\Sigma_{a=1}^{N-1} Q_a \otimes \epsilon_a$ with $Q_a \in \mathbb{R}^d$. The Q_a are then normalized Jacobi vectors. The basis corresponding to the original $N = 3$ (non-normalized) Jacobi vectors Q_1, Q_2 is

$$\epsilon_1 = \left(\frac{1}{m_1}, \frac{-1}{m_2}, 0 \right), \ \epsilon_2 = \left(\frac{-1}{m_1 + m_2}, \frac{-1}{m_1 + m_2}, \frac{1}{m_3} \right).$$

0.4 Reduction and the Shape Sphere

0.4.1 Reduction

The Galilean group acts by symmetries on the N-body problem. Whenever a Lie group G acts on a manifold $\mathcal{P}$ leaving a vector field X invariant, then X descends to a vector field $\bar{X}$ on the quotient space $\mathcal{P}/G$, provided that quotient is a manifold. We refer to the process of going from $(X, \mathcal{P})$ to $(\bar{X}, \mathcal{P}/G)$ as *reduction* and the quotient space $\mathcal{P}/G$ as the *reduced space.*[5]

The Galilean group does not act on the N-body phase $\mathcal{P} = \mathcal{P}(d, N)$ due to the boosts, which explicitly involve time. Nevertheless, the act of replacing $\mathcal{P}$ by $\mathcal{P}_0$ in Equation (0.53), achieved by using center-of-mass coordinates, is the moral equivalent of forming the quotient of N-body phase space by the subgroup of the Galilean group made up of boosts and translations. After getting rid of the boosts and translations, the orthogonal subgroup $O(d)$ of the Galilean group remains to act on the centered phase space $\mathcal{P}_0 = \hat{\mathbb{E}}_0 \times \mathbb{E}_0$. The quotient $\mathcal{P}_0/O(d)$ is the moral equivalent of $\mathcal{P}/G$ where G is the Galilean group.

However, the reduced space $\mathcal{P}_0/O(d)$ is not a manifold when we have more than two bodies, a shortfalling that makes reduction difficult to work with. But when $d = 2$ a simple trick turns the reduced space into a manifold: Get rid of reflections, thus replacing $O(2)$ by the rotation group $SO(2) \subset O(2)$. The resulting quotient space $\mathcal{P}_0(2, N)/SO(2)$ is now a manifold.

Why reduce? For the planar N-body problem the combined processes of fixing the center of mass, reducing by rotations, and fixing the angular momentum drops the dimension of phase space by 6, from $4N$ to $4N - 6$.

---

[5] Our definition of reduction does not agree with the usual definition found in symplectic geometry and geometric mechanics. See the notes in Section 0.6 and Appendix B on geometric mechanics regarding the relation between the two types of reduction.

When $N = 3$, this drop from 12 to 6 is a big deal. It allows us to visualize and formulate results that would be very difficult without reducing. When N is large, say $N = 10^8$, the drop by 6 is insignificant. Nevertheless, even here reduction can be a conceptual help.

In essence, the reduced phase space is the cotangent bundle of a manifold we call *shape space*. Two objects sitting in $\mathbb{R}^d$ have the same "shape" if there is a rigid motion taking one onto the other. We recall that a rigid motion is an element of the group $SE(d)$, the identity component of the group $E(d)$ of all isometries of $\mathbb{R}^d$. Alternatively, any rigid motion is the composition of a rotation (an element of $SO(d)$) and a translation (an element of $\mathbb{R}^d$).

Definition 0.31 Two configurations are said to be *oriented congruent* (or to have the same shape) if a rigid motion takes one to the other.

Shape space, denoted $Shape(d, N)$, is the space whose points are oriented congruence classes of configurations (labeled N-gons in d-space) endowed with the quotient metric.

The definition says that shape space is the quotient space $\mathbb{E}(d, N)/SE(d)$. Using the center-of-mass trick we have that $\mathbb{E}/\text{translations} = \mathbb{E}_0$ so that, equivalently, shape space can be defined as

$$Shape(d, N) = \mathbb{E}_0(d, N)/SO(d).$$

Write $\pi \colon \mathbb{E}_0 \to Sh$ for the projection. Using this second definition of shape space, the quotient metric d on shape space can be defined by

$$d(\pi(q), \pi(q')) = \inf_{g, h \in SO(d)} \|gq - hq'\|, \tag{0.64}$$

where $\| \cdot \|$ is the norm relative to the mass metric.

Th salient facts regarding reduction in the planar case are

$$Shape(2, N) = Cone(\mathbb{C}P^{N-2}) \tag{0.65}$$

and

$$\mathcal{P}_0(2, N)/SO(2) \cong T^*B \times \mathbb{R}, \text{ with } B = Shape(2, N) \setminus \Delta. \tag{0.66}$$

When $N = 3$ we have $N - 2 = 1$ and $\mathbb{C}P^1$ is diffeomorphic to a two-sphere, which we call the shape sphere. The cone over the two-sphere is $\mathbb{R}^3$ so that $Shape(2, 3) = \mathbb{R}^3$, which is wonderful for drawing pictures. See Section 0.4.2.

In Equation (0.65), $\mathbb{C}P^{N-2}$ denotes complex projective $N - 2$ space, endowed with its Fubini–Study metric. (See, for example, [23].) We will sometimes refer to this projective space as the shape sphere for the planar N-body problem, despite the fact that it is not a sphere. Its points represent oriented similarity classes of planar N-gons. To describe the cone over this shape sphere, we recall the cone, $Cone(X)$, over any metric space X. (Again, see [23].) Topologically, this cone is obtained by the standard cone construction of topology: Form the product $X \times [0, \infty)$ and then squash (identify) $X \times 0$ to a single point, the cone point, denoted by 0. If X's metric comes, like Fubini–Study, from a Riemannian metric ds_X^2 on X, then the metric on $Cone(X)$ is induced by the Riemannian metric $dr^2 + r^2 ds_X^2$ on $X \times (0, \infty)$, where r denotes the coordinate on $[0, \infty)$. With respect to this metric, the coordinate r is the distance from the cone point $r = 0$. In our N-body setting, 0 is the total collision shape and $r = \sqrt{I}$, where $I = \|q\|^2$ is the moment of inertia, as before. See Section 0.4.6 for a derivation of Equation (0.65).

In Equation (0.66) for the reduced phase space, the symbol $\Delta \subset Shape(2, N)$ denotes the image of the collision locus $\Delta \subset \mathbb{E}_0$ in the quotient shape space. Now the collision locus contains total collision, which corresponds to the cone point. The cone over a manifold X fails to be a manifold at the cone point when X is not a sphere. It follows that for all $N > 3$ the shape space fails to be a manifold at the cone point. This does not mess up the writing down of the ODEs because we have to delete the collision locus anyway to form phase space, and the collision locus contains the cone point, which is total collision.

The coordinate in the final $\mathbb{R}$ factor of Equation (0.66) represents total angular momentum J. By conservation of angular momentum, the reduced Newtonian vector field $\bar{X}$ on the reduced space is tangent to the level sets of J, and so we get a family of reduced flows on the spaces $T^* B = T^* B \times \{J_0\}$, parameterized by $J_0 \in \mathbb{R}$. These spaces are known as *symplectic reduced spaces.* The result of Equation (0.66) is a special case of Theorem B.7 (proven in Appendix B).

0.4.2 The Shape Sphere

When $N = 3$ the salient facts become simple and useful. For in this case $N - 2 = 1$ and $\mathbb{C}P^1 = \mathbb{S}^2(1/2)$ as a Riemannian manifold. Here $\mathbb{S}^k(\rho)$ denotes the sphere of radius ρ in $\mathbb{R}^k$. We call this sphere $\mathbb{S}^2(1/2)$ the *shape sphere*. The reader can find more details and derivations of the results of this section in Appendix C.

The cone over $\mathbb{S}^k(1)$ is isometric to $\mathbb{R}^{k+1}$ – this isomorphism is essentially spherical coordinates. We can scale spheres by dilations so $Cone(S^k(\rho))$ is homeomorphic to $\mathbb{R}^{k+1}$. However, this cone is not isometric to $\mathbb{R}^{k+1}$ unless $\rho = 1$. In all other cases, the cone metric becomes singular at the cone point. It follows that

$$Shape(2, 3) = \mathbb{R}^3, \tag{0.67}$$

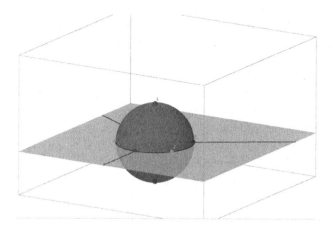

Figure 0.3 Shape space. The origin is triple collision. The plane represents the set of collinear triangles. The three rays represent the three binary collision sets. The sphere represents triangles of a fixed size and is identified with the shape sphere. The marked points on the sphere are the five central configurations and the three binary collision points.

with a spherically symmetric singular metric. See Figure 0.3. With regards to the reduced phase space, we then have that $\mathcal{P}_0(2,3)/SO(2)$ is $(\mathbb{R}^3 \backslash \Delta) \times \mathbb{R}^3 \times \mathbb{R}$. We find that Δ is the union of three coplanar rays through 0, in $\mathbb{R}^3$.

Under the identification (0.67) the shape projection is a homogeneous quadratic map

$$\pi \colon \mathbb{E} = \mathbb{C}^3 \to \mathbb{R}^3, \quad \pi(q) = w = (w_1, w_2, w_3),$$

whose fibers are the orbits of the group of rigid motions. In other words, $\pi(q) = \pi(q') \iff \exists \, g \in SE(2)$ such that $gq = q'$. See Equation (C.22) for an explicit formula for π. That equation represents the restriction of π to $\mathbb{E}_0 \cong \mathbb{C}^2$ using the normalized Jacobi vectors (Z_1, Z_2) described by Equation (0.63). If we further restrict π to the three-sphere $\{I = 1\} \subset \mathbb{C}^2$, we arrive at the famous Hopf fibration $\mathbb{S}^3 \to \mathbb{S}^2$ from topology. The Hopf fibration lies at the heart of the planar three-body problem.

We now describe some properties of π and the relationships between points or subsets of shape space and objects within the three-body problem. In what follows, $q = (q_1, q_2, q_3) \in \mathbb{C}^3$ denotes a configuration – the three vertices of a labeled triangle – and $\vec{w} = (w_1, w_2, w_3) = \pi(q) \in \mathbb{R}^3$ denotes the coordinates of its shape. The map π is homogeneous quadratic so that

$$\pi(\lambda q) = \lambda^2 \pi(q), \lambda \in \mathbb{R}. \tag{0.68}$$

See Figure 0.3 for a rendition of shape space. The origin $\vec{w} = 0$ of shape space represents triple collision. The distance from triple collision is given by

$$r = \|q\| = \sqrt{I} \quad \text{provided } m_1 q_1 + m_2 q_2 + m_3 q_3 = 0.$$

Here, I is the moment of inertia and is related to w by

$$2|w| = I, \text{ where } |w|^2 = w_1^2 + w_2^2 + w_3^2.$$

The coordinate w_3 is related to the signed area[6] A of the triangle q by

$$w_3 = \left(\sqrt{\frac{m_1 m_2 m_3}{m_1 + m_2 + m_3}} \right) A.$$

In particular, the plane $w_3 = 0$ defines the space of collinear triangles and contains triple collision. This collinear plane also contains the collision locus Δ, which is the union of three rays leaving the origin, the three binary collision rays $r_{ij} = 0$.

We can best understand shape space by further reducing by scaling to arrive at the *shape sphere*. The quadratic conditon (Equation (0.68)) implies that the dilates of any given shape form a ray through 0 in shape space. Such a ray hits the surface $I = 1$ at exactly one point.

Definition 0.32 The shape sphere is the unit sphere $I = 1$ of shape space.

Points of the shape sphere represent oriented similarity classes of (labeled) triangles.

The normalized potential (Equation (0.18)), induces a function on the shape sphere for which we use the same symbol,

$$\tilde{U} : \mathbb{S}^2 \to \mathbb{R} \cup \infty,$$

and which we call the shape potential. Indeed, U, the negative of the gravitational potential, is invariant under isometries of the plane, so induces a function $U = U(w_1, w_2, w_3)$ on shape space. Its normalization $\tilde{U} = rU$ is homogeneous of degree zero; that is, it is invariant under scalings (Equation (0.68)). Hence, it yields a function on the space of rays through 0 in $\mathbb{R}^3$; that is, on the shape sphere. See Figure 0.4 for a depiction of contours of $\tilde{U}$ on the shape sphere.

---

[6] The signed area A of the triangle q is given by

$$\frac{1}{2}(q_2 - q_1) \wedge (q_3 - q_1) = A,$$

where the wedge product $\mathbb{R}^2 \times \mathbb{R}^2 \to \mathbb{R}$ is the two-dimensional wedge product as interpreted in Equation (0.25). The absolute value of A is the usual area and A switches sign upon reflection of the labeled triangle q.

Figure 0.4 Contour plot for the shape potential $\tilde{U}$. The binary collisions are poles of $\tilde{U}$. The dotted contour passes through all all three Euler points in case the three masses are equal, and the projection of the figure eight solution to the shape sphere lies very close to this contour. Courtesy of Richard Moeckel.

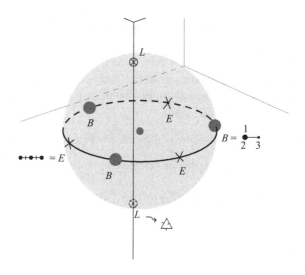

Figure 0.5 The shape sphere. The points marked B are the binary collision configurations. The points marked E are the Euler configurations. The points marked L are the Lagrange configurations.

We have drawn the shape sphere in Figure 0.5. Its equator, $w_3 = 0$, represents the collinear configurations. The marked points in that figure are the critical points of $\tilde{U}$. The binary collision points B are poles: $\tilde{U}(B) = +\infty$ and are the maxima. They lie on the collinear equator and interleaved between them are the three Euler configurations E that are saddle points of $\tilde{U}$. At the north and south poles are the Lagrange points L whose shapes are equilateral triangles. There is a "right-handed Lagrange point" and a "left-handed Lagrange point." When the masses are equal the Lagrange points sit

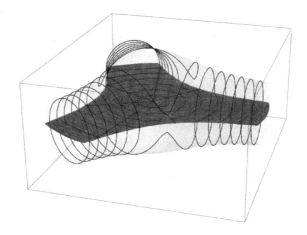

Figure 0.6 Contour level set $U = c$ drawn in shape space. It looks like the surface of a plumbing fixture whose three pipes are centered about the binary collision rays. The Hill region for energy $E = -c < 0$ projects onto the domain $U \geq c$ bounded by this contour surface. The planar domain, which is shaded inside the Hill region, is the Hill region for the collinear three-body problem. Courtesy of Richard Moeckel.

at the north and south poles of the shape sphere and the E and B points form a regular hexagon circumscribed by the equator.

Figure 0.6 depicts a level surface of the shape potential U within shape space. In order to plot the contours of the shape potential U or those of $\tilde{U}$ we need formulae for the side lengths r_{ij} in terms of the shape coordinates w_i. To get these formulae recall that the shape sphere is the sphere $|w| = 1/2$ of radius $1/2$ in the w-coordinates. Write B_{ij} for the binary collision point B in which masses i and j coincide. Normalize the shape sphere by multiplying vector points $\mathbf{w}$ on the sphere by 2 so as to expand it into the unit sphere and thus define

$$\mathbf{b}_{ij} = 2B_{ij}.$$

The $\mathbf{b}_{ij}$ have unit w-length, lie on the collinear plane $w_3 = 0$, and on the ij collision ray. The desired formulae are then

$$r_{ij}^2 = \frac{m_i + m_j}{m_i m_j}(\|\mathbf{w}\| - \mathbf{w} \cdot \mathbf{b}_{ij}). \tag{0.69}$$

Kinetic energy and total energy The Saari decomposition of kinetic energy, Equation (0.56), written in shape variables, is

$$K = \frac{1}{2}\frac{\|P\|^2}{M} + \frac{1}{2}\frac{J^2}{I} + \frac{1}{2}\frac{\|\dot{w}\|^2}{I}.$$

Here, $P = P(\dot{q})$ and $J = J(q,\dot{q})$ are the linear and angular momenta (Equations (0.22) and (0.23)), while $\dot{w} = \frac{d}{dt}\pi(q(t))$ is the shape velocity. The radial term $\frac{1}{2}\dot{r}^2$ and pure shape term $K_{p.s.}$ of Equation (0.56) sum to make the last term

$$K_{shape} = \frac{1}{2}\frac{\|\dot{w}\|^2}{I}, \quad \text{where } I = 2\|w\|. \tag{0.70}$$

We call K_{shape} the shape kinetic energy. In writing the expression we use the standard $\mathbb{R}^3$ inner product for w and $\dot{w}$ so that $\|w\|^2 = w_1^2 + w_2^2 + w_3^2$. We may assume that the linear momentum $P = 0$ since we will work in the center-of-mass frame. Then the total kinetic energy of a path $q(t)$ with constant angular momentum J and shape projection $w(t)$ is

$$K = K_{shape} + \frac{1}{2}\frac{J^2}{I},$$

while its total energy is

$$E = K_{shape} + \frac{1}{2}\frac{J^2}{I} - \frac{1}{r}\tilde{U}(s). \tag{0.71}$$

The shape kinetic energy K_{shape} corresponds to Riemannian metric away from triple collision, namely

$$ds_{shape}^2 = \frac{\|dw\|^2}{2\|w\|}, \tag{0.72}$$

where $\|dw\|^2 = dw_1^2 + dw_2^2 + dw_3^2$. When written in spherical coordinates, $(w_1, w_2, w_3) = \|w\|(\sin(\varphi)\sin(\theta),\ \sin(\varphi)\cos(\theta),\ \cos(\varphi))$ on shape space the metric can be rewritten in the useful form

$$ds_{shape}^2 = dr^2 + \frac{r^2}{4}(d\varphi^2 + \sin^2\varphi\, d\theta^2).$$

The distance function on shape space is induced by this Riemannian metric. For a derivation of these formulae for the shape metric see the proof of Proposition C.24 in Appendix C.

If ∇ denotes the Levi-Civita connection on shape space induced by the shape metric then we can rewrite the three-body equations for solutions that have fixed angular momentum J in the shape space form

$$\nabla_{\dot{\mathbf{w}}}\dot{\mathbf{w}} = -\nabla\left(V + \frac{1}{2}\frac{J^2}{I}\right) - \frac{J}{\|w\|^2}\mathbf{w} \times \dot{\mathbf{w}}. \tag{0.73}$$

The final term involves the standard vector cross product $\mathbf{w} \times \dot{\mathbf{w}}$ of the vector w and its time derivative $\dot{w}$. I write the two vectors in bold face here simply to clarify that this is a vector cross product. I will call these the *reduced equations* for the three-body problem after the notion of symplectic reduction. I derive the reduced equations in Section B.5.

When $J = 0$ the reduced equations simplify to

$$\nabla_{\dot{\mathbf{w}}}\dot{\mathbf{w}} = \nabla U.$$

In this guise, the reduced equations have been quite useful in proving new results. See for example [148]. In the hope that a reader might extend their use into the realm $J \neq 0$, I rewrite the reduced equations in purely Euclidean terms where they could easily be used for numerical integration. In order to do the rewriting, use the Lagrangian for the shape space metric, Equation (0.73), to compute its geodesic equations and thus find that $\nabla_{\dot{\mathbf{w}}}\dot{\mathbf{w}} = \ddot{\mathbf{w}} - \frac{\langle w, \dot{w} \rangle}{\|w\|^2}\dot{\mathbf{w}} + \frac{1}{2}\frac{\|\dot{w}\|^2}{\|w\|^2}\mathbf{w}$. Also, use the standard yoga of "raising indices" to see that $\nabla = \frac{1}{\varphi}\nabla_{Euc}$, where ∇_{Euc} denotes the standard Euclidean gradient operator and where $\varphi = \frac{1}{2|w|}$ is the conformal factor relating the shape space metric to the flat Euclidean metric in Equation (0.72). We arrive at the reduced equations in Euclidean form:

$$\ddot{\mathbf{w}} = \frac{\langle \mathbf{w}, \dot{\mathbf{w}} \rangle}{\|w\|^2}\dot{\mathbf{w}} - \frac{1}{2}\frac{\|\dot{w}\|^2}{\|w\|^2}\mathbf{w} + 2\|w\|\nabla_{Euc}\left(U - \frac{J^2}{4\|w\|}\right) - \frac{J}{\|w\|^2}\mathbf{w} \times \dot{\mathbf{w}}.$$

0.4.3 Area Law

Suppose we have a periodic shape curve $w \colon \mathbb{R} \to \mathbb{R}^3 \setminus \{0\}$ and we know that w is the shape projection of a smooth centered curve $q \colon \mathbb{R} \to \mathbb{C}^2 \setminus \{0\}$. If T is the period of w, then q is relative periodic with the same period T so that

$$q(t + T) = gq(t), g \in \mathbb{S}^1 = SO(2)$$

for some g that we call the monodromy of the loop w.

Theorem 0.33 (Area law) *If we know a closed shape curve $w(t)$ and if we also know that its lift $q(t)$ has angular momentum zero, then the holonomy g of the loop w is $g = exp(i\Delta\theta)$ where*

$$\Delta\theta = \frac{1}{2}\int_D dA,$$

where $D \subset \mathbb{S}^2$ is the disc in the shape sphere that bounds the radial projection of $w(t)$ to the shape sphere, and where dA is the usual spherical area form, the rotationally symmetric two-form whose integral over the whole sphere is 4π.

See [144] or [147] for a proof of Theorem 0.33. The corresponding curvature form, $\frac{1}{2}dA$, which is one-half the solid angle form on shape space, is encoded in the last term of Equation (0.73) and corresponds to the curvature two-form F that appears in Theorem B.7 and Equation (B.15).

To illustrate the area law, consider the isosceles subproblem of the three-body problem. Take $m_1 = m_2$ and consider the space of isosceles triangles

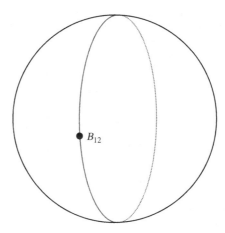

Figure 0.7 The 1-2 isosceles great circle on the shape sphere passes through the 1-2 binary collision point B_{12}.

$r_{13} = r_{23}$. This locus of triangles forms a great circle in the shape sphere that bounds a hemisphere. See Figure 0.7. The spherical area of a hemisphere is 2π, and $\frac{1}{2}(2\pi) = \pi$ so the area law asserts that the isosceles loop generates a monodromy of 180 degrees. In other words, any curve in centered configuration space that has zero angular momentum and projects onto the isosceles great circle, traversed once, has the property that the initial and final triangle of the curve are related to each other by a 180 degree rotation.

We can see this 180 degree rotation by hand. Imagine masses 1 and 2 as dumbbells you are holding in your hands so as to form an isosceles triangle with the third vertex, mass 3, being your head. Start with the dumbbells together, hanging down directly below your waist. You are forming the 1-2 binary collision shape. Now raise your arms out and up over your head symmetrically at the same rate, keeping your body as the symmetry axis of the resulting isosceles triangles. This is a zero angular momentum motion. Stop when the two dumbells touch again, now directly over your head. You are again forming a 1-2 binary collision shape, but this final degenerate triangle is the 180 degree rotation of the initial one. The shape projection of your motion traverses the isosceles great circle once. See Figure 0.8.

0.4.4 The Case of Equal Masses and Its Subproblems

When the masses are equal we can permute them to get additional symmetries of the N-body problem. Thus the permutation group S_3 on three letters forms

Figure 0.8 An isosceles three-body motion that projects onto the 1-2 isosceles great circle.

an additional group of symmetries of the equal-mass three-body problem, leaving both the kinetic and potential energies invariant. The figure eight solution relies on this symmetry. Combined with reflection $(q_1, q_2, q_3) \mapsto (\bar{q}_1, \bar{q}_2, \bar{q}_3)$, the group S_3 generates an action of the 12-element dihedral group D_6 of a hexagon on shape space and on the shape sphere. On the shape sphere this D_6 action is generated by reflections about four great circles, the three isosceles great circles $r_{ab} = r_{ac}$ and the collinear equator $w_3 = 0$.

These four circles correspond to four planes through the origin in shape space. Solutions tangent at any non-collision instant to any one of the planes remains within that plane. In this way, we can see that the equal-mass zero angular momentum planar three-body problem contains within it four subproblems. These are the three isosceles subproblems and the collinear three-body subproblem. These subproblems are two-degree-of-freedom problems so are simpler to analyze and understand than the full problem.

In the equal mass case, the Euler and binary points are equally spaced on the equator while the Lagrange points are placed at the geometric north and south poles.

0.4.5 Morse Theory and Dynamics

Figure 0.9 depicts the Morse theory associated to the shape potential. This picture governs many of the global features of planar three-body dynamics.

Recall that a function is called *Morse* if all its critical points are nondegenerate, meaning that its Hessian at these points is a nondegenerate quadratic form. Morse theory [126] reconstructs the topology of a manifold from that of the sublevel sets $f \le c$ of a Morse function f on the manifold, doing so by paying careful attention to how these sets change as one passes through critical values of the function.

The shape potential is Morse. In Figure 0.9 the super-level sets $\tilde{U} \ge c$ are indicated by shaded domains, with different panels given for different values of c, and that value decreasing as the panels descend. To understand what these

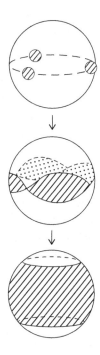

Figure 0.9 Bifurcations of the normalized Hill shape region as a function of the Dziobek parameter $|E|J^2$ when $E < 0$. The top panel is for $|E|J^2 \gg 1$. Bifurcations occur in the shape as the value of $\tilde{U}$ at the Euler configurations are crossed, and again, as the Lagrange value is passed. The bottom panel is for $|E|J^2$ small. (Figure a redrawn summary of pictures in [132].)

pictures tell us about dynamics, fix the values of the energy E and angular momentum J, thus defining a subvariety,

$$\mathcal{P}_{E,J} \subset \mathcal{P}_0,$$

invariant under the planar three-body flow.

Proposition 0.34 *Consider the projection $\mathcal{P}_0 = \mathcal{P}_0(2,3) \to \mathbb{S}^2$ of the centered planar three-body phase space to its shape sphere, namely the map that sends a phase point to the shape of its configuration. Under this projection, the energy–momentum level set $\mathcal{P}_{E,J}$, $E < 0$, projects onto the super-level set $\tilde{U} \geq c \subset \mathbb{S}^2$ where $c = |E|J^2$ is the Dziobek value given in Equation (0.31). If $E \geq 0$ then the level set projects onto all of $\mathbb{S}^2$.*

The proposition provides constraints on the possible shapes that might arise during the course of evolution. If $-EJ^2 := c \gg 1$ corresponding to the top panel of Figure 0.9 then the shape region of the proposition consists of three small disjoint discs, one centered at each of the three collisions. A small neighborhood of B_{23}, the collision of masses 2 and 3, is characterized by the two inequalities $r_{23} < \epsilon r_{13}, r_{23} < \epsilon r_{12}$ for ϵ small. It follows that for c in this range the dynamics always behave as that of a "tight binary:" two close masses, spinning fairly quickly about each other, with the third mass remaining relatively far away throughout the evolution. Which pair is tight cannot change with time since the discs are disconnected. (This provides for a kind of weak stability known as *Hill stability*. See the notes in Section 0.6.)

Definition 0.35 The Hill region for energy E and angular momentum J is the projection of $\mathcal{P}_{E,J}$ to configuration space $\mathbb{E}_0$. The Hill shape region is its projection to shape space. The normalized Hill shape region is its further projection to the shape sphere.

Thus, Proposition 0.34 describes the normalized Hill shape region in terms of $\tilde{U}$.

The middle panel of Figure 0.9 depicts a bifurcation occuring at the value of c for which $\tilde{U} = c$ passes through the three Euler points. The critical values of $\tilde{U}$ are the Euler points and the Lagrange points: the central configurations of the Tour (Chapter -1). These critical values are precisely those for which the topology of $\mathcal{P}_{E,J}$ changes, and are simultaneously the values of $c = -EJ^2$ for which $\mathcal{P}_{E,J}$ is not a smooth manifold. These critical values, generalized to higher N, are the concern of Question 1, the subject of Chapter 1.

0.4.6 Reduction for the General N-Body Problem

Here we derive the reduction representation (0.65) for the planar N-body shape space.

We start out with the full configuration space $\mathbb{E}(2, N) = \mathbb{C}^N$. Its quotient by the group $\mathbb{C}$ of translations is achieved by going to center-of-mass coordinates, and yields $\mathbb{E}_0(2, N) \cong \mathbb{C}^{N-1}$, with the origin representing total collision. The group of rotations and scalings forms the group $\mathbb{C}^*$ of nonzero complex scalars that acts on this $\mathbb{C}^{N-1}$ by scalar multiplication. One of the standard definitions of $\mathbb{CP}^{N-2}$ is that it is this quotient space $(\mathbb{C}^{N-1} \setminus 0)/\mathbb{C}^*$. Returning to the land of labeled N-gons, points of this projective space represent oriented *similarity classes* of labeled N-gons.

To understand the metric cone structure described by Equation (0.65), first quotient $\mathbb{C}^{N-1}$ by the scalings $q \mapsto \lambda q, \lambda > 0$ to identify $(\mathbb{C}^{N-1} \setminus 0)/(\text{scalings})$ with the sphere $\mathbb{S}^{2N-3} = \{r = 1\} \subset \mathbb{C}^{N-1}$. The remaining $\mathbb{S}^1 = \{e^{i\theta} : \theta \in \mathbb{R}\} \subset \mathbb{C}^*$ now acts on this sphere, so that the quotient $\mathbb{S}^{2n-3}/\mathbb{S}^1$ is another representation of the projective space. It is in this guise that we obtain the Fubini–Study metric on the projective space, as the quotient metric.

To get the metric cone structure for shape space, note that spherical coordinates yield $\mathbb{C}^{N-1} = Cone(\mathbb{S}^{2N-3})$ with distance from the cone point being $r = \|q\|$. Now scalings commute with rotations, and rotations act by isometries so that $Cone(\mathbb{S}^{2N-3})/\mathbb{S}^1 = Cone(\mathbb{S}^{2N-3}/\mathbb{S}^1)$, which is to say that, metrically speaking, shape space is the claimed cone over projective space.

Upon going to center-of-mass coordinates, the phase space $\mathcal{P}_0$ of the planar N-body problem becomes $T^*(\mathbb{C}^{N-1} \setminus \Delta)$. We then have the composed shape projection

$$\pi \colon \mathcal{P}_0 = T^*(\mathbb{C}^{N-1} \setminus \Delta) \to \mathbb{C}^{N-1} \setminus \Delta \to \mathbb{CP}^{N-2} \setminus \Delta.$$

Now, the N-body flow leaves invariant the functions of total energy, E, and angular momentum, J, so that $\mathcal{P}_0$ is foliated by invariant subvarieties of codimension two $\mathcal{P}_{E_0, J_0} = \{E = E_0, J = J_0\} \subset \mathcal{P}_0$. Consider their images under shape projection,

$$\Omega_{E,J} := \pi(\mathcal{P}_{E,J}) \subset \mathbb{CP}^{N-2}.$$

We ask the Morse-theory inspired question: How do the domains $\Omega_{E,J}$ depend on E and J? And how does this dependency affect dynamics on the $\mathcal{P}_{E,J}$?

Inspired by the usual notion of "Hill region" in classical mechanics, we will call $\Omega_{E,J}$ the *normalized Hill shape region* for the given choice of E and J. We will describe these regions in terms of the normalized potential $\tilde{U}$ defined earlier (Definition 0.18). We recall that $U = \frac{1}{r}\tilde{U}$ where $r^2 = \|q\|^2$. This function $\tilde{U}$ is invariant under rotations and scalings so can be viewed as a continuous function on normalized shape space:

$$\tilde{U} \colon \mathbb{CP}^{N-2} \to \mathbb{R} \cup \infty.$$

Here we added in ∞ to the range of the potential so that $\tilde{U}(\Delta) = \infty$. (Note that $\tilde{U}$ is analytic away from Δ.) Recall Equation (0.31), the scale-invariant Dziobek constant.

Proposition 0.36 *The normalized Hill shape regions coincide with superlevel sets of the normalized potential $\tilde{U}$ as follows. If $E J^2 < 0$, then*

$$\Omega_{E,J} = \left\{ s \in \mathbb{CP}^{N-2} : +\infty > \tilde{U}(s) \geq \sqrt{2|E|J^2} \right\}.$$

If $E J^2 \geq 0$ then $\Omega_{E,J} = \mathbb{CP}^{N-2} \setminus \Delta$.

Figure 0.10 depicts the normalized Hill shape region for the collinear four-body problem.

Remark 0.37   It can be useful to include Δ in the Hill region at times so that $\Omega_{E,J} \cup \Delta = \{s \in \mathbb{CP}^{N-2} : \tilde{U}(s) \geq \sqrt{2|E|J^2}\}$ for $E < 0$.

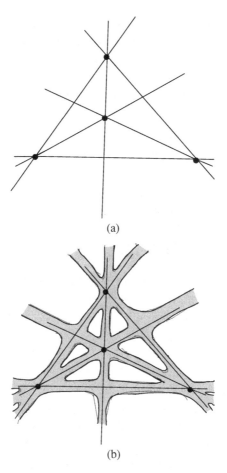

(a)

(b)

Figure 0.10 (a) The collision locus for the four-body problem. The plane represents $\mathbb{RP}^2$, the normalized shape space for the collinear four-body problem. The six lines are the six binary collision lines. The four marked points are the four triple collisions. (b) The Hill shape region is a tubular neighborhood of the collision locus.

Proof of Proposition 0.36 Recall that shape space is the quotient of configuration space by rigid motions and is coordinatized by $(r,s) \in (0,\infty) \times \mathbb{CP}^{N-2}$, with rs playing the role that $w \in \mathbb{R}^3$ played in describing three-body shape space. The energy decomposition (0.71) remains valid over shape space, and K_{shape}, the shape kinetic energy, is everywhere non-negative. Now freeze $J = J_0$ and set $V_{eff} = V_{eff}(r,s;J_0) = \frac{1}{2}\frac{J_0^2}{r^2} - U$ so that the energy decomposition (0.71) becomes $E = K_{shape} + V_{eff}(r,s;J_0)$. Fix $E = E_0 = const$. Then we must have that $V_{eff}(r,s;J) \leq E_0$ since $K_{shape} \geq 0$. Conversely, since the shape momenta vary over a vector space, we can always arrange that $K_{shape} =$

$E_0 - V_{eff}$ provided $E_0 - V_{eff} \geq 0$. This proves that $\Omega_{E_0, J_0} = \{ s \in \mathbb{CP}^{N-2} :$ $\exists r > 0 \quad V_{eff}(r, s, J_0) \leq E_0 \}$. Now freeze s as well as E_0, J_0 and look at the condition $V_{eff}(r, s, J_0) \leq E_0$ as a constraint on the positive radial variable r. Multiplying through by r^2 and rearranging yields $E_0 r^2 + r \tilde{U}(s) - \frac{1}{2} J_0^2 \geq$ 0. The left-hand side of this inequality is a quadratic polynomial in r whose discriminant is $\Delta = B^2 - 4AC = \tilde{U}(s)^2 + 2 E_0 J_0^2$. If $E_0 \geq 0$, the polynomial is eventually positive provided we take r sufficiently large, and hence every $s \in \mathbb{CP}^{N-2}$ lies in Ω_{E_0, J_0}. Conversely, if $E_0 < 0$ then this polynomial tends to negative infinity with increasing r and there may or may not be solutions to the inequality. By plotting the parabola that is the graph of $r \mapsto E_0 r^2 +$ $r \tilde{U}(s) - \frac{1}{2} J_0^2$, we see that there are positive solutions to the inequality if and only if the roots to this quadratic polynomial are real, which is to say if and only if $\Delta \geq 0$. The latter holds if and only if $\tilde{U}(s) \geq \sqrt{2 E_0 J_0^2}$, establishing the claimed characterization of Ω_{E_0, J_0}. QED

Corollary 0.38 *The normalized shape Hill regions change topology only at values $c = -E J^2$ that are critical values for $\tilde{U}$. The corresponding critical points are the only possible points at which $\{\tilde{U} = c\}$ might fail to be a smooth hypersurface.*

Remark 0.39 The corresponding *critical points* of $\tilde{U}$ are in bijection with central configurations which are in turn in bijection with relative equilibria. Our first open question asks whether the number of these critical points is finite.

Proof of Corollary 0.38 The proof of the corollary is application of the basic ideas of Morse theory. See [126]. Set $\Sigma_c := \{\tilde{U} \geq c\} \subset \mathbb{CP}^{N-2}$ and look at how these sets change as a function of c. We have $c_1 < c_2 \implies \Sigma_{c_1} \supset \Sigma_{c_2}$, $\Sigma_0 = \mathbb{CP}^{N-2}$, and $\Sigma_\infty = \Delta$. As we drop down from $c = +\infty$ to $c = 0$, the subsets are diffeomorphic as long as we do not encounter any critical values of $\tilde{U}$. In other words, if the interval $[a, b]$ is free of critical values, then all the sets $\Sigma_c, c \in [a, b]$ are diffeomorphic. So, every time the Dziobek parameter $c = -E J^2$ crosses a critical value of $\tilde{U}$, we expect a topological change in $\Omega_{E, J}$ and hence in $\mathcal{P}_{E, J}$. That the hypersurfaces are smooth when c is not critical is just the implicit function theorem. QED

0.4.7 The Shape Sphere for General d and N

Much of what we've said and derived in this chapter regarding reduction for the planar N-body problem carries over with little change for the N-body problem in $d > 2$ dimensions.

To start off with, recall that in a general metric space the sphere of radius r about a point p is the locus of points a distance r from p. Now take the metric space to be shape space, $Shape(d, N)$, and the point p to be the total collision in which all masses are located at the same point in $\mathbb{R}^d$.

Definition 0.40 The shape sphere for the N-body problem in d dimensions is the locus in shape space consisting of the sphere of radius one about total collision. This shape sphere will be denoted $ShapeS(d, N)$.

The shape sphere represents oriented similarity classes of labeled N-gons in $\mathbb{R}^d$. The shape sphere can be identified with the quotient space of the sphere $\{r = 1\}$ in $(\mathbb{E}_0(d, N) \cong \mathbb{R}^d)^{N-1}$ by the action of $SO(d)$. It has a metric inherited from the metric on the sphere in $\mathbb{E}_0$. Metrically speaking, we still have the relation

$$Shape(d, N) = Cone(ShapeS(d, N)).$$

When $d > 2$, the $SO(d)$ action on the sphere $\{r = 1\} \subset \mathbb{E}_0$ is not free and so the shape sphere fails to be a manifold, and this is true even at some non-collision points. (Specifically, the action fails to be free exactly at the configurations whose affine span in $\mathbb{R}^d$ has dimension less than $d - 1$.) Since $ShapeS(d, N)$ is not a manifold even away from collisions it becomes more difficult (and less useful) to make sense of the reduced phase space in this case. It can be done but so far without much practical use in solving problems. See for example [109].

0.5 Limitations

0.5.1 Finite Size, Harmonic Functions, and $1/r$

Planets and stars are balls of matter, not points. If the mass distribution of such a ball is spherically symmetric then we can replace the entire ball with a point mass placed at its center and whose value is the total mass of the ball. That single point mass will exert the same force on a test mass exterior to the ball as the total force exerted by all the infinitesimal masses comprising the ball. This fact is a wonderful theorem of Newton. See Newton [163, proposition LXX1, theorem XXX1], Chandrasekhar [24, chapter 15], or Feynman [64, volume 1, chapters 13–14]. This theorem follows from the fact that the fundamental solution to the Laplacian in $\mathbb{R}^3$ is a constant times $1/r$.

Moving from $\mathbb{R}^3$ to $\mathbb{R}^d$, the fundamental solution for the Laplacian on $\mathbb{R}^d$ is c/r^{d-2}. As a consequence, the same Newton theorem holds for spherically symmetric mass distributions in $\mathbb{R}^d$, provided we use as a pair potential $f_{ab}(r) = c/r^{d-2}$. For this reason people often object to studying power law potentials with arbitrary exponents α. They insist that you couple the dimension d with the exponent α of $f_{ab} = c/r^\alpha$ according to

$$\alpha = d - 2.$$

Exercise 0.41   Suppose we take the pair potential between two bodies in $\mathbb{R}^d$ to be of the form $m_a m_b f(r_{ab})$ so that the force on body a is $-m_a \nabla_a \Sigma m_b f(r_{ab})$. Replace the finite collection $m_1, m_2, \ldots, m_N$ of masses by a mass distribution $\rho(b)db$ so that the force of the mass distribution on a body x exterior to that mass distribution is, up to a constant, minus the gradient of $W(x) = \int_B f(|x - b|)\rho(b)db$. Suppose that the distribution is spherically symmetric so that $\rho(b)db = \rho(|b|)db$, with db being the Lebesgue measure in d-dimensions. Prove that $\nabla W(x) = c\nabla f(r)$ for some constant c if and only if $f(r) = c_2 r^{d-2}$. See [173] for a nice exploration of this exercise for power law potentials when $d = 3$.

If mass distributions for bodies are not spherically symmetric then the ODEs for their motions resulting from applying the principles of Newton's mechanics are not those of the standard N-body problem. For example, in planning satellite orbits one cannot simply use Kepler's equation with the center of mass of the Earth as the "Sun." The fact that the Earth is not perfectly spherical must be taken into account to accurately place and keep satellites in specified orbits. And the fact that the Earth and Moon are not perfectly rigid bodies is essential to understanding the tidal interactions coupling them and affecting the Earth's oceans.

0.5.2 Relativitistic Limitations

Special relativity tells us the speed of a massive object must be less than the speed c of light. But as our point masses approach each other, the forces between them increase so that as they tend to collision, their velocities tend to infinity, in clear violation of the dictates of special relativity. To be concrete, consider N equal masses with mass m engaged in a zero energy solution. From $H = K - U$ we get $K = U$. Now $K = \frac{m}{2}\Sigma|v_a|^2$ while $U = \Sigma Gm^2/r_{ab}^2$. If v_{max} is the maximum velocity of the bodies and r_{min} is the minimum of the interbody distances r_{ab}, then $Nmv_{max}^2 \geq 2K$ while $U \geq Gm^2/r_{min}^2$ so that $2Nmv_{max}^2 \geq Gm/r_{min}^2$, or $v_{max}^2 \geq Gm/2Nr_{min}^2$. The relativistic constraint

$c^2 \geq v_{max}^2$ yields $r_{min} \geq \sqrt{G/2N}(m/c)$. Imposing special relativity limits the approach to collision in gravity.

This discussion ignores the relativistic coupling of mass with velocity as measured in a fixed inertial frame. As the velocity of a massive particle increases so does mass, increasing without bound as the speed of light is approached. We will not even bother trying to figure out what to do with that massive can of worms in relation to any N-body-type equations.

A more serious philosophical objection to Newtonian gravity raised by special relativity is that forces cannot be transmitted instantaneously. This instantaneous transmission is the property of "action at a distance." In the standard N-body problem (Definition 0.5), a tiny change in the position of one body instantaneously affects the forces on all bodies, and hence their motions, even those light years away. It seems to be impossible to write down N-body-type ODEs that are invariant under the Lorentz transformations of special relativity. One can imagine trying to pose N-body problems using differential delay equations, rather than ODEs, staggering the time at which forces are applied according to the speed of light. This has not been done in a useful and consistent way. (See, however, the chapter notes in Section 0.6.) All attempts to consistently formulate versions of the N-body problem invariant under the transformations of special relativity rather than of Galilean relativity force us from the land of ODEs into the land of PDEs. Dust off your Sobolev estimates.

The need to find some marriage between Newton's N-body equations and special relativity led to general relativity. We refer the reader to the beauiful short book by Einstein [52]. The general relativistic two-body problem is that of two black holes (or asymptotically Schwarzschild solutions) moving under each other's gravitational fields and cannot be solved in closed form. Numerical solutions to this two-body problem proved crucial in interpreting the LIGO observations[7] of gravitational waves from a pair of black holes merging.

0.6 Chapter Notes

On Section 0.1

Chenciner [29] inspires my notation and set-up.

Topologists often call the collision locus Δ the fat diagonal.

---

[7] LIGO, or the Laser Interferometer Gravitational-Wave Observatory, is a large-scale physics experiment and observatory that has detected gravitational waves resulting from black hole mergers.

For general pair potentials $f_{ab}(r)$, people usually require that $f'_{ab}(r) \to 0$ as $r \to \infty$ and $|f'_{ab}(r)| \to \infty$ as $r \to 0$. The first condition asserts that forces decay as particles diverge from each other, and suggests that, eventually, solutions tend to move along lines. The second condition tells us that the forces blow up at collision.

We can make sense of the limiting case $\alpha = 0$ of the power law N-body problems by setting

$$f_{ab}(r) = -m_a m_a \ln(r); \alpha = 0.$$

This is justified by noting that

$$\lim_{\alpha \to 0} \frac{1}{\alpha}(r^{-\alpha} - 1) = -\ln(r),$$

as is seen by differentiating the function $x \mapsto r^{-x}$ at $x = 0$. It is also justified by noting that $d = 2$ corresponds to $\alpha = 2$ under the dimension-exponent relation $\alpha = d - 2$. The fundamental solution for the Laplacian on the plane is a multiple of $\ln(r)$.

How incomplete is the N-body flow? When a solution fails to exist for all time, then either it suffers a collision or it suffers one of the more mysterious non-collision singularities described in the Tour in Chapter -1, at the end of Section -1.4. Saari [180, 181] proved that the set of inital conditions yielding collision solutions has measure zero. This is true for any N and any $d > 1$ and an open interval of α-power laws including standard gravity ($\alpha = 1$). When $N = 3$ all singularities are collision singularities, a result that goes back at least to Sundman [202]. When $N = 4$, Saari [182] showed that the non-collision singularities form a set of measure zero. (Much more recently, this set was shown to be non-empty. See again the end of Section -1.4 of the Tour.) Whether the non-collision singularity set continues to have measure zero when $N > 4$ is an open question. (Experts weigh on the side of "yes.") See Knauf [97].

We can go quite far in formulating the N-body problem on a Riemannian manifold X in place of $\mathbb{R}^d$. To make sense of Newton's equations (Equation (0.4)) on X take $q_a(t) \in X$. Interpret the acceleration $\ddot{q}_a$ of Equation (0.4) to mean the covariant derivative $\nabla_{\dot{q}_a} \dot{q}_a$ along the curve $q_a(t)$.

We will need to choose forces F_{ba}. We could continue to use pair potentials $f_{ab} = f_{ab}(q_a, q_a)$ to define these forces. See Equation (0.12). What pair potentials should we take? One choice could simply be $f_{ab} = m_a m_b / r_{ab}$, where $r_{ab} = d_X(q_a, q_b)$ is the X-distance between q_a and q_b. Or we could take the more general $f(r_{ab})$ where $f: \mathbb{R}^+ \to \mathbb{R}$ is any pre-chosen function.

If we want the closest analogue to the standard N-body problem on a Riemannian manifold X, then we would formulate forces arising as ∇f where f solves the Poisson equation $\Delta f = \rho$ on X with ρ consisting of delta function sources. Here Δ is the Laplace–Beltrami operator for X. Thus we would take a pair potential $f_{ab}(q_a, q_b) = m_a m_b K(q_a, q_b)$ where $K(x, y)$ is Green's function for the Laplacian on X. Green's function K is the solution to $\Delta_x K(x, y) = \delta_y(x)$, where δ_y is the Dirac delta distribution at $y \in X$ and where Δ_x denotes the Laplace–Beltrami operator of (X, ds_X^2) acting on the x-variable.

This Green's function-based choice of pair potential is problematic when X is compact. No Green's function $K(x, y)$ exists! This is because there is no solution f to $\Delta f = \rho$ unless the total mass $\int_X \rho = 0$. This difficulty can be circumvented either by replacing the mass density $\rho = \delta_y$ by $\delta_y - c1$ where $c = 1/vol(X)$. See Aubin [13, chapter 4], particularly theorem 4.13, for definitions and some information on Green's functions on manifolds. In case X is the round sphere, this choice is equivalent to insisting that masses come in "antimatter pairs" with a repulsive antimass at the antipodal point $-q_a$ to each attractive mass location q_a, thus guaranteeing total mass $\Sigma m_a = 0$. For example, for the unit three-sphere one finds that $f_{ab} = m_a m_b \cot(r_{ab})$, which has poles of type $1/r$ or $1/|\pi - r|$ when $r = 0$ or at the antipodal point $r = \pi$, one representing an attracive force, and the antipodal a repulsive force. See Diacu et al. [43, section 3.3] for the analytic expressions.

Simply connected constant curvature universes X, namely hyperbolic space, the sphere, and Euclidean space, have Green's functions of the form $f_{ab}(q_a, q_b) = m_a m_b f(r_{ab})$. One might argue that these spaces, and only these spaces, are the proper arena for the N-body problem. Lobachewsky phrased the two-body problem on hyperbolic space (see [42] and references therein). The N-body problem (mostly the Kepler problem) has been studied to some extent on these spaces. See [43], particularly Section 0.1, for history around constant curvature universe N-body problems.

When we replace $\mathbb{R}^d$ by one of these non-Euclidean space forms its Galilean group shrinks. The boosts disappear! They disappear because $Iso(X)$ has no distinguished subgroup of translations, and so there seems to be no way to "go into a moving reference frame." The symmetry group of the N-body problem on such X's is $Iso(X) \times Iso(\mathbb{R})$, with no boosts. Although $dim(Iso(X)) = dim(Iso(\mathbb{R}^d))$ for these X's, the dimension of their Galilean group – the group of symmetries of the N-body problem – has dropped by $d = dim(X)$ from that of Euclidean space.

Because of this drop in symmetries, the two-body problem on the non-Euclidean space forms appears to be non-integrable. For this result for the sphere, see Borisov et al. [21]. It seems certain that the problem is also not integrable on hyperbolic space.

On Section 0.2

For detailed introductions to the two-body problem and Kepler's problem see Pollard [171], Singer [196], Albouy [3], or Landau and Lifshitz [101]. For the general central force problem see Landau and Lifshitz [101].

The Kepler problem on the sphere or hyperbolic space is also completely solvable. It arises as the limiting case of the two-body problem on these spaces when we let one of the two masses go to zero. Alternatively, place a massive immovable "sun" at some fixed point of X and have it attract "test masses." The solutions to these curved Kepler problems obey versions of Kepler's 1st and 2nd laws. In particular, if one adopts a synthetic definition of conic on these spaces, the solutions travel conics with the Sun as one focus. See [43] for some details.

On Section 0.3

Albouy advocates a way to quotient out by translations that does not involve the center of mass. He has christened his alternative method the method of *dispositions*. See [3, 8, 138]. Albouy's method is based on framing all objects as much as possible in terms of intrinsic linear algebra and has proved useful in attacking the open question of Chapter 1.

On Section 0.4

In 1970, Smale [197, theorem A] identified the planar N-body reduced phase space as we have done in Equation (0.66). In 1983, Moeckel (see [130]) drew clear pictures of the shape sphere that he used to explain and organize results. In 1987, Iwai, in a series of papers [86, 87, 88], stated explicit theorems identifying shape space in the context of both the classical and quantum three-body problem. In 1970, Easton [51] used a version of the *unoriented* shape sphere instead of our oriented shape sphere. (By "unoriented shape sphere" I mean configurations modulo similarities rather than our configurations modulo oriented similarities.) Easton used the squared side lengths r_{ab}^2 as coordinates for his unoriented shape space and identified it with one octant $(x, y, z \geq 0)$ of the standard two-sphere.

The general shape sphere as defined here appeared in statistics in the works of Kendall [93] and his students. Kendall writes "My interest in shape theory was prompted by a statistical topic on the fringes of archeology. When one looks at Stonehenge ..." Imagine comparing data from different archeological digs.

Regarding quotients by compact Lie groups, if G is a compact group acting smoothly on a manifold Q and if G acts freely on Q, then Q/G is also a manifold. To say that G acts freely means that $gs = s \implies g = e$, where e is the identity. The trick of replacing the full isometry group by its subgroup, the group of orientation-preserving isometries, ensures that the group acts freely away from total collision. More generally, whether or not Q/G is a manifold depends on whether or not the isotropy type of points of Q is constant. The isotropy subgroup of $q \in Q$ is the subgroup of G that does not move q under the action. Two points have the same isotropy type if their isotropy groups are conjugate. The more "generic" the point the smaller its isotropy type. Free actions are ones for which the isotropy type of all points is equal to the trivial subgroup $\{e\}$. See, for example, Bierstone [17] for more complete and involved information on quotients of manifolds by the actions of compact Lie groups.

The Morse-theoretic outlook on the bifurcations of the Hill shape region as the Dziobek parameter is varied comes from Moeckel's wonderful paper [132]. See Marchal [116, pp. 323–340] for a discussion of Hill stability.

On Section 0.5

Section 0.5.1
Instead of point masses, we can consider a continuous mass distribution ρ on $\mathbb{R}^3$. The resulting force on a point mass m placed at position q is $\mathbf{F}(q) = mG\nabla f$, where f solves the Poisson equation $\Delta f = \rho$ with appropriate decay conditions at infinity. The force due to another point mass m' and q' corresponds to the case where ρ is m' times the delta function at q'.

Section 0.5.2
In trying to overcome the inconsistencies between Newtonian mechanics and special relativity, Feynman, in his thesis [62], set up a quantum mechanics for Newton's equations with a fixed differential delay, which we might think of as the delay due to the finite speed of light. See also Feynman and Wheeler [63], who argue that one must apply a "Jacob's ladder" of retarded and advanced potentials – forward and backwards delays as it were – in order to formulate Maxwell's equations as an N-body variational principle.

PART TWO

The Questions

1

Are the Central Configurations Finite?

Open Question 1 Is the set of central configurations for the planar N-body problem, viewed modulo symmetries, a finite set?

1.1 Who Cares?

When I teach dynamics I give students explicit vector fields to analyze for homework. I teach them to start by finding the equilibria, the zeros, of their vector field. Once found, a linear analysis comes to the fore. A whole involved and well-known game unfolds.

Newton gave us homework in 1687 in the form of Equation (0.4). But his N-body system has no equilibria: N stars cannot just sit in space, unmoving. They attract each other! It seems that we cannot start the game.

The next best thing to an equilibrium is a relative equilibrium: an equilibrium modulo symmetries. Upon forming the quotient of phase space by the symmetry group of Galilean motions (Section 0.4), the relative equilibria project to equilibria for the quotient dynamics. We can use these quotient equilibria as replacements for equilibria, and can begin Newton's homework assignment on this quotient – the reduced phase space.

Definition 1.1 A relative equilibrium for the N-body problem in d-dimensional Euclidean space is a solution to the problem having the form

$$q(t) = g(t)q_*,$$

where $q_* \in \hat{\mathbb{E}} = (\mathbb{R}^d)^N \backslash \Delta$ is a fixed collision-free configuration and where $g(t) \in SO(d)$ is a one-parameter group of rotations of $\mathbb{R}^d$. A *relative equilibrium configuration* q_* – also called a *balanced configuration* – is a configuration through which passes a relative equilibrium.

75

For the planar N-body problem the central configurations are the same as the relative equilibrium configurations (see Proposition 1.3). All of the relative equilibria for the planar three-body problem were found by Euler and Lagrange. See Section $-1.2.1$. Each relative equilibrium configuration generates an entire family of solutions parameterized by solutions to Kepler's problem. Within this family one solution ends in total collision. These total collision solutions are those associated to central configurations.

Definition 1.2   A central configuration for the N-body problem in d-dimensional space is a configuration q for which the solution with initial conditions $(q, v) = (q, 0)$ – that is, all velocities zero – evolves by shrinking to total collision.

If a configuration q_* has center of mass zero, then it is a central configuration if and only if the solution $q(t)$ to the N-body equations having initial conditions $(q, v) = (q_*, 0)$ is of the form

$$q(t) = r(t)q_*, r(t) \in \mathbb{R}. \qquad (1.1)$$

We sometimes refer to such solutions as *dropped central solutions*.

The symmetries of Newton's equations imply that if q_* is a central configuration then so is any rotation or scaling of q_*. Thus central configurations come in continua – the orbits of any given central configuration under rotation and translation – and cannot form a finite set. When counting central configurations we must count each such orbit as a point if we are to have a chance of getting a finite number of them. We can now state our open question.

Completed Open Question Is the projection of the set of planar N-body central configurations to the planar N-body shape sphere $\mathbb{CP}^{N-2}$ a *finite* set?

The open question continues to be open, of course, for the N-body problem in d-space, $d > 2$, with the same restatement, now in terms of points of the shape sphere for the N-body problem in d-space as defined in Section 0.4.7.

1.1.1 Relations Between Central Configurations and Relative Equilibria

We repeat:

Proposition 1.3 *For the planar N-body problem the set of central configurations coincides with the set of relative equilibrium configurations.*

For the proof of Proposition 1.3, see Section 1.3.1.

For the N-body problem in d dimensions, $d > 2$, the relation between central configurations and relative equilibrium configurations is as given in Proposition 1.4.

Proposition 1.4 *Write RE to stand for the set of relative equilibrium configurations, and CC for the set of central configurations for the N-body problem in d-space.*
(A) If d is even then $CC \subset RE$.
(B) If d is odd then $CC \backslash RE \neq \emptyset$.
(C) If $d \geq 4$ then $RE \backslash CC \neq \emptyset$.

For the proof of (A) see Section 1.6.2. For the proof of (B) see Proposition 1.13, where we show that the regular simplex in any dimension is a central configuration. In odd dimensions the simplex cannot be rotated so as to yield a relative equilibrium. For example, the regular tetrahedron cannot be a relative equilibrium for the four-body problem in three-space since any one-parameter subgroup of rotations has an axis of rotation and hence is the rotation of some plane. The entire simplex does not sit in the plane, so one of the bodies is left doing circles above or below the plane, which contradicts the laws of gravity: the result cannot solve the four-body problem. For (C) see Section 1.6.

Remarks on Terminology
1. In Chapter 0 we gave another definition of relative equilibrium: a solution under which the mutual distances remain constant. See also Proposition 0.23. It is easy to show that these two definitions of relative equilibrium are equivalent.
2. Albouy and Chenciner [8] wrote a seminal work on expressing the N-body problem solely in terms of mutual distances. There they introduced the phrase "balanced configuration" as a synonym for "relative equilibrium configuration," part of the idea behind the word "balance" being that gravitation and centrifugal forces must "balance" in order to be able to spin such a configuration and get it to be a solution.

1.1.2 More Reasons to Care
Here are more reasons to care about central configurations.

(1) Central configurations correspond to the *actual equilibria* of a partial compactification of the N-body problem known as the McGehee blow-up. See Proposition 1.18.

(2) Central configurations are the only roads to total collision. See Proposition 1.19.

(3) Relative equilibria coincide with the critical points of the energy–momentum map. As such, they indicate bifurcations in the topology of the energy–momentum level sets. See Corollary 0.35 and Sections 1.1.3 and 1.3.

(4) Central configurations and relative equilibria generate the only solutions to the N-body problem for which we have explicit formulae. See Lemma 1.10 in Section 1.3.

1.1.3 Central Configurations as Critical Points

Recall the normalized potential

$$\tilde{U} = \sqrt{I}U : \hat{\mathbb{E}}_0 \to \mathbb{R}, \tag{1.2}$$

as stated in Definition 0.18 and Equation (0.48). Here U is, as usual, the negative of the total N-body potential, and $I(q) = \|q\|^2$ is the moment of inertia. *The critical points of $\tilde{U}$ are precisely the centered (meaning center-of-mass zero) central configurations* (see exercise 1.9); $\tilde{U}$ is invariant under scaling and rotations, and so descends to a function on the shape sphere as defined in Section 0.4.7, which we will denote by the same name, $\tilde{U}$. See Figures 0.4 and 0.9 for depictions of contours of $\tilde{U}$ when $N = 3$. The open question asks: *Does $\tilde{U}$, viewed as a function on the shape sphere, have a finite number of critical points?*

1.2 What's Known?

We saw the central configurations for the planar three-body problem in the tour of Chapter -1, namely the three collinear configurations discovered by Euler and the two equilateral configurations of Lagrange. These make up the row $N = 3$ in Table 1.1, from Moczurad and Zgliczyński [128]. Albouy [4] established the count listed for $N = 4$. As unlabeled shapes, the configurations he established are depicted in Figure -1.20. Table 1.1 lists the number of similarity classes of central configurations for the equal mass case up to $N = 7$ as a function of N. The rightmost column of the table lists the number of labeled oriented similarity classes of central configurations in the equal mass planar N-body problem as a function of the number, N, of bodies. Since the masses are equal, we can permute the mass labels of a central configuration and we will get another central configuration. The permutation group acts on the

Table 1.1. *Counting central configurations in the equal mass case. See [128].*
The middle column is the number of unlabeled distinct configurations. The
last column counts the number after labeling.

N	unlabeled #	labeled #
2	1	1
3	2	5
4	4	50
5	5	354
6	9	3,624
7	14	53,640

Table 1.2. *Established bounds for the number of planar central*
configurations.

N	#	authors	year
2	1	Newton; Kepler	1667
3	5	Euler; Lagrange	1767; 1772
4	$\leq 50^*$	Hampton–Moeckel; Simó	2006; 1978
5	generically finite	Albouy–Kaloshin	2012

set of equal-mass labeled oriented similarity classes of central configurations.
If we form the quotient of this set by the action of the permutation group,
we arrive at a new, smaller set whose cardinality is indicated by the middle
column, "unlabeled #," listing the count of distinct shapes of N-gons in the
plane, ignoring the mass labels at the vertices.

Table 1.2 summarizes known finiteness results for the planar N-body
problem.

$N = 4$ The asterisk by the result "50" for $N = 4$ in Table 1.2 is there
to indicate that the number 50 has not yet been rigorously established. Simó
[191], Moczurad and Zgliczyński [128], Moeckel [138], and others have done
careful numerics that indicate that 50 is the actual least upper bound, although
the rigorously established upper bound is 8,472 (see Hampton and Moeckel
[75]).

$N = 5$ In the last row of Table 1.2, "generically finite" means that the number
of central configurations mod symmetry is finite provided the vector of masses
$\vec{m} = (m_1, \ldots, m_5)$ comes from a generic set, specifically from a Zariski-dense
subset of the positive part of $\mathbb{R}^5$.

The following counterexample (Theorem 1.5) discovered by Gareth Roberts
[175] underlines the difficulty of the open question.

Theorem 1.5 *In the five-body problem if we allow one of the five masses to be negative then the answer to the open question is no. Specifically, if the mass distribution is* $(4, 4, 4, 4, -1)$ *then there exists a curve's worth of inequivalent central configurations within the planar five-body normalized shape space.*

Roberts' example is not a counterexample to the conjectured finiteness since for gravity we insist that all masses be positive. Understanding Roberts' example in a deep way was essential for Albouy and Kaloshin [7] in obtaining their $N = 5$ result.

1.2.1 A Few General Results

Definition 1.6 Write $\#(N, m, d)$ for the number of central configurations for the N-body problem with mass distribution m, and the bodies moving in d-space.

Moulton's Collinear Result

$$\#(N, m, 1) = \frac{1}{2} N!$$

Moulton [160] established this fact by showing that there is exactly one linear central configuration for each ordering of the N masses on the line. See Proposition 1.11 for a restatement and a proof. There are $N!$ such orderings, hence the count. If we view these collinear central configurations within the plane instead of the line then an extra symmetry acts: rotation of the line by 180 degrees within the plane, which acts by $O(1) = \mathbb{Z}_2$ on the line, flipping the ordering of the masses on the line. It is standard to count a linear central configuration and its flip as the same. We thus get $N!/2$ linear Moulton central configurations for any given mass distribution. When $N = 3$ these are the $3!/2 = 3$ central configurations found by Euler.

Xia's Open Finiteness Results

Xia [216] has established, for any N, the existence of an open set of mass distributions $m = (m_1, \ldots, m_N)$ such that

$$\exists\, m_1, \ldots, m_N : \#(N, m, 2) = (N - 2)! \left((N - 2)2^{N-1} + 1\right),$$

thus establishing finiteness for an open set of mass distributions. Xia's construction is iterative. He assumes that m_1 and m_2 are close to equal and then imposing conditions of the form $m_i \gg m_{i+1}$ on successively introduced masses. Think of a binary star around which circles a planet about which circles a moon about which circles an asteroid . . . .

Palmore's Morse Estimates

As described in Section 1.1.3 (see also Exercise 1.9) central configurations can be characterized as the critical points of the function $\tilde{U}$ on shape space (see Equation (1.2)). This function depends parametrically on the mass distribution $\boldsymbol{m} = (m_1, \ldots, m_N)$.

Proposition 1.7 (Palmore [168]; see Moeckel [138, proposition 25]) *Suppose that the normalized potential $\tilde{U}$ is a Morse function on the planar N-body shape sphere ($\mathbb{CP}^{N-2}$) for some mass distribution. Then, the answer to the open question is "yes" for that mass distribution. Moreover, the Morse lower bound*

$$\frac{3}{4}N! - \frac{1}{2}(N-1)! \leq \#(N, \boldsymbol{m}, 2)$$

holds for the number of central configurations.

For example, for $N = 10$ the Moulton number is 1,814,400, while the Morse lower bound of Palmore is 2,540,160.

Avoiding Collisions

Lemma 1.8 (Shub) *For each fixed positive mass distribution there is a neighborhood of the collision locus which is free of central configurations.*

For a proof of Shub's lemma, see [138, p.29, proposition 15] or the original work [190].

1.3 Central Configurations as Critical Points

In this section we show that central configurations are the critical points of the normalized potential. We then explore some consequences of that fact.

We start by viewing Equation (1.1) from Definition 1.2 as an ansatz for a solution to Newton's equations and plug the ansatz into the condensed form (Equation (0.38)) of the N-body equations. We find that

$$(\ddot{r})q_* = \frac{1}{r^2}\nabla U(q_*), \tag{1.3}$$

where we use that ∇U is homogeneous of degree -2. Take the *mass* inner product of both sides of this equation with q_* to arrive at

$$\ddot{r} = -\frac{\mu}{r^2}, \qquad (1.4)$$

where

$$\mu = \frac{U(q_*)}{I(q_*)}. \qquad (1.5)$$

In arriving at Equation (1.5) we've used $r^2 = I(q) = \langle q, q \rangle$ and $-U(q) = \langle q, \nabla U(q) \rangle$, the latter following from Euler's homogeneous function theorem applied to U. Equation (1.4) is the one-dimensional Kepler problem (see Section 0.2). If $r(t)$ solves Equation (1.4), then $\ddot{r} r^2 = -\mu = const.$, so Equation (1.3) yields

$$\nabla U(q) = -\mu q; \quad \text{for} \quad q = q_*. \qquad (1.6)$$

We will call this equation the *central configuration equation*.

These steps are reversible. Suppose that $q = q_*$ solves the central configuration equation (1.6). Form the one-dimensional Kepler problem (1.4) with Kepler constant μ given by Equation (1.5) and take the solution $r(t)$ having inital conditions $r(0) = 1, \dot{r}(0) = 0$ to arrive at a dropped central solution as per Definition 1.2. Thus central configurations are the solutions to the central configuration Equation (1.6).

We can rewrite the central configuration, Equation (1.6), as the Lagrange multiplier equation

$$\nabla U(q) = -\lambda \nabla I(q), \qquad \lambda = \mu/2, \qquad (1.7)$$

since $\nabla I = 2q$. Equation (1.7) says that q is a critical point of the function U restricted to the sphere $I = I(q_*)$.

We have just shown that central configurations are critical points of the function U upon restricting it to a sphere $I = const.$ Here are two additional characterizations of central configurations as critical points. Characterization A is the original characterization in terms of $\tilde{U}$.

Exercise 1.9   (A) Show that the central configurations are the critical points of the normalized potential $\tilde{U} = \sqrt{I} U$. (See also Definition 0.18.)

(B) Show that critical points of the function $U + I$ are central configurations and that any central configuration can be scaled so as to be a critical point of $U + I$.

(C) Modify the derivation so that it works for the power law potentials. Use $\tilde{U} = I^{\alpha/2} U$ where α is the exponent of the power law.

1.3.1 The Planar Case: Proof of Proposition 1.3

By making slight variations in the derivation just given of Equations (1.7) and (1.6) we will prove Proposition 1.3, which says that central configurations relative equilibrium configurations are the same thing in the planar N-body problem.

Identify $\mathbb{R}^2$ with $\mathbb{C}$ in the usual way, so that (x, y) is sent to $(x + iy)$ and rotation by θ becomes multiplication by $\lambda = e^{i\theta}$. Then $\mathbb{E} \cong \mathbb{C}^N$. Replace the real scalar $\lambda(t)$ in our ansatz with the complex scalar $\lambda(t) \in \mathbb{C}$. Observe that $\nabla U(\lambda q) = \frac{\bar{\lambda}}{|\lambda|^3} \nabla U(q)$ for $\lambda \in \mathbb{C}$. Now go through precisely the same manipulations as we went through above to see that our modified ansatz with $\lambda(t) \in \mathbb{C}$ satisfies Newton's equations if and only if Equation (1.6) is satisfied together with the two-dimensional Kepler problem $\ddot{\lambda} = -\mu \frac{\lambda}{|\lambda|^3}$. The Kepler constant μ is still given by Equation (1.5). Since our planar Kepler problem admits the circular solution $\lambda(t) = e^{i\omega t}$ with frequency ω satisfying $\omega^2 = \mu$, we have proved the proposition. QED

A moment's reflection shows we have proven more.

Lemma 1.10 *In the planar N-body problem a complex one-dimensional family of solutions called a homographic family and having the form $q(t) = \lambda(t)q_*$ passes through each central configuration $q_* \in \mathbb{C}^N$. Here, $\lambda(t)$ is any solution to the planar Kepler problem $\ddot{\lambda} = -\mu\frac{\lambda}{|\lambda|^3}$. The complex parameter can be taken to be the initial Keplerian velocity $\dot{\lambda}(0)$ at time 0 of the initial conditons $(\lambda(0), \dot{\lambda}(0))$ to our Kepler problem. Take $\lambda(0) = 1$ to ensure the family passes through q_* at time $t = 0$.*

The homographic family comprises the exact solutions a central configuration gives rise to (see the 4th "reason to care" in Section 1.1.2.). If we label the central configurations by their shapes $[s] \in \mathbb{CP}^{N-2}$ this family corresponds to a three-parameter family of solution curves (modulo time translation) to the planar N-body problem for each central configuration shape$[s]$. As parameters we can use the energy, angular momentum, and orientation of the solution in the plane.

1.3.2 Implications of Criticality

We describe three consequences of the fact that central configurations are critical points of a function on normalized shape space.

Collinear Central Configurations

Proposition 1.11 (Moulton) *Considered modulo scaling and translation, the collinear N-body problem has exactly $N!$ central configurations, one for*

each choice of orderings of placements of the masses on the line. When viewed as planar central configurations, these yield $N!/2$ collinear central configurations, modulo rotations, translations, and scalings.

Proof of Proposition 1.11 (On the Moulton configurations) We make use of the function $U + I$ of part (B) of Exercise 1.9. The configuration space of the collinear N-body problem is $\mathbb{R}^N$; $\mathbb{R}^N \setminus \Delta$ consists of $N!$ components, corresponding to the $N!$ possible orderings of N points on the line. Each component is convex. The restriction of $U + I$ to any component is easily checked to be convex. (See [138, p. 33–35] for the convexity computation.) $U + I$ blows up on Δ it blows up on the boundary of any one component. A convex function on a convex domain that blows up on the domain's boundary has a unique minimum in the interior of the domain. Rotation by 180 degrees in the plane acts by $O(2)$ on the line where it reverses the ordering of the masses on the line, resulting in the division by 2 when we go from the line to the plane, moving from the count of $N!$ to the count of $N!/2$. QED

Rank and Regular Simplices

In order to state the next consequence we will use the definition of the *rank* of a configuration.

Definition 1.12 The rank of a configuration $q = (q_1, \ldots, q_N) \in \mathbb{E}(d, N)$ is the dimension of the affine span $\mathbb{A} = \mathbb{A}(q) \subset \mathbb{R}^d$ of its vertices $q_1, \ldots, q_N \in \mathbb{R}^d$.

Recall that $\mathbb{A}(q)$ is the smallest affine subspace of $\mathbb{R}^d$ containing the N points q_a and can be characterized as the collection of all points P expressible in the form $P = \Sigma t_a q_a$ as $(t_1, \ldots, t_N)$ varies over $\mathbb{R}^N$. We can find an isometry of $\mathbb{R}^N$ whose fixed point set is $\mathbb{A}$. It follows that any solution to the N-body problem with initial condition $(q, 0)$ lies in $\mathbb{A}$. The rank of q is an integer that is less than or equal to both d and $N - 1$.

Proposition 1.13 (simplex) *The only central configuration for the N-body problem that has rank $N - 1$ is the regular $N - 1$ simplex, the unique-up-to-similarities configuration for which all side lengths r_{ab} are equal. The regular simplex is a central configuration regardless of the mass distribution $m_a, a = 1, \ldots, N$, and the exponent α of the power law, for any power law N-body problem, including the gravitational one.*

Proof of Proposition 1.13 (On the Regular simplex) We give the proof for the general power law case of $U = \Sigma m_a m_b / r_{ab}^{\alpha}$.

We may suppose that q is centered and that $d \leq N - 1$, since the space in which the dropped solution $q(t)$ with initial condition $(q, 0)$ evolves is the affine space $\mathbb{A}$ whose dimension is $rank(q) \leq min(d, N - 1)$. That is, we may assume that $\mathbb{R}^d = \mathbb{A}$. If the configuration q has rank $N - 1$ we have that $d = N - 1$ and no element of the linear isometry group $O(d)$ leaves q fixed. It follows that the rotational orbit $SO(d)q$ of q admits a neighborhood U on which $O(d)$ and, consequently, $SO(d)$ acts freely. We may take U to be an $SO(d)$-invariant neighborhood. Then $U/SO(d) \subset \mathbb{E}/SO(d)$ has dimension $\binom{N}{2}$, has the structure of a manifold, and the $\binom{N}{2}$ side lengths r_{ab} form a set of coordinates for $U/SO(d)$. We have $U = G\Sigma m_a m_b / r_{ab}^\alpha$ while $I = \frac{1}{M}\Sigma m_a m_b (r_{ab})^2$. Writing out the Lagrange multiplier equation $dU = -\frac{\mu}{2}dI$ in these variables we get

$$-\alpha\Sigma m_a m_b r_{ab}^{-\alpha-1} dr_{ab} = \frac{-\mu}{2}\frac{2}{M}\Sigma m_a m_b r_{ab} dr_{ab},$$

or $\Sigma m_a m_b \left(\alpha r_{ab}^{-(\alpha+1)} - \frac{\mu}{M}r_{ab}\right) dr_{ab} = 0$. Since the dr_{ab} are a basis of one-forms on $U/SO(d)$ we can cancel the common factor of $m_a m_b$ and are left with the $\binom{N}{2}$ equations $\alpha r_{ab}^{-(\alpha+1)} - \frac{\mu}{M}r_{ab} = 0$. This leaves us with $\frac{M\alpha}{\mu} = r_{ab}^\alpha$ so that all the r_{ab} must have the same length $(\frac{M\alpha}{\mu})^{1/\alpha}$. QED

Morse Theory and Lower Bounds

Recall that a critical point of a smooth function is called nondegenerate if its Hessian – the matrix of its second partial derivatives – is nondegenerate. A Morse function is a smooth function all of whose critical points are nondegenerate. Morse theory (see [126]) relates the topology of a manifold to the number and nature of critical points of a Morse function on the manifold. Using Morse theory, Palmore proved Proposition 1.14.

Proposition 1.14 *If, for a given choice* $\boldsymbol{m} = (m_1, \ldots, m_N)$ *of masses,* $\tilde{U}$ *is a Morse function on the normalized shape space for N-bodies in d-space, then the answer to the open question of this chapter is "yes" for these masses:* $\#(N, \boldsymbol{m}, d) < \infty$. *Moreover, in the case* $d = 2$ *of the planar N-body problem, we have Palmore's lower bound*

$$\#(N, \boldsymbol{m}, 2) \geq \frac{3}{2}N! - 2(N - 1)!.$$

Sketch of proof of Proposition 1.14 (On Morse and Palmore's estimate)

By the Morse lemma [126], the critical points of a Morse function are discrete. Hence, if $\tilde{U}$ is Morse, then its critical points are finite number *provided* they do not accumulate on the collision locus. (With the collison locus included the normalized shape space is compact.) That they do not so accumulate is due to the content of Shub's lemma mentioned above.

To prove Palmore's estimate, one combines the Morse inequalites with Moulton's theorem. The Morse inequalities are inequalities between weighted sums of critical points of a Morse function and sums of Betti numbers of the underlying manifold. Moulton asserts that $\tilde{U}$ has exactly $N!/2$ critical points within the collinear configurations $\mathbb{RP}^{N-2} \setminus \Delta \subset \mathbb{CP}^{N-2} \setminus \Delta$. These are all absolute minima for the restriction of $\tilde{U}$ to the collinear space while orthogonal to this space the Hessian is negative definite so that the Hessian of $\tilde{U}$ has index $N-2$, with $N-2$ positive, and negative signs for its diagonalized Hessian. One puts this data into the Morse inequalities to achieve the Palmore result. (In order to apply Morse theory we need Shub's lemma asserting that the central configurations do not accumulate on the collision locus.) For details see Moeckel [138, p. 45, proof of proposition 25], or the original reference [168].

Remark 1.15 A generic smooth function is Morse. We expect $\tilde{U}$ to be Morse for most mass distributions. Being Morse is an open condition. Xia [216], as described in Section 1.2.1, established that the set of masses for which $\tilde{U}$ is Morse is non-empty. One expects the subset of the mass simplex $\{m_i > 0, \Sigma m_1 = 1\}$ for which $\tilde{U}$ fails to be Morse to be an algebraic subvariety of this simplex. Establishing the algebraicity of this set would establish generic finiteness.

1.3.3 Relative Equilibria as Critical Points

The energy–momentum map for the planar N-body problem is the map $(E, J): \mathcal{P}_0 \to \mathbb{R}^2$ whose components are the energy H and angular momentum J.

Lemma 1.16 *A point in the planar N-body phase space is a critical point for the energy–momentum map if and only if it is a relative equilibrium.*

Proof Let ω be the symplectic form so that the Hamiltonian vector field is $X_H = \omega^{-1} dH$. By definition of momentum map, $\frac{\partial}{\partial \theta} = \omega^{-1} dJ$ where $\frac{\partial}{\partial \theta}$ is the infinitesimal generator of rotation. (See Exercise A.21.) For the sake of clarity we have used H for E, the energy viewed on the cotangent side. We then have, by linearity $dE(z_0) = \lambda dJ(z_0) \iff X_H(z_0) = \lambda \frac{\partial}{\partial \theta}|_{z_0}$, which in turn says that the time t dynamical evolution of the state z_0 is the same as rotation of the configuration by $exp(i\lambda t)$. QED

1.4 As Rest Points: McGehee Blow-Up

We go into detail regarding the reasons listed in Section 1.1.2 for studying central configurations. (We discussed the fourth reason in Lemma 1.10.)

1.4.1 Equilibria Out of Blow-Up

Write

$$r = \|q\| := \sqrt{I(q)},$$

so that $r = 0$ corresponds to total collision, and set

$$q = rs; \langle s, s \rangle = 1.$$

We call s the normalization of q and (r, s) "spherical coordinates" on configuration space. The McGehee blow-up is the change of variables

$$q = rs, \tag{1.8}$$

$$v = r^{-1/2}y, \tag{1.9}$$

$$dt = r^{3/2}d\tau, \tag{1.10}$$

that defines an invertible *transformation* $(q, v; t) \mapsto (r, s, y; \tau)$ of Galilean space–time provided $r > 0$. It changes both the dependent $((q, v))$ and independent (t) variables of Newton's equations. We assume that $q \in \mathbb{E}_0$ is centered and $v \in \mathbb{E}_0$ so the linear momentum is zero. Then the variable s lies on the unit sphere

$$s \in \mathbb{S} = S^{d(N-1)-1} = \{r = 1\} \subset \mathbb{E}_0.$$

Exercise 1.17 is a straightforward computation.

Exercise 1.17 Write $'$ for $\frac{d}{d\tau} = r^{3/2}\frac{d}{dt}$. Show that McGehee's transformation transforms Newton's equations (0.38) to the equations

$$r' = rv,$$
$$s' = y - vs, \tag{1.11}$$
$$y' = \nabla U(s) + \frac{1}{2}vy,$$

where $v = \langle s, y \rangle$. These equations are the McGehee blown-up equations. In the blown up variables, the energy is given by $H = r^{-1}(K(y) - U(s))$, and the angular momentum by $r^{1/2}J(s, y)$.

McGehee blow-up creates equilibria for Newton's equations on the previously unapproachable collision manifold $r = 0$.

Proposition 1.18 *When rewritten in McGehee variables the vector field defining the N-body equations extends to define an analytic vector field (1.11) valid on the collision manifold $r = 0$. Equilibria appear on the collision manifold and are in 2:1 correspondence with normalized central configurations. Each dropped central solution (see Equation (1.1) and the sentence following it) is associated to a pair of equilibria, being heteroclinic between the pair. See Figure 1.1.*

This proposition enables us to return to Newton's homework. We have equilibria! We can linearize about them. Rick Moeckel did this, and thus completed a good chunk of Newton's homework for the case $N = 3$. In so doing Moeckel obtained some of the deepest results achieved in the three-body problem in the last 50 years. See [129, 130, 132, 133, 135].

Proof of Proposition 1.18 A glance at the equations shows that they make sense when $r = 0$ and that $r = 0$ is an invariant submanifold. To find their equilibria set the left-hand sides of the three parts of Equation (1.11) to zero. Get $rv = 0$, $y = vs$, and $\nabla U(s) = -\frac{1}{2}vy$. We already know any equilibria must lie on $r = 0$, which dispenses with the first equation. The last two equations combine to give $\nabla U(s) = -(\frac{1}{2}v^2)s$, which is the equation for s to be a central configuration with eigenvalue parameter $\mu = \frac{1}{2}v^2$.

The energy equation yields

$$rE = K(y) - U(s), \tag{1.12}$$

which, upon setting $r = 0$, yields $\frac{1}{2}v^2 - U(s) = 0$. It follows that $v = \pm\sqrt{2U(s)}$ at the equilibria, which established the claimed 2:1-ness. (Recall $U > 0$ on the sphere $\mathbb{S}$.) The negative root represents collisions, since $r' < 0$ nearby according to the first part of Equations (1.11). The positive roots are time-reversed collsions: explosions.

Finally, fix any normalized central configuration s_0. Set $v_0 = \sqrt{2U(s_0)}$. When reparameterized in McGehee variables, the corresponding dropped solution has the form $(r, s, y) = (r(\tau), s_0, \lambda(\tau)s_0)$ and tends to $(0, s_0, v_0 s_0)$ in forward time and to $(0, s_0, -v_0)$ in backward time and so is heteroclinic between the positive and negative pairs of equilibria associated to s_0. QED

1.4.2 Saari Decomposition and the Homographic Family

When re-expressed in terms of the McGehee variables, the Saari decomposition (0.55) becomes

$$\frac{1}{2}\|y\|^2 = \frac{1}{2r}\langle J, \mathbb{I}(s)^{-1}J\rangle + \frac{1}{2}v^2 + \tilde{K}_{p.s.}, \tag{1.13}$$

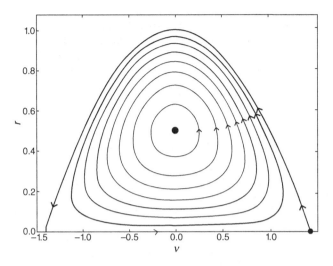

Figure 1.1 A central configuration family with fixed energy projected onto the $v-r$ plane. The arch and "floor" $r = 0$ comprise the rest cycle.

where $\tilde{K}_{p.s.} = \frac{1}{2}\langle (proj)y, (proj)y\rangle$ and where $proj = proj_s$ is the orthogonal projection onto the pure shape space part of the decomposition at the point $q = rs$. (The translational part $\frac{1}{2M}|P|^2$ of the Saari decomposition is not present because $P = 0$ when $y \in \mathbb{E}_0$.)

When the normalized kinetic energy of Equation (1.13) is evaluated along one of the homographic solutions of Lemma 1.10, the point $[s_0]$ on the normalized shape space remains constant so that $proj(y) = 0$ and the pure shape part of the kinetic energy is zero. Moreover, the motion is planar so that the angular momentum term there becomes $\frac{1}{2r}|J|^2$ as per the planar version of the Saari decomposition (0.56). It follows that this normalized kinetic energy is $K(y) = \frac{1}{2r}|J|^2 + \frac{1}{2}v^2$. Plugging this into Equation (1.12) for the energy E in McGehee coordinates we find the relation

$$rE = \frac{v^2}{2} + \frac{J^2}{2r} - U(s_0), \tag{1.14}$$

satisfied by homographic solutions having normalized shape s_0, energy E, and angular momentum J. For each fixed E, J this relation defines a curve in the $v-r$ plane. In Figure 1.1 we plot these curves for fixed $E < 0$, thus portraying a one-parameter family of curves in the $v-r$ plane, the parameter being J. This family forms a disc of solutions within the McGehee phase space. The fixed center, a stable equilibrium, corresponds to the relative equilibrium – the circularly rotation homographic solution. The bounding top arch corresponds to $J = 0$ and represents the dropped homographic solution. This visualization, and, in particular, the return leg – the heteroclinic connection lying on the collision manifold $r = 0$ and joining the two equilibrium points – will become essential in Chapter 3, where we discuss braids.

1.4.3 A Liapunov Function

On the collision locus, the *derivative* of the size change $v = \langle s, y \rangle$ acts like a Liapunov function. This will be useful later.

Proposition 1.19 $v' \geq 0$ *everywhere on the collision manifold* $\{r = 0\}$. *Moreover* $v' = 0$ *for a point on the collision manifold if and only if that point is an equilibrium.*

From $v = \langle s, y \rangle$ we get that $v' = \langle s', y \rangle + \langle s, y' \rangle$. Using Equations (1.11) we find

$$v' = \langle y - vs, y \rangle + \left\langle s, \nabla U(s) + \frac{1}{2}vy \right\rangle$$

$$= \langle y, y \rangle - v\langle s, y \rangle + \langle s, \nabla U(s) \rangle + \frac{1}{2}v\langle s, y \rangle$$

$$= |y|^2 - v^2 - U(s) + \frac{1}{2}v^2$$

$$= \frac{1}{2}\|y\|^2 - U(s) + \frac{1}{2}\|y\|^2 - \frac{1}{2}v^2$$

$$= rE + \left(\frac{1}{2}\|y\|^2 - \frac{1}{2}v^2 \right)$$

$$= rE + \left(\tilde{K}_{p.s.} + \frac{1}{2r}\langle J, \mathbb{I}(s)^{-1}J \rangle \right).$$

We used Euler's theorem for homogeneous functions to arrive at the third line and the energy relation (1.12) to arrive at the fourth line. To arrive at the final line we used Equation (1.13).

When we restrict this differential relation to the collision manifold $r = 0$ the term rE disappears, energy E being constant. The term involving angular momentum is positive and blows up when $r = 0$ unless $J = 0$. This blow-up is in keeping with Proposition 0.25, which asserts that any solution tending to total collision must have angular momentum zero. When $J = 0$ and $r = 0$ the equation for v's evolution becomes

$$v' = \tilde{K}_{p.s.} \geq 0.$$

Thus, if $v' = 0$ at some point we must have $K_{p.s.} = 0$ and $J = 0$ at that point. The condition $J = 0$ says the rotational component of y is zero and then $K_{p.s.} = 0$ implies that $proj(y) = y - vs = 0$. From the McGehee equations we get $s' = y - vs = 0$. The point $(0, y, s)$ is an equilibrium point of the collision flow. QED

Remark 1.20 The McGehee transformation can be viewed as an almost successful attempt to make Newton's equations scale invariant. The equations are not invariant under scaling, so the idea does not completely succeed. What McGehee did was to isolate all scale information into the single radial variable $r = \|q\|$. Scaling by $\lambda > 0$ acts by $q \mapsto \lambda q, \lambda \in \mathbb{R}$, so that r and $q \in \mathbb{E}$ are homogeneous of degree 1. The variable s is scale invariant. Under scaling $U \to \lambda^{-1} U$. The guiding principle for the remainder of the McGehee transformation can be taken to be that the total energy $E = K(v) - U(q)$ must also scale homogeneously. Since K is homogeneous of degree $+2$ in velocities v, the guiding principle requires that $v \mapsto \lambda^{-1/2} v$ so that K is also homogeneous of degree -1. (For α-force laws we would get $v \mapsto \lambda^{-\alpha/2} v$.) We can effect this scaling of v by insisting that time scale according to $t \mapsto \lambda^{3/2} t$, for then $v = \frac{dq}{dt} \mapsto \frac{\lambda dq}{\lambda^{3/2} dt} = \lambda^{-1/2} v$. This leads directly to the transformation Equations (1.10) of McGehee.

1.4.4 Roads to Total Collision

See Section 1.1.2, reason 2. Central configurations are the only roads in or out of total collision. See Figure 1.2. Here is the precise statement.

Theorem 1.21 *Suppose a solution $q(t)$ suffers a total collision as $t \to t_c$. Let $s(t) = \frac{1}{r(t)} q(t)$ be its normalized configuration curve. Then*

(A) *the set of accumulation points of $s(t)$ as $t \to t_c$ is a closed connected subset of the set of normalized central configurations.*

(B) *(no infinite spin) In the planar case and when the number of central configurations modulo symmetries is finite for the given mass distribution then the limit $\lim_{t \to t_c} s(t)$ exists.*

For $N = 3$, this result was established a long time ago by Sundman [201]. Part (B) for $N > 3$ is a new result, proved while this book was being prepared. See the end of the chapter notes in Section 3.9.

Before embarking on the proof of this theorem, the reader may want to refer back an earlier discussion of total collision, Proposition 0.25, where it

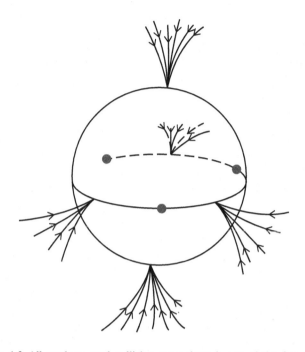

Figure 1.2 All roads to total collision enter through one of the five central configurations for the three-body problem. This is the $N = 3$ case of Sundman's theorem, Theorem 1.21.

was shown that solutions suffering total collision have angular momentum zero.

Proof of part (A) of Theorem 1.21 Re-express the solution in McGehee coordinates $r, s, y; \tau$ so that as $t \to t_c$ we have $\tau \to \infty$ and $r \to 0$. From $r' = rv$ and the fact that $r \to 0$ as $\tau \to \infty$, one expects that $v < 0$ for all τ large enough. We will verify this fact momentarily. In view of Proposition 1.18 characterizing the equilibria on the collision manifold in terms of central configurations, the first part of this theorem asserts that the ω-limit set of our solution curve $(r(\tau), s(\tau), y(\tau))$ is a non-empty subset of the negative branch ($v < 0$) of the equilibrium points of the collision-restricted flow.

We begin by showing that indeed $v < 0$ for all τ sufficiently large. In the first paragraph of the proof of Proposition 0.25, we used the Lagrange–Jacobi identity for $I = r^2$ to show that $\ddot{I} < 0$ in a one-sided neighborhood (t_2, t_c) of the collision time t_c. From $I = r^2$ we get $I' = 2rr' = 2r^2v$ and so $\dot{I} = 2r^{1/2}v$. It follows that, when expressed in Newtonian time, $v(t) < 0$ on this same one-sided neighborhood. In McGehee time τ, the one-sided neighborhood becomes a half line (τ_0, ∞), establishing the claim.

To establish this first part, it is enough to show that our collision orbit eventually lies in a bounded set of the McGehee-extended phase space. For this would imply that the ω-limit set is a non-empty compact invariant subset. See Zehnder [220, p. 138, proposition IV.8]. We showed in Proposition 1.19 that on the collision manifold, $\nu' \geq 0$ with equality if and only if we are at an equilibrium. If the ω-limit set contained a point where $\nu' \neq 0$, we could flow backward from that point, reaching points where $\nu \to -\infty$, and violating compactness of the ω-limit set. Consequently, all points in the ω-limit set are points where $\nu' = 0$ and hence equilibrium points.

It remains to show that the orbit eventually lies in a bounded subset. For this we invoke another theorem of Sundman, as described in Siegel and Moser [195, p. 69–70], expressed in Newtonian time. Use a time translation to shift the collision time from t_c to $t = 0$. Seigel and Moser showed that along any total collision orbit, $I(t) := r(t)^2 \sim t^{4/3}$ and $U \sim t^{-2/3}$. (Their proof is for $N = 3$ but carries through for general N.) It follows that $\tilde{U} = rU = O(1)$ is bounded as $t \to 0$. Now the McGehee energy is $\tilde{H} = rH = \frac{1}{2}K(y) - U(s)$, and $\tilde{H} \to 0$ as $r \to 0$. In McGehee variables, $U = \tilde{U}$ because $\|s\| = 1$ so the U term is bounded. Thus $K(y)$ is bounded and hence y is bounded. This completes the proof of part (A) of the theorem. QED

Proof of part (B) of Theorem 1.21 See [141]. The proof relies on the Lojasiewicz gradient inequality and the center manifold theorem.

1.5 Words on the $N = 4$ and $N = 5$ Proofs

The papers of Hampton and Moeckel [75] and Albouy and Kaloshin [7] that established $N = 4$ and $N = 5$ finiteness start by rewriting the condition for being a central configuration as a system of rational equations. They choose different sets of variables and hence begin with different sets of equations. In the first paper, all variables are invariant under $SE(2)$ and among their variables are the r_{ab}. The second paper uses the gauge-fixing condition $y_2 - y_1 = 0$ to get rid of the rotational freedom. Here $q_a = (x_a, y_a)$ are the positions. They take their variables to be the $x_{a+1} - x_a, y_{a+1} - y_a$, and r_{ab}^{-2}. In both cases, one can clear denominators so that the equations become polynomial and hence, allowing the variables to be complex, their equations define a complex algebraic variety V in some $\mathbb{C}^M$. The masses occur as coefficients for these polynomials.

An algebraic variety V in $\mathbb{C}^M$ is finite if and only if it is compact. Both papers now proceed by contradiction, assuming their variety V is non-compact and using the fact that if an algebraic variety $V \subset \mathbb{C}^M$ is non-compact then the image of its projection onto at least one of the M coordinate axes of $\mathbb{C}^M$ is non-compact. Since both V and the projection map are algebraic, non-compactness of the image implies that this image must be this entire coordinate axis minus a finite set of points. Call the coordinate of the axis z_i where $(z_1, \ldots, z_M)$ are standard coordinates on $\mathbb{C}^M$. Assuming V is infinite, it follows that they can let $z_i \to \infty$ along V and from here they arrive at contradictions.

The first paper achieves its contradictions using a beautiful result of Bernstein [16]. Each Laurent polynomial f in M variables has associated to it a Newton polyhedron $P_f \subset \mathbb{R}^M$. This P_f is the convex hull of the *support* of the polynomial, this support being a finite collection of integer vectors $vert(P_f) \subset \mathbb{Z}^M$, each vertex corresponding to one nonzero monomial in the polynomial, the vertex itself being the ordered list of exponents defining that monomial. Bernstein fixes the convex hulls P_i for each equation $f_i = 0$ and then shows that for a generic collection of polynomials f_i having these polyhedra, the number of solutions to the system of equations is constant and equal to the Minkowski mixed volumes of the convex hulls of the collection of Newton polyhedra. He also gives a specific criteria for finiteness involving subsystems associated to the faces of the polyhedra (his theorem B) and this is the workhorse to arrive at the contradiction. Some of this hard work is integer convex geometry.

The second paper achieves its contradictions in a more hands-on way. They pay particular attention to Robert's example (see Theorem 1.5 in this chapter), figuring out what makes it tick, taking care that in each of their many cases some Robert's-type phenomenon does not occur.

1.6 Relative Equilibria in Four Dimensions

Four dimensions supplies enough room to support relative equilibrium configurations that are not central configurations. These play a role in Chapter 2. Their existence is based on the normal form

$$\Omega = \begin{pmatrix} 0 & -\omega_1 & 0 & 0 \\ \omega_1 & 0 & 0 & 0 \\ 0 & 0 & 0 & -\omega_2 \\ 0 & 0 & \omega_2 & 0 \end{pmatrix}$$

for skew-symmetric 4×4 matrices. Identify $\mathbb{R}^4$ with $\mathbb{C}^2$ and this is the matrix for the real linear transformation $(z_1, z_2) \mapsto (i\omega_1 z_1, i\omega_2 z_2)$.

Recall that for three bodies the only non-collinear central configuration is the equilateral triangle configuration.

Proposition 1.22 *Every isosceles triangle is a relative equilibrium configuration for the equal mass three-body problem in $\mathbb{R}^4$. In the corresponding relative equilibrium solution, the two Jacobi vectors travel in orthogonal two-planes at different frequencies, the ratio of these frequencies being a function of the ratio of the side lengths of the isosceles triangle as described by Equation (1.22).*

These non-central relative equilibria are described in Albouy and Chenciner [8] who call the corresponding configurations "balanced." See also [31, 32] and Moeckel's scholarpedia article [137].

Proof of Proposition 1.22 The proof is achieved by a direct computation. Take $m_1 = m_2 = m_3 = 1$. Then we can write the equations defining the equal mass three-body problem (compare Equations (0.4) and (0.1)) as

$$
\begin{aligned}
\ddot{q}_1 &= \frac{q_2 - q_1}{r_{21}^3} + \frac{q_3 - q_1}{r_{31}^3}, \\
\ddot{q}_2 &= \frac{q_1 - q_2}{r_{21}^3} + \frac{q_3 - q_2}{r_{32}^3}, \\
\ddot{q}_3 &= \frac{q_1 - q_3}{r_{13}^3} + \frac{q_2 - q_3}{r_{23}^3}.
\end{aligned} \tag{1.15}
$$

Suppose the triangle to be isosceles with vertex 3 so that $r_{23} = r_{13}$. Set

$$a = r_{23} = r_{13} \text{ and } c = r_{12}.$$

The two Jacobi vectors corresponding to the choice of vertex 3 are $\xi_1 = q_1 - q_2$ and $\xi_2 = q_3 - \frac{1}{2}(q_1 + q_2)$. If we subtract the second part of Equation (1.15) from the first and use the isosceles condition, we get

$$
\ddot{\xi}_1 = -\left(\frac{2}{c^3} + \frac{1}{a^3}\right)\xi_1. \tag{1.16}
$$

If we add the first equation to the second and subtract that result from the third part of Equation (1.15), we get

$$
\ddot{\xi}_2 = -\frac{3}{a^3}\xi_2. \tag{1.17}
$$

These pair of equations suggests the ansatz

$$
\begin{aligned}
\xi_1(t) &= e^{i\omega_1 t}\xi_1(0), \\
\xi_2(t) &= e^{i\omega_2 t}\xi_2(0).
\end{aligned} \tag{1.18}
$$

The ansatz solves Newton's equations provided that

$$\omega_1^2 = \left(\frac{2}{c^3} + \frac{1}{a^3} \right),$$

$$\omega_2^2 = \frac{3}{a^3},$$

(1.19)

and that a and c remain constant. To guarantee this constancy, recall that $\xi_1(0), \xi_2(0) \in \mathbb{C}^2 = \mathbb{R}^4$. Take the two vectors to lie in Hermitian orthogonal complex subspaces,

$$\xi_1(0) = (z_1, 0),$$

$$\xi_2(0) = (0, z_2),$$

(1.20)

so that at every instant $\xi_1(t)$ and $\xi_2(t)$ are perpendicular. Then

$$c = r_{12} = |\xi_1| = |z_1|,$$

while

$$a = \sqrt{|z_1|^2 + \frac{1}{4}|z_2|^2},$$

which are indeed constant. The equation for a is seen by noting that

$$q_3 - q_1 = q_3 - \frac{1}{2}(q_1 + q_2) - \frac{1}{2}(q_1 - q_2)$$

$$= \xi_2 - \frac{1}{2}\xi_1,$$

(1.21)

and, similarly,

$$q_3 - q_2 = \xi_2 + \frac{1}{2}\xi_1,$$

and then using the orthogonality of ξ_1, ξ_2 and the fact that $|\xi_i| = |z_i|$ to compute

$$a^2 = \frac{1}{4}|z_1|^2 + |z_2|^2.$$

Summarizing, the solution ansatz defined in Equations (1.18) and (1.20) yields a relative equilibrium solution to Newton's equations provided the frequencies satisfy relations (1.19) with $c = |z_1|, a = \sqrt{\frac{1}{4}|z_1|^2 + |z_2|^2}$.

If we divide the two frequency relations we get

$$\frac{\omega_1^2}{\omega_2^2} = \frac{1}{3}\left(\frac{2a^3}{c^3} + 1 \right).$$

(1.22)

Note, in particular, that $|\omega_1| = |\omega_2|$ if and only if $a = c$, which is to say, if and only if the triangle is equilateral (and thus central). QED

1.6.1 Balancing in Higher Dimensions

Albouy and Chenciner [8] get more general. (See also [31].) They replace the old ansatz for a central configuration (see Equation (1.1)) with $q(t) = g(t)q_*$ where $g(t) = exp(t\Omega)$ and $\Omega \in so(d)$ is a constant skew-symmetric matrix. Plug this ansatz into Newton's equations, $\ddot{q} = F(q)$. Use $F(gq) = gF(q)$. We get

$$- Sq_* = F(q_*); S = -\Omega^2 = \Omega\Omega^T. \tag{1.23}$$

These are the equations for a relative equilibrium configuration in any dimension. In these equations, S is an arbitrary symmetric positive semi-definite all of whose nonzero eigenvalues have even multiplicity. Each eigenvalue pair corresponds under a square root to a two-by-two rotational block of an Ω put into normal form. See [32] for investigations of this system.

Albouy and Chenciner call Equations (1.23) the "balancing equations" rather than "equations for relative equilibria." See the chapter notes in Section 3.9 concerning their nomenclature.

1.6.2 Almost Complex Structures and a Loose End

An almost complex structure on $\mathbb{R}^d$ is a skew-symmetric transformation $\mathbb{J}: \mathbb{R}^d \to \mathbb{R}^d$ that satisfies $\mathbb{J}^2 = -I$. When we identify $\mathbb{C}$ with $\mathbb{R}^2$ in the standard way, the complex multiplication by i defines the almost complex structure on $\mathbb{R}^2$. If $\mathbb{R}^d$ admits an almost complex structure then $d = 2k$ is even. Conversely, every even-dimensional Euclidean space admits a complex structure, and the structure yields an isomorphism $\mathbb{R}^d \cong \mathbb{C}^{d/2}$ under which $\mathbb{J}$ corresponds to complex multiplication by i. Armed with our almost complex structure $\mathbb{J}$ on our even-dimensional $\mathbb{R}^d$, we can now tie up a loose end and prove part (A) of Proposition 1.4.

Proof of (A) of Proposition 1.4 In proving the planar Proposition 1.3, we used a variation of the ansatz (1.1), namely the ansatz $q(t) = exp(i\omega t)q_*$. Replace i by $\mathbb{J}$ to obtain the solution ansatz $q(t) = exp(\omega\mathbb{J}t)q_*$ where ω is a scalar. The proof of Proposition 1.3 goes through verbatim upon making this replacement, yielding part (A) of Proposition 1.4 in even dimensions: central configurations yield relative equilibria. QED

As an alternative proof, set $\Omega = \omega\mathbb{J}$ in the balancing equations, Equations (1.23). Verify that the balancing equations become the equations for central configurations.

Example 1.23 The regular tetrahedron is a central configuration for the four-body problem in $\mathbb{R}^3$. See Proposition 1.13. The regular tetrahedron cannot be made into a relative equilibrium for the problem: It cannot be spun about an axis in space and turned into a solution to Newton's equations. However, $\mathbb{R}^4 = \mathbb{C}^2$ admits an almost complex structure. By taking the standard embedding $\mathbb{R}^3 \subset \mathbb{R}^4$ we promote the regular tetrahedron from a central configuration to a relative equilibrium configuration for the four-body problem in $\mathbb{R}^4$. A similar trick works for the regular N simplex for any even N.

1.7 Further Questions

1. Rick Moeckel posed the following question. See Albouy, Cabral, and Santos [6, problem 15].

Is it true that there is a neighborhood U of equal masses in the space of mass distributions $m = (m_1, m_2, \ldots, m_N) \subset \mathbb{R}_+^N$ for the planar N-body problem with the property that no relative equilibria is linearly stable for masses m in U? (Strictly speaking, I mean "spectrally stable" when I say "linearly stable." The notion is defined and investigated in Chapter 2.) More specifically, if we order the masses so that m_1 is the largest mass, does there exist a constant $c < 1$, independent of N, such that if a relative equilibrium solution is spectrally stable then we must have $m_i < cm_1$, for all $i > 1$?

In the three-body problem, the answer is "yes." The Euler relative equilibria are never linearly stable. And the Lagrange relative equilibria are only stable when one mass is dominant. See Figure 1.3, where the shaded region indicates the set of masses for which the Lagrange relative equilibrium is linearly stable.

2. Prove that part (B) of Theorem 1.21 holds for the spatial N-body problem.

3. Generalize the Albouy–Chenciner isosceles relative equilibria to non-regular tetrahedra spinning in six-dimensional Euclidean space, using $\mathbb{R}^6 = \mathbb{C}^3$ (and spinning each of the three Jacobi vectors within its own two-plane $\mathbb{C}$, these three two-planes being orthogonal).

1.8 Chapter Notes

The Problem

Chazy [25] is perhaps the first to have explored Open Problem 1. Aurel Wintner [215] stated the problem in 1941. See the final part of the first section of a paper by Albouy and Kaloshin [7] for precise passages and a detailed history. Smale lists this as problem 6 in his collection [198] of problems for the twenty-first century, his answer to Hilbert's list. Albouy, Cabral, and Santos [6] list it as problem 9 in their N-body problem list.

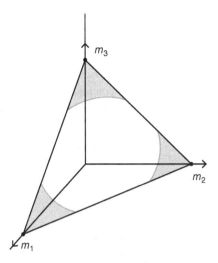

Figure 1.3 The triangle represents the set of all mass ratios and the simplex $m_i >$ $0, m_1 + m_2 + m_3 = 1$. The shaded region indicates the set of masses for which the Lagrange relative equilibrium is linearly stable.

On Section 1.1

See Moeckel [137, 138] for in-depth discussions of central configurations. See also Hampton [74] for an overview and a lively discussion. Notably, Hampton looks into the central configurations for the α-power law N-body problems and argues that characterizing this set in the limit $\alpha \to \infty$ may yield important information.

On Section 1.4

For the theorem regarding the "only roads in to total collision" and McGehee blow-up generally, I recommend the paper "A l'infini en temps fini" [29] by Chenciner. McGehee developed his blow-up in the beautifully written paper [121]. His blow-up method has become a central tool in mathematical celestial mechanics and was essential in Xia's discovery of non-collision singularities [217].

Sundman's theorem and the problem of Infinite spin See Siegel and Moser [195, p. 69–90, sections 12 and 13] for a careful proof of Theorem 1.21 when $N = 3$. The problem addressed by Theorem 1.19, part (B), has been dubbed the problem of "infinite spin." Wintner asks this question [215, p. 283, section 368]. Saari [187] called the question of infinite spin the "Painlevé–Wintner" problem. The proof of part (B) when the shape potential is Morse seems to have been fairly well known and essentially due to Chazy. While preparing this book, Moeckel and I [141] were able to get rid of the assumption that the critical points of the shape potential were nondegenerate and thus establish (B). Several papers (e.g. [53, 187]) have claimed to have proven that infinite spin collision solutions do not exist, that is, they claim to have proved part (B) of the theorem. The essence of their error is an erroneous implication that "having zero rotational part" implies having finite spin. See the appendix to [141] for details.

On Section 1.6

Regarding the terms "balanced" versus "relative equilibrium." Albouy and Chenciner [8] use "configuration équilibrée" which they translated as "balanced configurations." My American educated mind learned the phrase "relative equilibrium" early on and cannot forget it, so whenever I would hear these two talk about balanced configurations I could not remember what on earth they were talking about. By translating "équilibrée" as equilibrium and reattaching "relative" I found I could regain the meaning. Apologies to the two Alains for my change in their phraseology.

2

Are There Any Stable Periodic Orbits?

Open Question 2 Are there any stable periodic orbits?

2.1 Really?!

How could such a basic question remain unanswered? But if "stable" means "Lyapunov stable" then this question is open *for the planar three-body problem.*

We recall the definition of Lyapunov stable.

Definition 2.1 An orbit for a dynamical system is Lyapunov stable if every neighborhood U_1 of that orbit contains a neighborhood U_2 such that all solutions starting in U_2 remain in U_1 for all future times. (See Figure 2.1.)

We must interpret the open question in the center-of-mass frame, otherwise the answer becomes "no." Take any periodic solution and apply an arbitrarily small Galilean boost. We obtain a new solution diverging from our starting solution despite having arbitrarily close initial conditions.

Recall that a relative periodic solution is one for which the relative distances $r_{ab}(t)$ are periodic functions of t. Such an orbit is periodic or quasiperiodic in the center-of-mass frame. Reformulating the question then:

Are there any Lyapunov stable relative periodic orbits for the planar three-body problem when viewed in the center-of-mass frame?

2.1.1 Asymptotic versus Lyapunov Stability

Asymptotic stability is stronger than Lyapunov stability. It is the kind of stability that often first comes to mind and is the type of stability preferred by engineers.

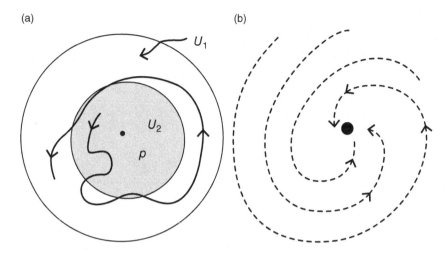

Figure 2.1 (a) A Lyapunov stable fixed point. (b) An asymptotically stable fixed point.

Definition 2.2   An orbit c for a dynamical system is asymptotically stable if every orbit in some neighborhood of c converges to c and if, moreover, c is Lyapunov stable.

It may seem that the added condition that the orbit of Definition 2.2 be Lyapunov stable is redundant. It is not. See the notes in Section 2.9.

As an example, the origin $(x, v) = (0, 0)$ is a Lyapunov stable fixed point for the harmonic oscillator $\dot{x} = v, \dot{v} = -x$, but it is not asymptotically stable. We can make the origin be asymptotically stable by adding a bit of friction to the system. The damped harmonic oscillator $\dot{x} = v, \dot{v} = -x - \epsilon v$ does have the origin as an asymptotically stable equilibrium. Typically, some dissipation is required in order to get any asymptotic stability in physical systems.

Small neighborhoods V of an asymptotically stable orbit c shrink onto the orbit: $\lim_{t \to +\infty} \varphi_t(V) = c$. In particular, the volumes of the $\varphi_t(V)$ eventually decrease with increasing t. As a consequence, Hamiltonian systems cannot have asymptotically stable orbits since the flow of such a system conserves phase space volume. (See Exercises A.3 and A.15 in Appendix A.) The N-body problem, being Hamiltonian, never has asymptotically stable orbits.

2.1.2 The Oldest Question in Dynamical Systems

(See Herman [81] and the chapter notes (Section 2.9.1).) *Arbitrarily close to the initial conditions of any relative periodic solution for the planar three-body problem are there initial conditions whose solutions are unbounded?*

Recall that a three-body orbit is unbounded if at least one of its mutual distances $r_{ab}(t)$ is an unbounded function of t. (See Definition 0.23 and Proposition 0.24.) In terms of the Hill region picture at negative energy for the planar three-body problem as depicted in Figure 0.6, unbounded solutions are those whose projections to shape space wind down one of the three arms of our plumbing fixture, heading off to infinity down that arm. A "yes" answer to the oldest question would give a dramatic "no" to our open question.

2.2 What's Known?

I know two methods for establishing Lyapunov stability in Hamiltonian systems. One is the energy–momentum method (or its close cousin, the energy-Casimir method). See Section 2.2.2. The other is the KAM method. See Section 2.4.5. Neither method works for the N-body problem in the plane or in space when $N > 2$. The KAM method fails for dimensional reasons. The KAM method can only guarantee Lyapunov stability for systems with two degrees of freedom. The planar three-body problem has three degrees of freedom after reduction. All these other problems have more degrees of freedom.

2.2.1 Weaker Types of Stability

Lyapunov stability seems out of reach for relative periodic solutions in the planar or spatial three-body problem. Having almost nothing to say regarding the main open problem for Lyapunov stability, we spend the bulk of this chapter discussing other forms of stability that can actually be achieved. These other types are linear – or more accurately – spectral stability, KAM stability, and Nekhoroshev stability. We've already written down the acronym KAM four times. KAM stability with its associated theorems tends to dominate conversations around stability in celestial mechanics. This chapter is no exception as more than half the chapter ends up devoted to the topic.

2.2.2 Four Dimensions Stabilize!

Allow bodies to move in four-space and they surprise us.

Theorem 2.3 (See Albouy and Dullin [5], and Dullin and Scheurle [48].) *The three-body problem in four dimensions admits Lyapunov stable orbits. The isosceles relative equilibria described at the end of Chapter 1 (see Proposition 1.22) are Lyapunov stable for equal and near-equal masses.*

Sadly, Theorem 2.3 seems to have no bearing on the planar or spatial three-body problems.

The proof relies on the constancy of energy and angular momentum, together with some understanding of the topology of energy–momentum level sets. The method is known as the *energy–momentum* or *energy–Casimir* method. To explain, the method begins with an equilibrium point p of a Hamiltonian system, so a critical point of the energy. If p is a nondegenerate minimum (or maximum) with energy level $H(p) = 0$, then, by Morse theory, the Hamiltonian has the form $H = \Sigma(\xi_i^2)$ relative to some coordinates ξ_i on phase space centered at p. The level sets $H = \epsilon$ are then invariant spheres surrounding p and collapsing to p with ϵ. The connected components of the sublevel sets $0 \leq H < \epsilon$ form a system of invariant balls $\mathbb{B}(\epsilon)$ shrinking to p as $\epsilon \to 0$, thus providing a system of invariant Lyapunov neighborhoods of p.

Now, suppose that the periodic orbit in question is a relative equilibrium and so a critical point of the energy–momentum map. Such a point corresponds to a single point in the symplectic reduced space obtained by fixing the angular momentum map and dividing out by the rotations that fix that value of the momentum. See Appendix B. Suppose that this point is a nondegenerate minimum of the reduced energy – the energy on that reduced space. Then the same logic as in the previous paragraph holds and we are guaranteed that the relative equilibria is Lyapunov stable within its reduced space. If we are at a generic value of momentum, then the reduced space varies smoothly and one can show that the relative equilibrium is in fact Lyapunov stable. The reduced invariant balls lift to invariant Lyapunov neighborhoods of the form $\mathbb{S}^1 \times \mathbb{B}(\epsilon)$, shrinking to the orbit as $\epsilon \to 0$.

This is what Albouy and Dullin [5] and Dullin and Scheurle [48] achieve with their theorem: The reduced energy has a nondegenerate minimum corresponding to the isosceles relative equilibrium provided that spinning isosceles triangle is not equilateral. The "not equilateral" condition implies that the rank of the angular momentum is 4 and is essential to their proof. In two- and three-dimensional space the maximal rank of J is 2.

2.2.3 Instability

Arnold Diffusion

Arnol'd diffusion is the catch-all phrase for mechanisms for escaping arbitrarily small neighborhoods of KAM stable orbits. A "no" answer to Open Question 2 would imply that every KAM stable periodic orbit for the N-body problem admits Arnol'd diffusion. A "yes" answer to the Oldest Question (2.1.2) would go further and imply that each of these Arnol'd diffusions actually diffuses all the way out to infinity. Arnol'd diffusion is believed to occur for generic Hamiltonian perturbations of Hamiltonian systems. But the vector field defining the N-body problem is not generic. Generic forays into the empire of Arnol'd diffusion do not offer much hope in providing a "no!" answer to Open Question 2 or to Herman's "oldest question."

Strong Force Instability

If we use the $\alpha = 2$ power law potential (Equation (0.14)) instead of the Newtonian potential, then the Lagrange–Jacobi identity (Equation (0.49)) asserts that $\ddot{I} = 4H$ where H is the energy. It follows that periodic orbits only exist when $H = 0$ and $\dot{I}(0) = 0$. If either condition is violated, then I either increases without bound or goes to zero, indicating total collision. Since we can change the energy (and $\dot{I}(0)$) with small perturbations we've answered the oldest question with a resounding "no!" in this case: A small perturbation can make a zero energy initial condition positive, and hence, arbitrarily nearby to any periodic solution there is an unbounded solution.

2.2.4 And Our Solar System?

Is our own solar system stable? This existential question has driven centuries of work in celestial mechanics. We recommend Laskar [104] and [102, section 3]. See also the first sections of [38], and the 2023 wikipedia entry [214] for engaging treatments. Starting in 1989 and continuing through the 1990s, Laskar demonstrated that the dynamics of the planetary orbits in our solar system are chaotic, with a Lyapunov time scale of five million years. This implies mixing and indeterminacy of *some* phase space coordinates on that timescale, notably of the eccentricities and inclinations of the inner planets. In truth, there is no one single periodic orbit to speak of when imagining long-term solar system dynamics. Instead of asking is this or that periodic orbit approximating the solar system stable in this or that sense, one should ask questions such as "will the Earth remain at roughly the same distance from the Sun over the next 3 billion years?" Due to the intrinsically chaotic nature of the dynamics, questions of this type must be approached statistically. Laskar ([102], in particular, pp. 150–152) showed that over a timescale of a few billion years or so Mercury might collide with the Sun or Venus.

2.2.5 Angular Momentum as Stabilizer

A physical principle underlies the appearance of Lyapunov stable orbits for the three-body problem in dimension $d = 4$ and its apparent lack of Lyapunov

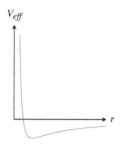

Figure 2.2 A graph of the effective potential for the Kepler problem as a function of distance.

stable orbits in dimensions $d = 2$ or 3. That principle asserts that solutions seek out states of lowest energy, but that angular momentum is "more conserved" than energy.

Proposition 2.4 *In the two-body (Kepler) problem, at fixed nonzero angular momentum, the energy is bounded below and the solutions that minimize energy are the circular orbits.*

This proposition holds regardless of the dimension d within which the bodies move. We contrast this situation with that of the three-body problem.

Proposition 2.5 *In the spatial or planar three-body problem, at fixed angular momentum, the energy is unbounded below. However, for the three-body problem in four-space, at fixed generic (= rank 4) value of the angular momentum, the energy is bounded below.*

The proof of Proposition 2.4 is summarized by graphing the *effective potential* $V_{eff}(r)$, which is the sum of the angular part of the kinetic energy with the usual potential energy. We have $V_{eff}(r) = \frac{1}{2}\frac{J^2}{r^2} - \frac{\mu}{r}$, where J is the angular momentum and $\mu > 0$ the Kepler constant. See Figure 2.2.

The global minimum of V_{eff} occurs at the radius r of the circular orbit whose angular momentum is J. Proposition 2.4 is held responsible for the near-circular nature of the orbits of the planets. The idea is that during planetary formation, dissipative forces are involved, but these forces conserve angular momentum. Over millions of years, the orbits of the planets around the Sun dissipated energy while conserving angular momentum and thus settled down to near-circular orbits.

In the spatial or planar three-body problem an angular momentum–preserving dissipative energy cascade has no lower bound: The infimum of energy at fixed angular momentum is negative infinity, corresponding to two of the bodies forming a fast tight binary orbit whose distance from the third tends to infinity. However, in the four-dimensional three-body problem if we initialize with a four-dimensional angular momentum – one of rank 4 – then the infimum of energy at this fixed angular momentum is finite. An energy cascade would bring us down to the global energy minimum on the angular momentum level set – presuming its existence. This global minimum does exist when the masses are equal or near equal and is the Lyapunov stable relative equilibria of Theorem 2.3.

2.3 Spectral Stability, Mandatory 1's, Return Map, and Reduction

This section is inspired by chapter 6 of Meyer [123].

In this section we describe spectral stability, a weak form of linear stability. In Section 2.4 we describe KAM and Nekhoroshev stability. The implications among these types of stability are

$$\text{Nekhoroshev} \implies \text{KAM} \implies \text{spectral}. \tag{2.1}$$

When celestial mechanicians say an orbit is stable, typically they mean spectrally stable.

Spectral stability has to do with the spectrum of a linear operator A, the monodromy, associated to a periodic orbit. (See Equation (2.3).) Being linear, A always has the zero vector, corresponding to the orbit itself, as a fixed point. We say that the orbit is *linearly stable* if the zero vector is Lyapunov stable for A. Then

$$\text{linear} \implies \text{spectral} \tag{2.2}$$

stability. Linear stability *never occurs* in the N-body problem due to the mandatory 1's and their consequent nontrivial Jordan blocks. In this section, we describe those mandatory 1's and the methods to get rid of them: slices and reduction.

2.3.1 Linearizing About an Orbit: Spectral Stability

Suppose $\gamma(t)$ is a periodic orbit for a vector field X on a manifold $\mathcal{P}$. Let Φ_t be the flow of the vector field so that $\gamma(t) = \Phi_t(z_0)$, where $z_0 = \gamma(0)$ and $\Phi_T(z_0) = z_0$ where T is the period of γ. The linearized flow map

$$A = d\Phi_T(z_0) \colon T_{z_0}\mathcal{P} \to T_{z_0}\mathcal{P} \tag{2.3}$$

contains crucial stability information regarding the orbit.

Definition 2.6 The linear map A defined by Equation (2.3) is called the monodromy matrix of the periodic orbit.

We can compute A by linearizing our nonlinear vector field $z \mapsto X(z)$ along the orbit. Write our original ODE as $\dot{z} = X(z)$ where, for simplicity, $\mathcal{P} = \mathbb{R}^m$ so that $X \colon \mathbb{R}^m \to \mathbb{R}^m$. Consider a family of solutions $z_\epsilon(t)$ to this ODE depending smoothly on some parameter ϵ. For example, ϵ might parameterize a curve of initial conditions. Setting $\delta z(t) := \frac{d}{d\epsilon}|_{\epsilon=0} z_\epsilon(t)$ and differentiating the ODE, we arrive at the linear time-dependent ODE

$$\dot{\delta z} = DX(z(t))\delta z(t),$$

which governs the variation of solutions and supplements our original ODE. If $\Phi_t \colon \mathbb{R}^m \to \mathbb{R}^m$ is the time t flow of our ODE, then $z_\epsilon(t) = \Phi_t(z(0) + \epsilon \delta z_0)$ is such a family of solutions. Differentiating with respect to ϵ shows that the linearization of the flow is given by

$$D\Phi_t(z(0))(\delta z_0) = \delta z(t),$$

where $\delta z(t)$ solves the linearized ODE and has initial condition δz_0. Thus $A(\delta z_0) = \delta z(T)$. This relation between the linearized ODE and linearized flow provides a useful numerical method for computing the monodromy matrix on the fly and is essential for estimating linear stability.

The following propositions help organize the theory.

Proposition 2.7 *If the monodromy matrix A associated to a periodic orbit has an eigenvalue with modulus greater than 1, then that orbit is* not *Lyapunov stable.*

We call a flow *reversible* if, whenever $\gamma(t)$ is a solution, so is $\gamma(-t)$.

Proposition 2.8 *If the flow is reversible, then whenever λ is an eigenvalue for A, so is λ^{-1}.*

Since Newton's equations are time-reversible, it follows from Propositions 2.7 and 2.8 that if an orbit is to have a chance of being Lyapunov stable then the eigenvalues of its monodromy must all lie on the unit circle. We call such orbits *spectrally stable*.

Definition 2.9 The periodic Hamiltonian orbit is called spectrally stable if the eigenvalues of its monodromy all lie on the unit circle.

Remark 2.10 A synonym for spectrally stable is elliptic. Occasionally we may abuse terminology and refer to spectral stability as linear stability. Apologies in advance. Spectral stability is thus a neccessary condition for Lyapunov stability.

Sketch proof of Proposition 2.7 An eigenvalue λ outside the unit circle guarantees the Lyapunov *instability* of 0 for the linear map A. Indeed, suppose for simplicity that $\lambda > 1$ is positive real and let v be its eigenvector. We get $A^n v = \lambda^n v \to \infty$ as $n \to \infty$. Since the linearized flow is a first-order approximation to the flow, orbits of the flow itself must move away from the orbit in the direction of v when perturbed in the direction v. Roughly speaking, the eigenspace for such an unstable eigenvalue λ corresponds to a growing mode for the dynamics. One can turn this rough speaking into a rigorous proof through the magic of the unstable manifold of the orbit, augmented by the use of its center-stable manifold.

Remark 2.11 Lyapunov has an older proof of Proposition 2.7 using his functions. See the chapter notes in Section 2.9.

Proof of Proposition 2.8 In the case of Newton's equations, the map $(q, v) \mapsto (q, -v)$ of phase space induces time reversal. Call this map $\hat{T}$. Time reversibility is then encoded in the equality $\Phi_t(\hat{T}\xi) = \hat{T}(\Phi_{-t}(\xi))$ where we've set $\xi = (q, v)$. Differentiating with respect to ξ as ξ approaches points of the orbit, with time t set to $t = T$, the period, and using $\Phi_{-t} = \Phi_t^{-1}$, we get $A\hat{T} = \hat{T}A^{-1}$. Now, if $Av = \lambda v$ then $A^{-1}v = \frac{1}{\lambda}v$, and it follows that $A\hat{T}v = \frac{1}{\lambda}\hat{T}v$. QED

We write the eigenvalues of the monodromy operator A of a spectrally stable orbit as

$$\sigma(A) = \{\lambda_j, \bar{\lambda}_j = exp(\pm i2\pi\omega_j)\} \quad \text{(spectrally stable case)} \quad (2.4)$$

and call the ω_j the *frequencies* of the elliptic orbit. The reason the eigenvalues come in complex conjugate pairs is due to Proposition 2.12.

Proposition 2.12 *If the flow is symplectic, then whenever λ is an eigenvalue for A then so are $\lambda^{-1}, \bar{\lambda},$ and $\bar{\lambda}^{-1}$.*

Proof of Proposition 2.12 If X is Hamiltonian then Φ_T is a symplectic map and so A is a linear symplectic map. The spectrum $\sigma(A) \subset \mathbb{C}$ of any linear symplectic map A enjoys the symmetries

$$\lambda \in \sigma(A) \iff \frac{1}{\lambda}, \bar{\lambda}, \frac{1}{\bar{\lambda}} \in \sigma(A). \quad (2.5)$$

See, for example, [11, section 42, p. 226]. Note in particular, that as in time-reversible symmetry, for every eigenvalue inside the unit circle there is one outside. QED

A Mandatory 1

From $\Phi_t^* X = X$ it follows that

$$AX = X, \qquad (2.6)$$

where these last X's mean $X(z_0)$. Thus 1 is always an eigenvalue of the monodromy matrix.

Spectral versus Linear Stability

In Section 2.3.2, we will see that the monodromy A of autonomous Hamiltonian systems like the N-body problem always has at least two 1's in their spectrum. This leaves open the possibility that A contains a nilpotent block: $M = \begin{pmatrix} 1 & 1 \\ 0 & 1 \end{pmatrix}$ in its Jordan normal form. Indeed, we will see that such a block *always* occurs in the monodromy for the N-body problem.

Due to the appearance of such blocks, we have been careful to use the phrase "spectrally stable" instead of "linearly stable." The nilpotent block matrix M is "spectrally stable," having spectrum consisting of $1 \in \mathbb{S}^1$, but is not linearly stable since $M^n = \begin{pmatrix} 1 & n \\ 0 & 1 \end{pmatrix}$ and consequently the orbit of the vector $(0, \epsilon)$ under M consists of the points $(n\epsilon, \epsilon)$ and tends to infinity with n. We have called the orbit, or its monodromy A, linearly stable if, when we look at $(A, T_{z_0}\mathcal{P})$ as a dynamical system, the fixed point 0 is Lyapunov stable. Linear stability in this sense *never occurs* in the N-body problem due to the nontrivial Jordan blocks.

However, as we amble away from the details, we may occasionally confuse spectral and linear stability in the text, and in life.

2.3.2 Symmetry-Induced 1's in the Spectrum

Having 1's in the spectrum of a linear map is an indication of possible instability, even though 1 lies on the unit circle. For example, the hypothesis of the standard version of the KAM stability theorem (see Theorem 2.15) require that there be no 1's in the spectrum of a linear operator closely related to the monodromy operator A. But several 1's always occur in the spectrum of the monodromy map A for any periodic orbit of the N-body problem due to Galilean symmetry. We will need to understand *how* these potentially destabilizing 1's arise from symmetries in order to dispense with them and proceed to a useful version of KAM stability.

We have already seen (Equation (2.6)) that the monodromy always has 1 as an eigenvalue. We will now show that for the Newtonian N-body problem, the monodromy A of *any* periodic orbit has a Jordan block of the form

$$\begin{pmatrix} 1 & -3/2 \\ 0 & 1 \end{pmatrix}$$

associated to energy-time, guaranteeing that 0 is not Lyapunov stable for the linear map A. (See Section 2.3.1.)

The eigenvector $X = X(z_0)$ of Equation (2.6) accounts for the first column of our claimed Jordan block, and we call it the direction of time evolution since $\dot{\zeta} = X(\zeta)$ is our dynamics. The vector $(1, 0)$ relative to the block structure represents X. To account for the second column of the block, we will show that the scaling symmetry of Newton's equations leads to a vector field D, the dilation vector field, having $AD = D + \frac{3}{2}X$, the information encoded in the second column of the block structure. The space-time dilational symmetry as described in Section 0.1.8 is $q(t) \mapsto \lambda q(\lambda^{-3/2}t)$. At the phase-space level this symmetry is $S_\lambda((q, p)) := (\lambda q, \lambda^{-1/2}p)$. If $\Phi_T(z_0) = z_0$, then the flow through $S_\lambda z_0$ is periodic with period $\lambda^{3/2}T$, which is to say that $\Phi_{\lambda^{3/2}T}(S_\lambda z_0) = S_\lambda z_0$. Now set $\lambda = exp(\epsilon)$ and differentiate this identity at $\epsilon = 0$ to find that $AD + \frac{3}{2}X = D$, where $D = D(z_0)$ is the infinitesimal generator of the dilation, evalutated at z_0. In the process, one computes from $D(z) = \frac{d}{d\epsilon}|_{\epsilon=0}S_{exp(\epsilon)}z$ that $D((q, p)) = (q, -\frac{1}{2}p)$.

Remark 2.13 In an exercise at the end of this chapter (see Exercise 2.25) I ask you to show that D and X are linearly independent at all phase points except for the initial conditons for dropped central configurations.

An analogue of the generalized eigenvector D *dual* to the time direction X exists for periodic orbits in any autonomous Hamiltonian system, regardless of whether or not X admits a scaling symmetry. This additional vector corresponds to the fact that we can usually continue the periodic orbit on to nearby energy levels. As long as the period of the continued orbit changes to first order with the change in energy level then the corresponding two-by-two matrix has such a nontrivial Jordan block.

We have just established an instance of a theorem in symplectic linear algebra: If 1 is a generalized eigenvalue of a linear symplectic map then it comes with even multiplicity. The theorem can be established by verifying that the generalized eigenspace for 1 is a symplectic subspace.

The remaining pairs for the centered N-body problem are rotation-angular momentum pairs. These pairs of 1's are a kind of linear shadow of Noether's theorem (see Section A.5). Recall that this theorem associates a conservation law to each infinitesimal symmetry of a Hamiltonian system. Write an infinitesimal symmetry as a vector field V and the corresponding conserved quantity as f. They *commute*: $df(V) = 0$. Assume that V is independent of X at the base point z_0 of our periodic orbit. Since the Hamiltonian flow commutes with the flow of V and preserves f, we have that $AV = V$ and $A^* df(z_0) = df(z_0)$. So A preserves the nonzero vector $V(z_0)$ and the nonzero covector $df(z_0)$ whose kernel contains V. Choose a basis $e_1, \ldots, e_n$ for our tangent space $T_{z_0}\mathcal{P}$ such that $e_1 = V$ and $df(z_0) = \theta_2$ where $\theta_1, \theta_2, \ldots, \theta_n$ is the dual basis to $e_1, \ldots, e_n$. Relative to this basis, A has the form

$$
A = \begin{pmatrix}
1 & * & \cdots & * & * \\
0 & 1 & 0 & \cdots & 0 \\
0 & * & * & \cdots & * \\
\vdots & \vdots & \vdots & \vdots & \vdots \\
0 & * & * & \cdots & *
\end{pmatrix},
$$

exhibiting 1 as a double root for A's characteristic polynomial, and the possibility of a corresponding two-by-two Jordan block. (Regarding the 2nd row of the matrix, recall that the matrix representation of A^T acting on the dual space is as the transpose matrix.) This Jordan block is distinct from the one exhibited for time-energy described above. So it yields another pair of 1's.

Concretely, consider the case of the planar N-body problem. We can rotate our periodic solution $q(t)$ by θ radians to form the new solution $R_\theta q(t)$ without changing its period T. It follows that $A\frac{\partial}{\partial\theta} = \frac{\partial}{\partial\theta}$ where $\frac{\partial}{\partial\theta}$ is the infinitesimal generator of rotation evaluated at the phase point corresponding to $z_0 = (q(0), \dot{q}(0))$. The N-body flow also preserves the angular momentum J, the Hamiltonian generator of rotation, from which it follows that $A^* dJ = dJ$ with the one-form dJ being evaluated at z_0. As in the previous paragraph, this establishes the two-by-two Jordan block for the monodromy associated to angular-momentum-rotation.

For the *centered* planar N-body problem there are only these four mandatory symmetry induced 1's: the time-energy pair and the rotation-angular momentum pair. For the centered spatial N-body problem there are typically six mandatory 1's: the time-energy pair and two pairs corresponding to rotation-angular momentum. As the dimension d increases, the number of mandatory 1's for the N-body problem in d-space grows. This the total number of mandatory 1's that can be expressed in terms of the dimension of $SO(d)$ and its co-adjoint orbits and is associated to the dimension of the symplectic reduced space at the value of angular momentum for the orbit. See Meyer [123, section 6.7] for details.

2.3.3 The Return Map and Reduction

The return map replaces the study of the flow near a periodic orbit with the study of a map near a fixed point. And it gets rid of the mandatory symmetry-induced 1's in the spectrum of the monodromy.

Let γ be a periodic orbit of the smooth vector field X. Choose an embedded disc $\mathbb{D}$ transverse to the orbit at the point $z_0 = \gamma(0)$. View the points of $\mathbb{D}$ as initial conditions for the flow of X and flow them forward until their orbits first pierce the disc again. In this we obtain the return map

$$R\colon (\mathbb{D}, z_0) - \to (\mathbb{D}, z_0),$$

which has z_0 as a fixed point. (The broken arrow notation indicates that the domain of the map R need not be all of $\mathbb{D}$, but rather only some neighborhood of $z_0 \in \mathbb{D}$.) Other names for the *return map* are Poincaré return map, Poincaré section, or simply section.

Since X is nonzero near z_0, the flow-box theorem[1] tells us that $\mathbb{D} \times I$ provides a model for the flow near z_0, with X tangent to the direction of the interval I. Recall that $AX = X$. It follows that A induces a linear map $T_{z_0}\mathcal{P}/\mathbb{R}X(z_0) \to T_{z_0}\mathcal{P}/\mathbb{R}X(z_0)$. The differential dR_0 realizes this quotient map, with $T_0\mathbb{D}$ realizing the quotient vector space. In this way we get rid of the first mandatory 1.

There were two 1's in A when X was a Hamiltonian vector field for a time-independent Hamiltonian H, these 1's corresponding to energy-time. We just got rid of one of them. To get rid of the other, choose $\mathbb{D}$ to lie within the energy level $\{H = h_0\}$ containing our orbit and transverse to the orbit within that energy level. This $\mathbb{D}$ has one dimension less than the disc of the previous paragraph. (We could, for example, take this new disc to be the intersection of our previous disc with the energy level set.) We will call such a $\mathbb{D}$ a *symplectic slice* since a bit of symplectic geometry establishes that $\mathbb{D}$ is a symplectic submanifold of phase space. See for example Section B.8.

Now $\Phi_T \circ H = H$ so that A also maps $T_{z_0}\{H = h_0\} \subset T_{z_0}\mathcal{P}$ to itself. Moreover, from $dH(X) = 0$ we have that $X(z_0) \subset T_{z_0}\{H = h_0\}$ so that A induces a linear map $T_{z_0}\{H = h_0\}/\mathbb{R}X(z_0) \to T_{z_0}\{H = h_0\}/\mathbb{R}X(z_0)$.

Definition 2.14 We call this linear map on $T_{z_0}\{H = h_0\}/\mathbb{R}X(z_0)$ induced by A the *energy-reduced monodromy*.

---

[1] This is essentially the same as the straightening theorem for vector fields. See https://en .wikipedia.org/wiki/Straightening_theorem_for_vector_fields or [2, p. 67].

We can view the construction of the energy-reduced monodromy in terms of symplectic reduction. See Section B.7 for more on this perspective. In reduction, we fix the value of the momentum map and quotient out the result by the group action. The Hamiltonian H is the momentum map for the $\mathbb{R}$-action induced by the Hamiltonian flow of our vector field X, so that the corresponding reduced space is, formally, the quotient space $\{H = h_0\}/\mathbb{R}$. This quotient space is rarely a manifold, but, by the flow-box theorem it is locally a manifold near points such as z_0 where $X(z_0) \neq 0$. Since the disc $\mathbb{D}$ is transverse to the flow and lies in the energy level set, $\mathbb{D}$ forms a realization of this local quotient. In particular, the symplectic form, restricted to the disc, is symplectic and the return map R is a symplectic map of a disc whose codimension is 2 within the full phase space.

The other mandatory 1's for the N-body problem are associated with rotation and angular momentum. We can get rid of them by working in a higher codimension disc that yields a local realization of the symplectic reduced space for energy–momentum. Let us see how to form this disc in the case of the planar N-body problem. Use the infinitesimal generator $\frac{\partial}{\partial\theta}$ of the circle action and the scalar angular momentum J and their associated mandatory pair of 1's. As we saw above, in addition to $AX = X$ and $A^*dH = dH$ we have $A\frac{\partial}{\partial\theta} = \frac{\partial}{\partial\theta}$ and $A^*dJ = dJ$. It follows that A induces a linear operator on the linear space $T_{z_0}\{H = h_0, J = J_0\}/Span(\frac{\partial}{\partial\theta}|_{z_0}, X(z_0))$. In keeping with Definition 2.14 we call this operator the *energy–momentum reduced monodromy*. (This linear space is a symplectic vector space whose dimension is 4 less than $2(N - 1)$ provided that $dH \wedge dJ \neq 0$ at z_0.) To form the corresponding return map, take a disc $\mathbb{D}$ that lies within $\{H = h_0, J = J_0\}$ and is transverse to both X and $\frac{\partial}{\partial\theta}$. Rotates of orbits nearby to our γ will then intersect the disc transversally and yield a return map whose linearization at the origin is the operator A.

More generally speaking, suppose we have conservation laws $f_1, \ldots, f_k$ as described towards the end of Section 2.3.2, conservation laws that fit together so as to form the momentum map $f = (f_1, \ldots, f_k)$ for some G action on our phase space (See Definition A.22). Suppose that f, like angular momentum, is equivariant, so that $f(gz) = g \cdot f(z)$ where $v \mapsto g \cdot v$ denotes the co-adjoint representation of G on its dual Lie algebra $\mathfrak{g}^* \cong \mathbb{R}^k$. Write $\mu = f(z_0) \in \mathfrak{g}^*$. Let G_μ be the Lie algebra of the stabilizer of μ. The standard Marsden–Meyer–Weinstein construction (Section B.6) of the symplectic reduced space tells us that this reduced space is $f^{-1}(\mu)/G_\mu$. As a local model for this reduced space we take a disc $\mathbb{D} \subset f^{-1}(\mu)$ that is transverse to the G_μ-orbits. Intersect this disc with our energy disc from the previous paragraph and we get a symplectic disc and a symplectic return map R as above. (Compare with Section B.8.) This is our final disc. The Poincaré return map is defined by returning to the disc modulo G_μ. The differential at the origin of the return map represents the linear map A modulo the generalized eigenspace of all of our mandatory 1's. It is this return map that is the subject of the KAM stability theorem in Section 2.4.

Instead of proceeding as above with transverse discs on the original phase space, it may sometimes be simpler to work on the symplectic reduced space, with its Hamiltonian flow. Suppose that the orbit through z_0 upstairs in the full phase space is not periodic, but rather is relatively periodic. Then downstairs on the reduced space it is represented by a bona fide periodic orbit. If z_0 is not a relative equilibrium, then the downstairs orbit is not a zero of the vector field. We then take our disc in the reduced space, *within the reduced energy level* there, and transverse to the reduced Hamiltonian vector field near the image of z_0. We thus get a symplectic return map R on a disc (diffeomorphic to the disc of the preceding paragraph) and proceed as before.

2.4 KAM Stability and Its Consequents

2.4.1 The Theorem

The acronym KAM stands for the three mathematicians Kolmogorov, Arnold, and Moser. The KAM theorem applies to certain spectrally stable periodic orbits. (For spectral stability, see Definition 2.9.) Unfamiliar terminology used in our statement of the theorem will be described immediately following the statement.

Theorem 2.15 (KAM; energy version) *Let R be the energy-fixed return map for a periodic orbit of a smooth autonomous Hamiltonian system. Suppose that:*

- *The derivative of R at the origin, which is the energy-reduced monodromy of Definition 2.14, is spectrally stable.*
- *(L) The eigenvalues of this monodromy satisfy the finite number of non-resonance inequalities specified by Equation (2.7).*
- *(T) The third-order Taylor expansion of R at the origin satisfies the twist condition as specified by Equation (2.13).*

Then that orbit lies in a tubular neighborhood inside its energy level that is partially filled by irrational Lagrangian tori invariant under the flow. These are the KAM tori. See Figure 2.3. The Lebesgue measure of the union of these tori is positive and their density tends to 1 as we approach the orbit.

The hypothesis of Theorem 2.15, conditions (L) (for linear) and (T) (for twist), are a finite number of open conditions on the third-order Taylor

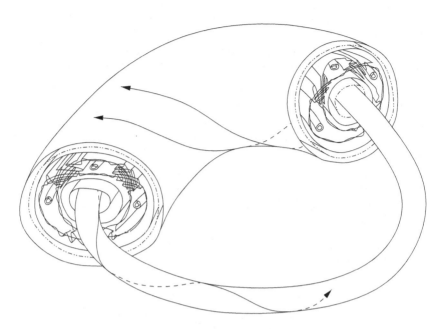

Figure 2.3 KAM tori surrounding a hidden KAM stable periodic orbit. Courtesy of Jacques Féjoz.

expansion of R at the origin. These hypotheses can be verified numerically and made rigorous using interval arithmetic if desired, allowing for numerical proofs of KAM stability. Condition (L) implies, among other things, that 1 is not in the spectrum of dR_0, hence all our attention is on getting rid of mandatory 1's.

Definition 2.16   A KAM stable orbit is an orbit satisfying the hypothesis of the KAM theorem, Theorem 2.15.

We also owe the reader explanations of terms used in the theorem's conclusion. Phase space $\mathcal{P}$ comes with a symplectic form ω and the N-body vector field is a Hamiltonian vector field relative to this form. (See Definition A.10.) Let $2m$ be the dimension of P. Then m is called the number of degrees of freedom of the system. We call a submanifold of a phase space Lagrangian if its dimension is m and if the restriction of ω to the submanifold is zero. (A submanifold on which $\omega = 0$ must have dimension m or less.) This gives meaning to the term Lagrangian tori. We say that a flow on a torus is *linear* if there exist angular coordinates θ_i for the torus such that the defining vector field is $\dot{\theta}^i = \nu^i = const$. If the list of frequencies $(\nu_1, \ldots, \nu_m)$ is independent over $\mathbb{Q}$, we say that the linear flow, or the torus itself, is *irrational*. Here, that flow is the restriction of the given vector field to the invariant torus. If M is a measurable subset of the phase space that contains our orbit γ, then the "density of M as we approach γ" is $\lim_{\epsilon \to 0} v(N_\epsilon(\gamma) \cap M)/v(N_\epsilon(\gamma))$ where v denotes volume, that is, the Lebesgue measure, and where $N_\epsilon(\gamma)$ means the ϵ neighborhood of the set γ. (The ϵ neighborhood can be taken relative to any distance function that gives the same topology as the manifold topology without changing the density.)

KAM stability is a probabalistic form of Lyapunov stability. Suppose that you start within ϵ of a KAM stable orbit. With high probability you will land on a KAM torus and then you stay on it for all time. The *probability* that you wander away from your ϵ-neighborhood tends to 0 with ϵ.

2.4.2 Applying KAM to the N-Body Problem

Due to the mandatory 1's described above and forced on us by Galilean symmetry, periodic orbits of N-body systems *never* satisfy hypothesis (L) of the KAM stability theorem, Theorem 2.15. As we described in Section 2.3.3, we can rid our orbit of these additional 1's by adapting the return map disc $\mathbb{D}$ to angular momentum as well as energy, and making sure that this disc is transvserse to the relevant subgroup (G_μ) of G. This process of adapting discs is locally equivalent to forming the *symplectic reduced space* associated to rotations and to the orbit in question. See Appendix B regarding symplectic reduction. The language of reduction makes it easier for us to state the appropriate KAM theorems.

Let γ be our orbit in question, the solution to a Hamiltonian system with a Lie group symmetry; that is, rotations. Let J be the momentum map for this symmetry (angular momentum for the N-body problem). Let $J_0 = J(\gamma(0)) = J(\gamma(t))$ be the constant value of the momentum map on the orbit. By the symplectic reduced space for γ we mean the symplectic reduced space $\{J = J_0\}/G_{J_0}$ as described in Section B.6, together with the projected orbit $\pi(\gamma(t))$ where $\pi: \{J = J_0\} \to \{J = J_0\}/G_{J_0}$ is the projection. Here $G_{J_0} \subset G$ is the co-adjoint stabilizer of $J_0 \in \mathfrak{g}^*$.

Definition 2.17 We will say that the relative periodic orbit γ is reduced KAM stable if its projection $\pi(\gamma)$ to its reduced space near $\pi(\gamma)$ satisfies the hypotheses (L) and (T) of Theorem 2.15 for the Hamiltonian flow induced on the reduced space.

The planar N-body problem has symmetry group $G = \mathbb{S}^1$, the group of rotations. (Remember: we fix the center of mass at the origin so can ignore translations.) Since the circle is Abelian we have that $G_{J_0} = \mathbb{S}^1$. The symplectic reduced space is $T^*Cone(\mathbb{C}P^{N-2} \setminus \Delta)$ as we described in the introductory chapter, Chapter 0, Equations (0.63) and (0.64). (The symplectic form on the reduced space is *twisted* by adding to the canonical form the pullback of the Fubini–Study form on the shape sphere $\mathbb{C}P^{N-2}$ and multiplying it by $-J_0$.) The reduced space has dimension $4(N-2) + 2$. The dimension of the energy–momentum adapted discs have two dimensions less (getting rid of the two mandatory 1's for energy and dilation, or time) so have dimension $4(N-2)$, or 4 for the planar three-body problem.

Theorem 2.18 (Reduced KAM stability; planar case) *Let $\gamma(t)$ be a relative periodic orbit for the* planar *N-body system that is reduced KAM stable. Then a rotationally invariant neighborhood of γ in the full phase space is partially filled with irrational invariant tori whose density tends to 1 as we approach the orbit. On these tori the relative distances r_{ab} are quasi-periodic and hence are bounded functions of time.*

Remark 2.19 I am certain some version of this theorem holds for the N-body problem in d-space, $d > 2$. See the chapter notes in Section 2.9.

Proof of Theorem 2.18 The reduced orbit $\pi \circ \gamma$ is periodic. Apply the KAM theorem 2.15 to this reduced orbit. One gets KAM stability at the reduced level, for that value of energy and momentum, with consequent slice map defined on a disc of dimension $4(N-2)$ within the reduced space. Now vary the initial conditions, allowing energy and momentum to vary. We can follow the reduced periodic orbit into all the nearby energy–momentum reduced spaces by the implicit function theorem since the assumption of the theorem excludes 1's from the spectrum of the reduced monodromy. Because of the planarity assumption, these reduced spaces do not jump in dimension and vary smoothly with the initial condition. Since the KAM conditions (L) and (T) are open conditions they persist into nearby values of angular momentum and energy, yielding KAM stable orbits within each reduced space, and, in particular, are of positive measure and density 1 nearby in these reduced spaces. Now pull back up to the total phase space to see that the rotationally invariant family of nearby relative periodic orbits are surrounded by tori (no longer Lagrangian) whose total measure is positive and whose density tends to 1 as we approach any orbit in the family. QED

2.4.3 Parsing the KAM Theorem

In this section we describe in detail conditions (L) and (T) of the KAM theorem, Theorem 2.15, and end with a recapitulation of the theorem.

Non-resonance: Condition (L)

Here, we describe condition (L), for linear, occurring in Theorem 2.15.

Identify the periodic orbit in question with the circle $\mathbb{S}^1$ and the transverse symplectic disc with a $2n$-dimensional disc $\mathbb{D} \subset \mathbb{C}^n = \mathbb{R}^{2n}$, where $2 + 2n$ is the dimension· of the phase space and the disc lies in an energy level. A tubular neighborhood of the orbit within the energy level set is diffeomorphic to $\mathbb{S}^1 \times \mathbb{D}^{2n}$, with $\mathbb{S}^1 \times 0$ corresponding to the orbit, and with $\mathbb{D}^{2n}$ symplectically diffeomorphic to the slice. Spectral stability of our orbit implies that we can coordinatize the $\mathbb{C}^n$ factor so that the linearized return map at the origin takes the form

$$dR_0(z_1, \ldots, z_k) = (\lambda_1 z_1, \ldots, \lambda_n z_n), \text{ where } \lambda_j = exp(2\pi i \omega_j) \in S^1.$$

(See condition (2.4)). Recall that dR_0 can be identified with the quotient map induced by the monodromy A upon modding out by the two-dimensional subspace corresponding to the energy–time 1's in its spectrum.

Definition 2.20 Condition (L), the *non-resonance* condition of Theorem 2.15, is the condition that there are *no* nonzero integer vector solutions $(k_0, k_1, \ldots, k_n)$ to the linear equations

$$k_1\omega_1 + k_2\omega_2 + \cdots + k_n\omega_n = k_0 \text{ having } 0 < \Sigma_{j=1}^n |k_j| \leq 4. \tag{2.7}$$

A linear equality among frequencies ω_j involving integer coefficients k_j is called a *resonance*. The size $\Sigma_{j=1}^n |k_j|$ of the integers involved is the *order* of the resonance. Thus condition (L) asserts that there are no resonances of order 4 or less. In particular, condition (L) excludes any of the λ_j from being 1.

Twist and Oscillators: Condition (T)

Here, we describe condition (T) of in Theorem 2.15. We will need to begin with some background on action-angle coordinates and the Birkhoff normal form.

Symplectic structure and tori

The standard symplectic structure on the disc $\mathbb{D}^{2n}$ referred to above is the one induced from $\mathbb{C}^n$. Write $z_j = q_j + ip_j$ with $i = \sqrt{-1}$. Then the q_j, p_j are Darboux coordinates so that $\omega = \Sigma dq_j \wedge dp_j$. Now $dq_j \wedge dp_j = \frac{1}{2i} d\bar{z}_j \wedge dz_j = dI_j \wedge d\theta_j$.

The linearized return map dR_0 leaves the n-dimensional torus $\{z \in \mathbb{C}^N : |z_j| = c_j, j = 1, \ldots, n\}$ invariant for each choice of the positive real constants c_j. These tori are Lagrangian relative to the standard symplectic structure. A subcollection of these tori will become, upon perturbation from dR_0 to R, the KAM tori of the KAM theorem.

Oscillators and Normal Forms. The harmonic oscillator with frequencies ω_j is described by the quadratic Hamiltonian $H_{osc}(z_1, \ldots, z_n) = \frac{1}{2}\Sigma\omega_j|z_j|^2$. The Hamiltonian vector field for the harmonic oscillator is linear with eigenvalues $\pm i\omega_j$ and its time 2π flow is dR_0.

Define new coordinates (I_j, θ_j) on $\mathbb{C}^n$ called *action-angle coordinates* by

$$I_j = \frac{1}{2}|z_j|^2, \theta_j = Arg(z_j),$$

so that

$$H_{osc} = \Sigma\omega_j I_j. \tag{2.8}$$

These are Darboux coordinates: $\omega = \Sigma dI_j \wedge d\theta_j$. Their Darboux nature is seen by noting that $dx \wedge dy = d(\frac{1}{2}r^2) \wedge d\theta$ expresses the planar area form in polar coordinates. Thus $dq_j \wedge dp_j = dI_j \wedge d\theta_j$, so the Darboux coordinates q_j, p_j play the role of Cartesian coordinates (x, y) while (I_j, θ_j) play the role of polar coordinates (r, θ). The tori described above are given by $I_j = (1/2)c_j^2$.

Condition (T) has to do with the second-order term in what is known as the Birkhoff normal form for R. In order to describe this form, let's see what happens when we replace H_{osc} by a Hamiltonian $H = h(I_1, \ldots, I_n)$ depending only on the actions I_j. A look at Hamilton's equations for such an H shows that its time t flow has the form $I_j \rightarrow I_j$ while $\theta_j \mapsto \theta_j + t\nu_j(I)$, where $\nu_j = \frac{\partial h}{\partial I_j}$. Thus the flow of this H leaves the tori invariant and is linear on each torus. We call any symplectic map obtained by setting $t = 1$ for a flow coming from such an H an *integrable oscillator map*. We call the map $I \mapsto \nu(I) = (\nu_1(I), \ldots, \nu_n(I))$, the *frequency map* associated to our integrable oscillator map.

The integrable oscillator maps relevant for condition (T) are those with Hamiltonian of the model form

$$H_{model} = H_{osc} + h_2, \tag{2.9}$$

where

$$h_2 = \frac{1}{2}\Sigma a_{jk}I_j I_k, \tag{2.10}$$

and where the a_{jk} are the entries of a symmetric matrix. In other words, their Hamiltonians are linear plus quadratic in the actions. The corresponding frequency map for this H_{model} is $\nu_j(I) = \omega_j + \Sigma a_{jk}I_k$. Let us write $R_{model} \colon \mathbb{D} \rightarrow \mathbb{D}$ for the time 1 map associated to H_{model}. Thus,

$$R_{model} \colon (I_j, \theta_j) \mapsto (I_j, \theta_j + \nu_j(I)) \text{ where } \nu_j(I) = 2\pi\omega_j + \Sigma a_{jm}I_m. \tag{2.11}$$

The Birkhoff normal form theorem applied to our return map R is an iterative algorithm that yields a formal series $H = h_1 + h_2 + h_3 + \cdots$ of integrable Hamiltonians and a successive symplectic change of variables such that the associated flow R_s of the partial sum $h_1 + \cdots + h_s$ agrees with R up to order $2s + 1$ in cartesian coordinates. Here, the h_k are homogeneous of degree k in the actions so that $h_1 = H_{osc}$. To proceed from R_{s-1} to R_s in the algorithm requires that non-resonance conditions of order $2s$ hold for the frequencies ν_j of R_0, that is, for the oscillator flow. See Arnol'd [11, Appendix 7B], particularly Theorem 2 on p. 388 regarding the Birkhoff normal form.

Under our non-resonance hypothesis (L) this algorithm yields a symplectic polynomial diffeomorphism fixing the origin and converting the *3rd jet* of our given return map R to that of an integrable oscillator map of the above quadratic type of H_{model}. We obtain a polynomial canonical change of variables that puts our return map R into the form

$$R = R_{model} + O(r^5), \tag{2.12}$$

where R_{model} is as in Equation (2.11). *The term h_2 of H_{model} is called the second-order Birkhoff invariant.* (See Birkhoff [19, p. 78–88] and Arnol'd [11, appendix 7], especially p. 388–389.) The r^5 of the error term refers to the r of $r^2 = 2\Sigma_j I_j = \Sigma |z_j|^2$. (Note: third-order polynomials in the original Cartesian q_j, p_j are $O(r^3)$.) Under assumption (L), the spectral invariants of the symmetric matrix a_{jk} become symplectic invariants – unchanged by symplectic conjugacies. In other words, they are independent of choice of Darboux coordinates, or equivalently, under the choice of symplectic disc $\mathbb{D}^{2n}$.

Definition 2.21 (Condition (T).) The twist condition (T) is the condition that the matrix a_{jm} arising in the second-order Birkhoff normal form h_2 be nondegenerate,

$$det(a_{jm}) \neq 0. \tag{2.13}$$

Physically, the twist condition asserts that the frequencies vary with amplitudes and that this variation is nondegenerate: The frequency map is a local diffeomorphism near $I = 0$.

Definition 2.22 The symmetric matrix a_{jk} in Equation (2.13) is called the torsion matrix.

Exercise 2.23 Show that relative to Cartesian coordinates the twist condition is a condition on the third-order Taylor expansion of R at the origin. Hint: Trace through the changes of variables $q_j = \sqrt{2I_j}\cos(\theta_j), p_j = \sqrt{2I_j}\sin(\theta_j)$ and look at the normal form in Cartesian coordinates.

Recapitulation of the KAM Theorem

The invariant tori guaranteed to exist are those for which the frequencies $\nu(I)$ are "sufficiently irrational." (See [11, 58]). The condition on frequencies of being sufficiently irrational selects out a frequency set $F \subset \mathbb{R}^n$ of positive measure. On these tori the Hamiltonian flow becomes, after a time reparameterization, the flow governed by $\dot{\theta}_0 = 1$, $\dot{\theta}_j = \nu_j(I)$, where θ_0 is the coordinate around the orbit circle S^1. The corresponding flow, restricted to such a torus, is the standard irrational linear flow on a torus, an ergodic flow all of whose orbits are dense in the torus.

The twist condition (2.13), condition (T), is a nondegeneracy condition on the frequency map. Condition (T) guarantees that the inverse image $K := \nu^{-1}(F)$ of this frequency set in action space continues to have positive measure. Here, K has the structure of a fat Cantor set in action space, that is, topologically a Cantor set, but one whose Lebesgue measure relative to $dI_1 dI_2 \cdots dI_n$ is positive. The set of invariant tori whose existence the KAM theorem guarantees is diffeomorphic to $S^1 \times K \times \mathbb{T}^n = K \times \mathbb{T}^{n+1}$, where the torus $\mathbb{T}^n$ is coordinatized by the θ_j, $j > 0$ now. We call these the *surviving tori*. The Lebesgue measure on the energy surface is $dI_1 \cdots dI_n d\theta_1 \cdots d\theta_n d\theta_0$, showing that the measure of the set of surviving tori is positive. Divide the measure of the set of surviving tori having $|I| \leq \epsilon$ by the measure of the subset of phase space having $|I| \leq \epsilon$ to form the relative measure of surviving tori. The assertion of "density 1" in Theorem 2.15 is the assertion that this relative measure tends to 1 as $\epsilon \to 0$.

2.4.4 KAM Heuristics

Irrational tori are hard to destroy. This principle drives KAM. The Birkhoff normal form R_{model} (Equation (2.11)) leaves each of the tori $I_j = const.$ invariant. Our return map R is a perturbation of this normal form. As we get closer to the orbit ($r \to 0$) the perturbation $R - R_{model}$ tends to zero. The principle tells us to expect that more and more of the normal form's irrational tori will survive as $r \to 0$.

Why are irrational tori harder to destroy through perturbations than rational ones? An irrational torus comes as a solid block, being the closure of any one of its orbits. In contrast, the orbits of a rational torus are already individually closed, which means that such a torus is a rather floppy, weakly held together ensemble of springs, an $(n - 1)$-parameter ensemble of already closed orbits.

The principle that irrational tori are sturdier than rational ones is most easily brought to fruition, that is, turned into proofs, in looking at the theory of circle diffeomorphisms and its relation to an area-preserving map R of the two-dimensional disc. When $n = 1$ in the discussion of the oscillator approximation above (see around Equation (2.8)) we just have one action $I = I_1$ and one angle $\theta = \theta_1$. Both R and the model map R_{model} are area-preserving diffeomorphisms of the disc that map the origin to itself. The map

R_{model} has the additional feature that it preserves I so we can view it as a one-parameter family of circle maps $f_I : \theta \mapsto R_I(\theta) = \theta + \nu(I)$ parameterized by I. The circle labeled by I is irrational or rational depending on whether or not $\nu(I) = \omega + aI$ is rational or irrational. The theory of circle diffeomorphisms associates to every circle diffeomorphism $f : \mathbb{S}^1 \to \mathbb{S}^1$ a rotation number ν. If ν is irrational then that theory supplies us with a change of variables that turns f into the model map $f_\nu : \theta \mapsto \theta + \nu$ of irrational rotation by ν. In other words, if we perturb such an f, keeping its rotation number fixed, we end up with the *same map* after a change of coordinates (homeomorphism of the circle). If ν is rational, there is typically no such change of variables: Maps with rational rotation number are not stable under perturbations.

Now we turn our perturbation on, replacing the map R_{model} with R. We would like to think of R as a one-parameter of circle maps. This is false – many of the rational circles for R_{model} get destroyed – but let us see where it takes us. The circle theory suggests that the R-dynamics on the irrational circles gets perturbed – one can conjugate to make it look like R_{model}. Essentially, one wants to fix the irrational ν and follow the circle and the map to a nearby invariant circle for R under which R is conjugate to the irrational translation map f_I with $\nu = \nu(I)$. The irrational circles near the origin wiggle a bit, but most remain. The circle theory suggests one of the key ideas in the proofs of KAM: Focus on the rotation number (frequency!) and not the action when imagining following the circles as we turn the perturbation on. Label invariant circles (tori in higher dimensions) according to rotation number (frequency vector), not according to action.

2.4.5 Arnol'd Diffusion: Dodging Sticky Tori

KAM tori, being Lagrangian submanifolds, have half the dimension of the phase space $\mathcal{P}$ within which they sit. Write $2(n + 1)$ for the dimension of this phase space, so that the tori have dimension $n + 1$, which is also known as the number of degrees of freedom of the mechanical system described by the dynamics. Since the KAM torii are Lagrangian and live inside the same energy level as their mother orbit, they form a family of $(n + 1)$-dimensional submanifolds embedded within that $(2n + 1)$-dimensional energy surface containing the orbit.

Two-Degree-of-Freedom Stability

If the number of degrees of freedom is 2 then $n = 1$. Each KAM torus is two-dimensional within its three-dimensional energy level set and separates the energy level into two components, one containing the orbit, thereby preventing escape from the orbit. Thus the concentric family of KAM tori force Lyapunov stability of the orbit they surround. An example is provided by the planar circular restricted three-body problem. Here, Lyapunov stability has been established by KAM for some orbits. See, for example, Meyer and Schmidt [125] and references therein.

Clever Diffusion

As soon as the number of degrees of freedom is greater than 2 the tori no longer block motion. For example, if $n = 2$ then the KAM tori are three-dimensional within a five-dimensional energy level set, and the complement of any torus will be a connected subset of the energy level set. A clever trajectory can wind around and dodge all the many KAM tori, eventually escaping any given neighborhood of the orbit. This escape process, this evading the Cantor net of KAM tori, is known as *Arnol'd diffusion* and is currently the only approach towards a "no" answer to our Open Question 2, or its unbounded variant, Herman's "oldest question."

2.4.6 Nekhoroshev Stability

Although escape from the KAM tori surrounding an orbit may happen, when it does, it takes a long long time.

Theorem 2.24 (Nekhoroshev) *If a periodic orbit is KAM stable and its torsion matrix a_{jk} is positive definite, then there are positive numbers a, b, C with the following significance. For all $\epsilon > 0$ sufficiently small any solution starting a distance ϵ from the orbit remains within ϵ^a of the orbit for times t with $|t| \leq C \exp(\epsilon^{-b})$.*

See the chapter notes (Section 2.9) regarding proofs of this theorem.

2.4.7 Breakdown of Tori and Diffusion Heuristics

Return to the area-preserving return map R of the two-disc as described in Section 2.4.4. We viewed R as a perturbation of its second-order Birkhoff normal form R_{model}. Recall that we can view the latter as a one-parameter family of circle diffeomorphisms $f_I : \mathbb{S}^1 \to \mathbb{S}^1$, the parameter I being the action $I = \frac{1}{2}r^2$. Circle maps with irrational rotation numbers $\nu(I)$ for I small correspond to the surviving KAM tori. What about circles with rational rotation number $\nu = p/q$ with $p, q \in \mathbb{Z}$?

The typical circle map with rational rotation number p/q has a finite even number of period-q orbits. The orbits are interleaved between stable and unstable so that as we go around the circle, the points forming these orbits alternate between being stable and unstable fixed points of the map f^q. The stable orbits form the ω-limit set of f^q while the unstable orbits are its α-limit set. Thus points lying between these periodic orbits, upon being iterated under f, accumulate onto one of the stable orbits in forward time and to an unstable one in backward time.

Now return to our return map R, viewed as a perturbation of the integrable map R_{model}. What happens to the rational invariant circles of R_{model} upon perturbation? They do not remain whole, that is, invariant, under R. Rather, the invariant circle for the rational rotation number p/q is destroyed, replaced by a finite number of periodic orbits of period q alternating between hyperbolic and elliptic. Being in the symplectic world, we have restrictions on the eigenvalues for R^q at these orbits. If an orbit has an unstable direction (eigenvalue outside the unit circle) then it also has a stable one (eigenvalue inside the unit circle). In this way, our rational invariant circle bifurcates into a finite number of elliptically stable periodic orbits that seed the elliptic islands and an equal number of hyperbolic orbits that seed a thin forest of "chaos."

The dynamics near the hyperbolic orbits, of necessity, must wander off into two dimensions and so cannot stick to any one curve. The original invariant rational circle has been destroyed. Typically, the stable and unstable manifolds of these hyperbolic orbits intersect transversally, leading to homoclinic or heteroclinic tangles and what is called *Hamiltonian chaos*. This chaotic behavior lives in the interstices between the surviving irrational KAM circles.

The elliptic periodic orbits provide us with opportunities to apply KAM again, now to R^q. Typically KAM succeeds. Most of the elliptically stable satellite orbits of our original orbit will be KAM stable and so will be surrounded by their own *elliptic islands*. The pattern repeats around each satellite orbit, leading to an overall fractal type of structure.

The model map used for the situation just described is the standard map pioneered by Chirikov [37] (see also references therein). See Figure 2.4.

2.5 Case Study: The Eight

Simó's article [192] inspired this section.

In the Tour, Chapter −1 of this book, we described the figure eight orbit and gave its initial conditions to 12 decimal places. See Figure −1.8 and Section −1.2.3 there. Simó found these initial conditions by starting out with the variational description of the eight as described in Chapter 3 (see Section 3.3.3). He then integrated these initial conditions forward using Newton's three-body equations and a trusted numerical integrator.

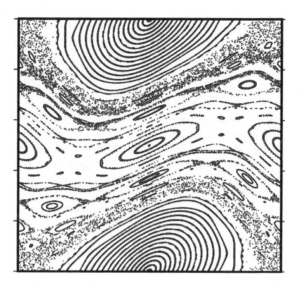

Figure 2.4 Some orbits for the standard map at an intermediate value of forcing, illustrating islands of stability (KAM circles) and ergodic-looking orbits diffusing around the islands. From [37].

Surprise, surprise! After hundreds of thousands of periods, the shape of the orbit looked just like it did after a single period. The figure eight was practically stable. In contrast, if one numerically integrates the initial conditions for the equal mass circular Lagrange orbit, a spectrally unstable orbit, then after a few periods the resulting orbit bears no resemblance to the circular Lagrange orbit.

Simó later proved that the eight was KAM stable by verifying the KAM hypotheses (L) and (T) as follows. The reduced space for the planar three-body problem at angular momentum zero (or any angular momentum) has dimension 6. Subtract 2 for the energy-time mandatory 1's to see that the relevant return map for investigating near-eight behavior is a symplectic map $R: (\mathbb{D}^4, 0) \to (\mathbb{D}^4, 0)$ of a four-disc. Its differential dR_0 has four eigenvalues. It follows that if we find eigenvalues $e^{i\omega_1}, e^{-i\omega_1}, e^{i\omega_2}, e^{-i\omega_2}$ with ω_1, ω_2 clearly distinct, then we have found all the eigenvalues for dR_0 and the orbit is elliptically stable.

Simó [192] investigated the stability of the eight and the dynamics of its surroundings with great care. He computed the eigenvalues of the differential of the return map to be $e^{i2\pi\omega_1}, e^{-i2\pi\omega_1}, e^{i2\pi\omega_2}, e^{-i2\pi\omega_2}$, where

$$\omega_1 = .008422724708131, \quad \omega_2 = .298092529004750.$$

(The eigenvalues are independent of the period or size chosen for the eight since scaled versions of the eight yield conjugate return maps, hence identical eigenvalues.) To verify KAM condition (L) one checks that the 64 numbers $k_1\omega_1 + k_2\omega_2$ with $k_i = -4, -3, -2, -1, 1, 2, 3, 4$ are not integers. We left $k_i = 0$ out of the sum set since it is obvious that $n\omega_i \neq \mathbb{Z}$ for $n = 1, 2, 3, 4$.

When linear algebra, implemented on a machine, tells us that one eigenvalue of the monodromy is $\lambda = e^{i\omega_1}$, how do we know that its true value is not actually $\lambda = (1/.9999...9)e^{i\omega_1}$, in which case the orbit would be spectrally unstable? There is no closed form expression for the eight. Approximate representations of its monodromy matrix A and linearized return map dR_0 must be found with the aid of numerical integration. We do not (and cannot) have exact closed form expressions for A or its eigenvalues.

We know that these four eigenvalues actually lie on the unit circle by invoking the spectral symmetry of Equation (2.5). If one eigenvalue $\lambda = re^{i\omega}$ were to lie outside but very close to the unit circle then it must arise as one of a quartet of eigenvalues: $\lambda, 1/\bar{\lambda} = \frac{1}{r}e^{i\omega}, \bar{\lambda}, 1/\lambda$. If the closeness of r to 1 was below machine precision, we could not tell λ from $1/\bar{\lambda}$ or $\bar{\lambda}$ from $1/\lambda$, but, taken together, $\lambda, 1/\bar{\lambda}, \bar{\lambda}, 1/\lambda$ would have to account for a four-dimensional eigenspace, exhausting our spectrum, and making the occurence of the other found pair of eigenvalues $e^{\pm i\omega_2}$ impossible. So it is enough to find four distinct eigenvalues lying within machine precision of the unit circle and splitting up into two complex conjugate pairs in order to *guarantee* that our orbit is linearly stable. See also Kapela and Simó [90, section 6].

Simó numerically approximated the Taylor expansion of the return map of the eight, thereby establishing the twist condition (T) and establishing that it is KAM stable [192]. In that paper you can find pictures of KAM tori around the eight. Later, Kapela and Simó [90] used interval arithmetic to rigorously establish the two KAM conditions (L) and (T), thereby guaranteeing the eight's KAM stability.

Domain of Stability

The eight is KAM stable, but how far can we push it and retain stability? We can push it by changing its initial conditions, keeping the masses the same, or by varying the masses away from equality. Simó investigated the results of both types of pushing. For the first, he took a certain two-dimensional slice of the Poincaré disc and integrated forward. As he integrated he kept track of the minimum and maximum of the interbody distances and threw orbits out as unstable when these distances went below or above some tolerance. He kept track of up to 30,000 crossings and then plotted the resulting *Julia set* of stable orbits. The result in indicated in Figure 2.5.

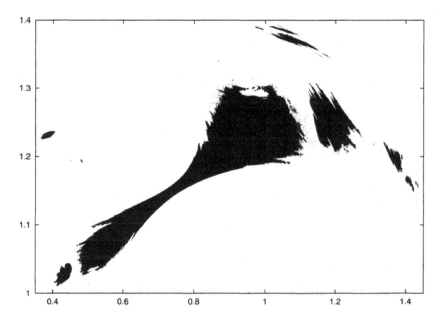

Figure 2.5 Stability zone around the eight. Black indicates stable orbits. The axes, $\dot{x}_3, \dot{y}_3$, parameterize a two-dimensional slice of initial conditions as described in the text. Courtesy of C. Simó [192].

The specific slice Simó took was as follows. He insisted the initial positions were collinear. After reduction, the set of phase points with angular momentum zero and energy $1/2$ is four-dimensional and can be parameterized near the eight by the initial velocities $\dot{x}_2, \dot{y}_2, \dot{x}_3, \dot{y}_3$ of bodies 2 and 3. His slice was determined by the eight relation $\dot{x}_2 = -\frac{1}{2}\dot{x}_3, \dot{y}_2 = -\frac{1}{2}\dot{x}_3$. In Figure 2.5 the axes are then the inital values of $\dot{x}_3, \dot{y}_3$. The black regions are stable according to the numerical tests. The initial condition for the eight is close to the center of the plot, in the narrow pinched section of the main black region.

In order to follow the eight orbit as masses vary, use the implicit function theorem. Let R denote the return map, now viewed as a function of the masses $m \in U \subset \Delta_2$ as well as the initial positions $x \in \mathbb{D}^4$, with the masses entering through Newton's defining equations. Here, $\Delta_2 = \{(m_1, m_2, m_3) : m_i > 0, m_1+m_2+m_3 = 3\}$ is the mass simplex and U is some small open subset near the point $m_* = (1, 1, 1)$ of equal masses. Thus $R \colon \mathbb{D}^4 \times U \to \mathbb{D}^4$. A periodic orbit is a fixed point of R, so a pair (x, m) with $R(x, m) = x$. Set up coordinates so that $x = 0, m = m_*$ corresponds to the figure eight. Rewrite the fixed point equation as $F(x, m) = 0$ where $F(x, m) = R(x, m) - x$. Thus $F(0, m_*) = 0$. The inverse function theorem asserts that we can follow the solution $(0, m_*)$ to

$F(x, m) = 0$ to a nearby family of solutions $(x(m), m)$ provided $\frac{\partial F}{\partial x}|_{0, m_*} \neq 0$. Here, the partial means the derivative with respect to all directions $x \in \mathbb{D}^4$ and saying that the partial derivative is not zero means that it is invertible. But $\frac{\partial F}{\partial x} = DR_0 - Id$ with $m = m_*$. This matrix is invertible *exactly* when 1 is not an eigenvalue of dR_0. The condition (L) on the reduced eight, substantiated by Simó's eigenvalue computations, guarantees this invertibility of $\frac{\partial F}{\partial x}|_{0, m_*}$. So, we can follow the orbit uniquely to get a nearby family of near-eights. Simó followed these mass-dependent eights, keeping track of their stability. He found that these periodic orbits became linearly unstable when he varied the masses by 10^{-5} away from the equal mass case.

Eights in the Universe

Are there figure eight triple star systems in our universe? Heggie [78] investigated this question numerically. He then threw pairs of equal mass binary systems at each other numerically to get a sense of how often figure eights might form in a galaxy. He writes that "a small fraction of one percent" of the experiments led to eights that lasted a few periods. If the masses are not equal, the creation of eights becomes even less likely, since, as Simó discovered, moving a fraction of a percent away from equal masses leads to instability. Heggie summarized his investigations by asserting in an email that "there are somewhere between one figure eight orbit per galaxy and one per universe."[2]

2.6 Planetary Systems

We presented KAM theory so as to apply to a given spectrally stable orbit in a fixed system. KAM theory did not arise this way, but rather it arose to answer families of problems that were explicitly perturbational, with a built-in small parameter ϵ.

A model example of the kind of situation KAM was first built for arises in our solar system, with its eight or so planets. Our Sun is a thousand times more massive than Jupiter, the most massive planet. Take as perturbation parameter ϵ the ratio of the planets' masses to that of the sun.

View the solar system as a special case of the general N-body problem. Select one of the N masses, m_1, to represent the Sun. Scale all the other

---

[2] Simó, C. Personal communication, December 12, 2000.

masses by ϵ, thus arriving at the one-parameter family of N-body problems given by the mass ratios $[m_1, m_2, \ldots, m_N] = [1, \epsilon\mu_2, \epsilon\mu_3, \ldots, \epsilon\mu_N]$. Work in the center-of-mass frame. When we take the limit $\epsilon \to 0$ the interplanetary forces disappear, so we get $N - 1$ uncoupled Kepler problems with the Sun sitting at the common origin. This limit problem is an exactly solvable integrable Hamiltonian system. Turn ϵ back on and we have our perturbation problem, so that the real solar system is viewed as a perturbation of an ideal limit where none of the planets interact with each other.

Near the regions in phase space of interest we can put the Hamiltonian for the system into a form reminiscent of our model Hamiltonian, Equation (2.9), namely,

$$H(I, \theta) = h(I) + \epsilon f(I, \theta, \epsilon). \tag{2.14}$$

Here, $I = (I_1, \ldots, I_m), \theta = (\theta_1, \ldots, \theta_m)$ are the list of action-angle coordinates for the $N - 1$ uncoupled Kepler problems so that m is the number of degrees of freedom of the entire system, so $m = 2(N - 1)$ for the planar problem. The θ_i are angular variables representing the instantaneous positions of the ith planet on its unperturbed orbit when $\epsilon = 0$. Poincaré referred to understanding the dynamics of systems of the form given in Equation (2.14) as the "general problem of dynamics" [169, volume 1, section 13]. Kolmogorov took up Poincaré's challenge and initiated KAM theory.

To get KAM off the ground, one starts with an $\epsilon = 0$ orbit whose uncoupled Keplerian orbits all have negative energy and nonzero angular momentum and so are bounded non-collision ellipses. We assume the ellipses are widely separated from each other to avoid encountering situations where the orbits of two planets come close, since if they did come close once ϵ is turned on, near collisions might happen and huge forces could result, swamping perturbational thinking. One assumes that the individual Keplerian periods are all irrationally related, so that the closure of the resulting unperturbed orbit is a torus of maximal dimension for the unperturbed system.

Even under these assumptions, severe problems arise in trying to apply KAM theory. The resulting tori are not Lagrangian but rather have dimension less than $m/2$. This failure has to do with additional 1's in the linearization that cannot be accounted for by symmetries. These extra 1's arise because, for the Kepler problem, all orbits of the same (negative) energy are periodic of the same frequency. (This extra degeneracy also goes under the name $1:1$ *resonance*.) An entire subfield of KAM theory has developed, known as *degenerate Hamiltonian KAM theory* due to these extra 1's that arise in the planetary problem. Some references on the planetary problem include Féjoz [58] and Chiercia [36].

2.7 Parabolic Infinity and the Oldest Question

Following McGehee [120], Robinson [176], and Guardia et al. [72] we can add a kind of boundary X_∞ at infinity to three-body phase space. When the energy is negative, the points of X_∞ correspond to ideal motions during which two of the three distances r_{ab} is infinite, while the remaining distance stays bounded. The three-body motion extends to X_∞, where it corresponds to Keplerian motions of the bound pair. (See Section 4.4.3 for this boundary when the energy is positive.)

The interior of X_∞ is comprised of the ω-limit points of the hyperbolic–elliptic orbits. An unbounded orbit is hyperbolic–elliptic if the limiting speed of the solo escaper is nonzero. In contrast, the boundary of X_∞ corresponds to the ω-limit points of parabolic–elliptic orbits: orbits for which the limiting speed of the solo escaper is zero.

Suppose that the answer to the oldest question is "yes" for energy E. And suppose that we are handed a KAM stable orbit $c(t)$. Then there must exist a sequence of initial conditions $\gamma_j(0), j = 1, 2, 3, \ldots$, converging to $c(0)$ as $j \to \infty$ but for which the corresponding solutions $\gamma_j(t)$ have ω-limit points lying on X_∞, as $t \to \infty$. These ω-limit points must be parabolic–elliptic, that is, in the boundary of X_∞, since the hyperbolic–elliptic points are open and $c(t)$ is not hyperbolic–elliptic.

Invert this logic. Suppose that the parabolic–elliptic points at inifinity admit nice stable manifolds, manifolds of "honest" orbits converging to them. Then the union of these stable manifolds would form the stable manifold of parabolic infinity. (In various instances the stable manifold of parabolic infinity has been established to exist. See the references at the beginning of this section.) A "yes" answer to the oldest question implies that this stable manifold of parabolic infinity enters every neighborhood of our KAM orbit $c(t)$ and, indeed, of every periodic orbit. So, one approach to the oldest question is to study the topological closure of the stable manifold of parabolic infinity. If this closure contains all of the periodic orbits then the answer to the oldest question is "yes." (And the converse seems to hold as well.)

2.8 An Exercise and More Questions

Exercise 2.25 Let $D(q, p) = (q, -\frac{1}{2}p)$ be the dilation vector field used to create the Jordan block in the second and third paragraphs of Section 2.3.2. Let $X(q, p) = (p, \nabla U(q))$ be the Hamiltonian vector field. Show that $X(q, p)$ and $D(q, p)$ are linearly dependent at a point with $q \notin \Delta$ if and only if q is a central configuration and $v = \mathbb{M}^{-1}p$ is a velocity associated with the corresponding dropped central configuration.

Two related questions are as follows:

1. Recall that the three-body problem is actually a family of problems, the parameters of the family being the mass ratios $m = (m_1 : m_2 : m_3)$, the value of the angular momentum J, and the value of the energy E. The figure eight orbit has zero angular momentum and equal mass ratios. Do there exist spectrally stable periodic orbits for the three-body problem for every value of the parameters of mass ratios, angular momentum, and negative energy?

2. Does Arnol'd diffusion occur for the figure eight orbit? Are there families of solutions to the equal mass zero angular momentum planar three-body problem that start arbitrarily close to the eight and "diffuse" all the way to infinity? One plausible scenario for such a diffusion would be to first diffuse out to one of the eight's three nearby companion hyperbolic eights that Simó found [192], and then to follow that hyperbolic eight's unstable manifold. If that unstable manifold connected the hyperbolic eight to infinity, we would have an escape route for our family to infinity.

2.9 Chapter Notes

On Section 2.1

The Oldest Question Herman [81] named this question "the oldest question in dynamical systems" and attributed it to Newton. Perhaps Herman was thinking of the following passage from the second paragraph of Book III, the General Scholium of Newton's *Principia* (see [163] or [164]), "Elegantissima haecce solis, planetarum et cometarum compages non nisi consilio et dominio entis intelligentis et potentis oriri potuit," which translates to: "This most elegant arrangement of the Sun, planets, and comets could not have arisen except by the design and control of an intelligent and powerful being."

Definition of Asymptotic stability It may seem that the condition that the orbit in question be Lyapunov stable is redundant in the definition we gave of asymptotic stability. It is not. Figure 2.6 depicts a dynamical system on the circle with a single fixed point c such that every orbit converges to c but for which c is not Lyapunov stable. Orbits converge to c in both forward and backward time.

This flow is a translation in disguise. Think of the circle as $\mathbb{R} \cup \infty$, covered with two charts denoted x and u. The x-chart covers $\mathbb{R}$ and the u-chart covers

Figure 2.6 Here c is a global attractor but is not Lyapunov stable.

∞ and on the overlap the coordinates are related by $u = 1/x$. We have depicted the translation vector field $\frac{\partial}{\partial x}$. Viewed in the chart at infinity, this vector field is $-u^2 \frac{\partial}{\partial u}$. The vector field is analytic with a single fixed point at infinity. All trajectories limit to the fixed point in both forward and backward time.

Every $\mathbb{S}^n = \mathbb{R}^n \cup \infty$ admits an analogous translation flow. Cover $\mathbb{S}^n$ with two coordinate charts x and u, with $u = 0$ corresponding to ∞. The overlap map is given by the inversion $u = x/\|x\|^2$. Take a nonzero constant v and consider the vector field that is equal to v in the x-chart. This vector field extends analytically to the whole sphere, where it vanishes quadratically at ∞. All trajectories limit in both forward and backward time to ∞. But ∞ is not Lyapunov stable. In Figure 2.7 we indicate the corresponding phase portrait for $n = 2$ in the u-chart so that ∞ has been placed at the origin.

On Section 2.2

See Meyer's book [123, chapter 6, p. 74] for a nice summary of various types of stability. See also Laskar's Scholarpedia [105] for more details concerning the stability of our solar system.

With reference to Section 2.2.4, the heavier outer planets, Jupiter and beyond, with their high angular momentum, seem to be in approximately stable orbits for longer than the expected age of the Sun.

Lagrange, Laplace, Poisson, and others developed perturbation theory, the method of averaging, and a host of other tools in order to attack the problem of the stability of the solar system. They derived *secular dynamics*, approximating ODEs on a phase space whose points represent $(N - 1)$-tuples of Keplerian conics about an attracting sun. Secular dynamics describes the slow deformations of the planetary Keplerian ellipses. Solutions can be developed in power series of small parameters: mass ratios to Sun, and eccentricities and inclinations of the planet's orbits. One of the big events in this line of research was when Laplace showed that "to first order" in such an

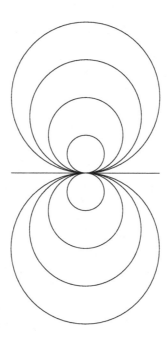

Figure 2.7 A Lyapunov unstable equilibrium point that is a global attractor.

approximation the semi-major axes of the planetary orbits don't change (see [103] and references therein).

On Section 2.3

Regarding the proof of Proposition 2.7   Here is the proof I have in mind. The direct sum of all eigenspaces for eigenvalues having modulus greater than 1 forms a subspace T^- of the tangent space to $\mathcal{P}$ at a point of the orbit. The unstable manifold theorem asserts that T^- is tangent to an immersed manifold sharing its dimension that is invariant under the dynamics and whose orbits converge, under the time-reversed flow, to the orbit at a rate governed by the size of the eigenvalues. Unwinding the precise meaning of this convergence shows that if we flow points of the unstable manifold forward in time then they will leave any sufficiently small fixed neighborhood of the orbit. (Some of these points might return to the neighborhood later!) For more details see most graduate books in dynamical systems, for example, Katok and Hasselblatt [91, p. 240, theorem 6.2.3] combined with the paper of Kelly [92].

Meyer and Hall [124, p. 228, paragraph 3] tell us that buried somewhere within Lyapunov's thesis [112] there is a proof of this instability result. Apparently Hale found it. See pp. 294–295 and, in particular, the long middle paragraph on p. 296 from the last chapter of [73] for Hale's version of Lyapunov's proof using Lyapunov functions.

I would prefer a proof that went along the following lines. For each generalized eigenvalue of the unit circle take its generalized eigenspace. Hypothetical Fact: There is an invariant manifold associated just to that eigenvalue on which the dynamics is conjugate to the linear dynamics restricted to that eigenspace. Such a fact seems quite close to the usual statements of the unstable manifold theorem for a periodic orbit or fixed point. See [112, p. 253, theorem II] for a fairly standard early statement of the unstable manifold theorem, but, again, without the needed center manifold for the proof I have in mind. I could not find a statement or a proof of this hypothetical fact in the literature.

Being unsatisfied with the proofs I found in the literature, I wrote up my own. It follows the lines suggested by Lyapunov. Find a function that is positive semidefinite about the point 0 that represents the orbit, and which increases upon iteration provided one starts inside some particular invariant set – in this case a cone having the origin as cone point. You can find this proof on my web page [156].

On Section 2.4

Regarding Theorem 2.15 I cobbled together this statement of the KAM theorem from appendix 8 of Arnol'd's mechanics book [11] (in particular, sections 5 and 6 of that appendix and its p. 411) and conversations with Simó and Chenciner.

Regarding Theorem 2.18 I could not find a clearly stated version of a reduced KAM theorem, so put this one together for the case of the circle. I have no doubt that there are versions for any compact Lie group. One twist will be that the reduced spaces vary in dimension as we vary the angular momentum J when the dimension d is greater than 2. Another twist will be that the reduced spaces can be singular varieties.

The "A" of KAM I took Arnol'd kayaking once and won't be forgetting it. We were crossing the San Francisco Bay, entering the strait between Angel's Island and the Marin Shore, orthogonal to the path of a sailboat race. They were large, 15 meters or more in length, traveling much faster than we could paddle. They had much too much inertia to dodge or brake for a pair of mosquito-like kayaks trying to cross their swift paths. Three zipped by us, one barely missing us, as we crossed into the edge of the pack. I had visions of my budding career about to go down in flames, as the mathematican who had killed Arnol'd. I told Arnol'd we needed to turn back for safety. He insisted on paddling on to Marin to complete the journey across the Bay. In desperation, I remembered Arnol'd's story of earlier difficulties with tides when attempting a swim across the San Francisco Bay (see the book "Swimming Against the Tides" [94]) and used his own story to convince him to turn around. He lived many more years.

Nekhoroshev I could not find a statement of this version (Theorem 2.24) of Nekhoroshev's theorem in the literature, nor did I attempt to prove it. So, I am being slightly dishonest in calling it a theorem. You would have to build your own proof. The closest things I could find to this version of the theorem were theorems valid only for equilibria of flows rather than fixed points of return maps. See Niederman [165] and Fasso et al. [55].

Most versions of the Nekhoroshev theorem have as their starting place a perturbed integrable flow with Hamiltonian having the form $H(I, \theta; \epsilon) = h(I) + \epsilon f(I, \theta)$. The positive-definiteness assumption on the torsion matrix becomes an assumption on the uniform positive-definiteness of the Hessian $\frac{\partial^2 h}{\partial I_i \partial I_j}$. I tried to massage the map version into the flow version but was not successful. Perhaps some reader will do this.

Proofs of Nekhoroshev-type theorems are based on estimating travel time near and along resonance webs – collections of hypersurfaces where the resonance conditions $\Sigma k_a \nu_a(I) = k_0$ hold with k_a, k_0 integer. The only known escape routes are along resonance webs. The analysis is deep and complicated and I won't try to touch it. I refer interested readers to Pöschel [172] and to the literature survey contained in p. 119–131 of the book by Dumas [49].

That hero of axiomatic quantum field theory, James Glimm, came up with a version of Nekhoroshev's theorem about six years before Nekhoroshev [68]. In Glimm's version the time-of-escape bound is polynomial in $1/\epsilon$ rather than "beyond all orders" $(O(exp(-k/\epsilon)))$. Glimm's methods and versions have not been synchronized and compared with Nekhoroshev's.

I would like to thank Francesco Fasso and Laurent Niederman for help understanding the Nekhoroshev theorem and the literature around it.

On Section 2.5

When Chenciner and I rediscovered Moore's figure eight orbit, Carles Simó became involved. He discovered that the orbit was KAM stable. And through that adventure it slowly began to dawn on me that KAM stability was essentially the only kind of stability available for the periodic orbits arising in the N-body problem.

Gareth Roberts [175] gave a separate proof of the spectral stability of the eight based on the hexagonal symmetry of the eight. (See Chapter 3 concerning this symmetry.) Using these symmetries, Roberts could reduce spectral stability computations to the analysis of a single two-by-two matrix culled from this one-twelfth and the symplectic realizations of the symmetry operations.

More celestial case studies The KAM stability of the circular Lagrange orbit has been studied in several papers. Most concern the Lagrange orbit in the context of the restricted three-body problem where it forms an equilibrium point when viewed in the rotating frame. See [39, 15, 125]. Yu [219] does address the KAM stability of the circular Lagrange orbit for the full three-body problem. There, the Lagrange orbit is a relative equilibrium and so forms a critical circle for the energy–momentum map, forming the singular locus for its level set of energy and momentum. It seems that, implicitly, Yu must work on a space with a conical singularity. It would be fun to see how this singularity is reflected in the analysis.

3

Is Every Braid Realized?

Open Question 3 Is every braid type on N strands realized by a periodic solution of the planar Newtonian N-body problem?

A special case of this question inspired much of my work and remains open.

Open Question 3.1 Is every *relative* braid type on three strands realized by a *relative* periodic solution of the equal-mass *zero angular momentum* Newtonian three-body problem?

3.1 Braiding Bodies

When three points move in the plane without colliding they generate a braid on three strands. Take time to be a third axis orthogonal to the plane. Then the three moving points $q_a \in \mathbb{R}^2 = \mathbb{C}, a = 1, 2, 3$, sweep out three non-intersecting curves $t \mapsto (q_a(t), t)$ in Galilean space–time $\mathbb{R}^2 \times \mathbb{R}$ whose union forms a traditional braid on three strands. Figure 3.1 shows the braid generated by the figure eight solution to the three-body problem while Figure 3.2 shows the braid made when points 1 and 2 form a tight binary system circling each other while 3 is far away. These space–time braids provide a means of visualizing free homotopy classes in $\mathbb{C}^3 \setminus \Delta$, the configuration space of the three-body problem.

By a *braid type* on N strands we mean a free homotopy class of loops for the planar collision-free N-body configuration space, $\mathbb{C}^N \setminus \Delta$. Braid types are in bijective correspondence with conjugacy classes of the fundamental group of $\mathbb{C}^N \setminus \Delta$. This group, denoted P_N, is known as the *pure braid group* or *colored braid group* on N strands and forms a finite index subgroup of the more famous Artin braid group B_N. (See Appendix E of this book or chapters 1 and 2 of [20] for basics on the braid groups.)

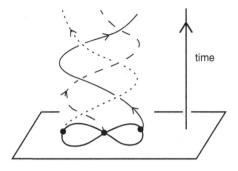

Figure 3.1 The braid generated by the figure eight orbit.

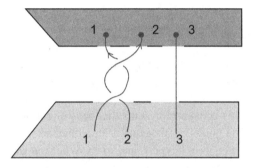

Figure 3.2 The braid A_{12} generated by a tight binary forms one of the generators of the pure braid group.

Open Question 3.1 stated a relative version of Open Question 3. Recall that a planar N-body solution is called *relative periodic* of period T if $q(t + T) = Rq(t)$ for some rotation matrix $R \in SO(2) = \mathbb{S}^1$. The rotation group in turn generates the center Z of the pure braid group by taking the base configuration $q_0 \in \mathbb{C}^N \setminus \Delta$ of $P_N = \pi_1(\mathbb{C}^N \setminus \Delta, q_0)$ and rotating it one full revolution by forming the loop $t \mapsto e^{2\pi i t} q_0$. Since the N-body configuration space modulo rotations is homotopic to shape space, P_N / Z is the fundamental group of N-body shape space.

Definition 3.1 A relative braid type on N strands is a free homotopy class in the collision-free planar N-body shape space, or, equivalently, a conjugacy class in the projective pure braid group P_N / Z. We sometimes refer to *relative braid types* as *reduced free homotopy classes*, shape space being the *reduction* of configuration space by the rotation group.

Open Question 3.1 is a special case of the question: Is every relative braid type on three strands realized by a relative periodic orbit? Eclipse sequences provide a concrete astronomical way to understand the $N = 3$ relative braid types. We describe these sequences next.

3.1.1 Eclipse Sequences

As our three bodies move in the plane, every once in a while they form a line. An *eclipse* has occurred. Label that eclipse by the label of the occluding mass. In this way any collision-free solution ticks off an *eclipse sequence* – a "word" in the three "letters" 1, 2, and 3. (In [148, 149] I called the eclipse sequence the *syzygy sequence*.) For example, the figure eight solution makes the eclipse sequence 123123 in one period. The eclipse sequence made during one period of a relatively periodic curve encodes that curve's relative braid type. (See Lemma 3.3.)

We can read off the eclipse sequence of a solution curve by projecting the curve to the shape sphere. Recall that the collision locus Δ projects to the three binary collision points $(12), (23), (31)$ that lie on the shape sphere's equator and that this equator represents the locus of collinear configurations. The binary collision points cut the equator into three arcs, labeled 1, 2, and 3. Arc 1 represents an eclipse of type 1, and so on. When our shape curve crosses arc 1, an eclipse of type 1 has occurred. When our shape curve crosses the arc labeled a an eclipse of type a has occured.

If the curve is relative periodic then its shape projection is closed, and hence crosses the equator an even number of times. (Crossings are transverse for non-collinear solutions.) Thus the eclipse sequence made in one period of a relatively periodic curve has an even number of letters, so we will say it is *even*. This suggests the following version of this chapter's Open Question 3.1.

Open Question 3.2   Is every even periodic eclipse sequence realized by a relative periodic solution of the planar three-body problem?

This is a stronger version of the question "*Is every relative braid type on three strands realized by a relative periodic orbit?*" For example, the free homotopy class of the eight is encoded by 11123123 in addition to the eight's sequence 123123. The two sequences differ by a stutter: the insertion of 11 into the eight's sequence. A *stutter* in an eclipse sequence is two consecutive appearances of the same letter. For example "1223" has a stutter. Crossing the same labeled equatorial arc 2 of the shape sphere twice in a row is a homotopically removable "mistake" that yields the stuttered 2. See Figure 3.3 for the picture of a homotopy eliminating such a stutter.

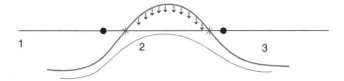

Figure 3.3 Eliminating a 22 stutter by homotoping a section of the curve to lie in one hemisphere of the shape sphere.

Reduced Periodic Signed Eclipse Sequences

It will be helpful to formalize the relationship between relative free homotopy classes and eclipse sequences described in the previous paragraph.

Definition 3.2 An eclipse sequence is called reduced when it has no stutters, and periodic reduced if, when written on a circle, it has no stutters.

The point of the *periodic reduced* definition is that since our shape curve is periodic, so is its eclipse sequence. We should think of its symbols $1, 2, 3$ as written cyclically on a circle. When written cyclically, we must require that the sequence continues to have no stutters if we want it to uniquely represent the free homotopy class. For example, the sequence 1231, viewed as a periodic eclipse sequence, is not stutter-free, since when wrapped around it is the same as 1123. The periodic reduced version of 1231 is therefore 23.

One final issue needs to be accounted for in order to get a unique eclipse sequence for each relative free homotopy class.

Signing the Sequence

Label the northern hemisphere of the shape sphere with a $+$ and the southern hemisphere with a $-$. Decorate our eclipse sequences accordingly so that we write 1^+ if the curve crosses arc 1 from the upper hemisphere to the lower hemisphere, or equivalently, if the triangle goes from positively oriented to negatively oriented. We write 1^- if it crosses arc 1 from the lower hemisphere heading into the upper hemisphere. In this way the solution ticks off a signed sequence such as $1^+ 2^- 3^+ 2^- \ldots$. The superscripts $+$ and $-$ must alternate. A choice of sign for the first letter determines the signs for all rest. It follows that there are exactly two signed eclipse sequences for a given unsigned eclipse sequences.

Lemma 3.3 formalizes how eclipse sequences encode relative braid types on three strands.

Lemma 3.3 *The following sets are naturally isomorphic:*

- *(A) the set of relative braid types on three strands,*
- *(B) the set of free homotopy classes for the shape sphere minus collisions,*
- *(C) the set of signed reduced periodic even-length eclipse sequences.*

Proof of lemma (A) $\Longleftrightarrow$ (B): A loop in the collision-free shape sphere $\mathbb{S}^2 \setminus \Delta$ represents a *relative* periodic curve: a curve that closes up modulo a rotation. (Recall Δ, the collision locus, consists of the three binary collision points.) Two different lifts $q(t), \tilde{q}(t)$ of the same shape curve differ by a time-dependent rotation: $\tilde{q}(t) = g(t)q(t)$. If each lift is closed, then the difference of their homotopy classes is represented by the rotational loop $g(t)q(0)$ generated by the action of the rotation group, and any such rotational loop lies in the center Z of the pure braid group. Since P_3 is the fundamental group of the collision-free configuration space, it follows that P_3/Z is the fundamental group of $\mathbb{S}^2 \setminus \Delta$. Now take conjugacy classes.

(B) $\Longleftrightarrow$ (C): We have seen that homotopies can be used to eliminate stutters so that each realized eclipse sequence can be reduced so as to be a signed reduced periodic even-length eclipse sequence. To show that all such sequences are realized, draw a closed squiggle on the sphere which avoids the collision points and has the given eclipse sequence. QED

Remark 3.4 (Lagrange: An empty eclipse sequence) The Lagrange relative equilibrium solution $(-1.2.1)$ consists of a rotating equilateral triangle. Suffering no eclipses, it generates the empty eclipse sequence. Projected to the shape sphere, the Lagrange solution is trivial homotopically: it is a constant loop. At the unreduced level, it realizes the generator of the center of P_3.

Remark 3.5 (Signed to unsigned and reflection) Recall that there are exactly two signed eclipse sequences for a given unsigned eclipse sequence. If we choose a relative periodic curve $q: \mathbb{R} \to \mathbb{C}^3 \setminus \Delta$ that represents one of these sequences, then to obtain a relative periodic curve representing the other one just apply a planar reflection to q. This works since such a reflection acts on the shape sphere by reflection about the equator. Due to this simple 2:1 nature, in what follows we do not bother with signs decorating eclipse sequences.

Remark 3.6 (Other representations) The fundamental group π_1 of $\mathbb{S}^2 \setminus \Delta$ is the free group on two letters. In order to see this, for each binary collision point in Δ choose a based loop circling that point once, counterclockwise. Label these three loops A, B, C. They generate π_1 and are subject to the single relation $ABC = 1$. It follows that the loops A and B alone generate the fundamental group. Set $A^{-1} = a$ and $B^{-1} = b$. Then we can write elements of π_1 as words in a, A, b, B subject to the relation $aA = 1 = bB$. Moore [157] and Li and Liao [110, 111] use this representation for braids on three strands.

Exercise 3.7 Show that modulo particle relabeling (permutations of the three masses) and reflection, there are less than 2^{M-2} distinct reduced eclipse sequences of length M.

3.2 Variational Methods: Our Motivation

The following success story motivates the questions of this chapter.

Theorem 3.8 *If X is a compact Riemannian manifold, then every free homotopy class of loops in X is realized by a periodic geodesic whose length minimizes the length among all closed curves in this class.*

Sketch proof Use the Riemannian metric to define the length $\ell(c)$ of a path c in X. Geodesics, almost by definition, extremize length: They are critical points of the length functional ℓ. Fix a free homotopy class α in X. Minimize the length among all loops $c \in \alpha$. Since minimizers are extremals, this minimum length loop yields the desired representative geodesic.

But does the minimizer exist? This is the heart of the issue.

We follow the *direct method in the calculus of variations* to achieve existence. Let $M(\alpha)$ be the infimum of the lengths $\ell(c)$ as c varies over all loops representing the given class α. Then, by the definition of infimum, there exists a *minimizing sequence* $\{\gamma_n, n = 1, 2, 3, \ldots, \infty\}$, which is to say, a sequence of smooth loops $\gamma_n \in \alpha$ such that $\ell(\gamma_n) \to M(\alpha)$ as $n \to \infty$. After reparameterizing the γ_n by arclength we can use compactness and the Arzela–Ascoli theorem (see Step 1 in Section 3.5.1) to establish that some subsequence of the γ_n tends, in the C^0 topology, to a curve γ_*, the alleged *minimizer*. By C^0 convergence, the curve γ_* represents α. Some functional analysis shows that γ_* is actually smooth, has length $M(\alpha)$, and satisfies the Euler–Lagrange equations, which is to say, the geodesic equations. QED

The sketched proof is a textbook example of *the direct method in the calculus of variations*. Much of this chapter concerns the successes and failures of using the method for the N-body problem. The method requires a variational principle: a functional on paths whose critical points solve Newton's N-body equations. The N-body problem admits several variational principles, the most useful of which has proven to be the action principle. The action principle works for any Newton's equations, that is, an equation of the form

$\nabla_{\dot{q}}\dot{q} = \nabla U(q)$ on any Riemannian manifold, and, in particular, it works for the N-body problem.[1]

Definition 3.9 The action A of a path $q : [0, T] \rightarrow \mathbb{E}$ is the integral $A(q(\cdot)) = \int_0^T L(q(t), \dot{q}(t))dt$ where $L(q, v) = K(v) + U(q)$. The function $L \colon T\mathbb{E} \rightarrow \mathbb{R}$ is called the Lagrangian.

A critical point of the action is a solution to Newton's equations, *provided* that critical point is collision-free. We will refer to this fact as the *action principle*, or sometimes as the *principle of least action*.

There are two points on which a warning should be given. First, critical points need not be minimizers, so refering to the action principle as the principle of least action is a bit of a misnomer. Second, the Lagrangian is $K + U$. The Hamiltonian (conserved energy) is $K - U$. Compare them. Do not confuse them.

We go through the direct method in the calculus of variations in some detail in section 3.5. There, the reader can see some of the fine points and subtleties and a bit around the "some functional analysis" assertion at the end of our sketch proof.

In the action-minimizing approach to Open Question 3 we fix the period T and the *braid type* of the competing curves and try to copy the compact Riemannian sketch proof. One would like to say the action-minimizer exists and is our desired curve.

Without imposing additional assumptions on competing curves, the direct approach fails due to the non-compactness of the N-body configuration space $\hat{\mathbb{E}} = \mathbb{C}^N \setminus \Delta$. There are two sources of non-compactness: $r_{ab} \rightarrow 0$ and $r_{ab} \rightarrow \infty$. We might call these the *infrared* ($r \rightarrow \infty$) and *ultraviolet* ($r \rightarrow 0$) divergences, following the terminology of quantum field theory. The more difficult of the two divergences to deal with is the ultraviolet, the collisions.

The difficulty that collisions impose for the direct method arises from the fact that there exist finite-time solutions that end in collision and have finite action. It follows from their existence that we can pass through collisions and so deform continuously from one braid type to another while the action also changes continuously. Perhaps the action is slightly lower in some adjoing free homotopy class. As a result, our action-minimizing sequences of paths will want to *jump* out of their fixed braid type by trying to reach this adjoining class and in doing so tend to a collision path. Newton's equations break down at collisions, and we cannot make much use of these collision minimizers.

---

[1] Another action principle is due to Jacobi and Maupertuis. See Appendix F. Here we fix the energy E and form a degenerate Riemannian metric, most of whose geodesics are solutions to Newton's problem at that energy E.

In 1977, W. Gordon [70] showed that this jumping phenomenon actually happens for the standard two-body problem.

Theorem 3.10 (W. Gordon [70]) *Consider the braid type in which the two bodies wind twice around each other counterclockwise in the standard two-body problem. (See Figure 3.2.) Fix the period T. Then any action minimizing sequence of loops having period T and this braid type tends to a collision path.*

In the proof of Gordon's theorem (see Section 3.6.1), the minimizing sequence tries to leave the "winding twice" class and jump to the nearby "winding once" class where the action is lower. In the process of trying to jump, the minimizing sequence must tend to a collision.

To get the direct method to work, we have had to either leave the world of the Newtonian $1/r$ potential or supplement topological braid type constraints with symmetry conditions that arise if the masses are all equal. Some of these successes are detailed in Section 3.3.

Remark 3.11 The Jacobi–Maupertuis (JM) length is a direct extension of the length functional in Riemannian geometry to mechanical problems. See Appendix F. The JM length is the path length for a degenerate Riemannian metric known as the JM metric, a metric that depends on the choice of energy E, and is defined on the Hill region $\{U \geq -E\} \subset \mathbb{E}$. In the interior of the Hill region, this metric is an honest Riemannian metric and its geodesics are reparameterizations of the energy E solutions to Newton's equations. It is tempting to try to use the JM metric to get interesting periodic solutions. So far most attempts have failed. Their failure is due to the fact that the JM metric degenerates to zero on the Hill boundary $\{U = E\}$. Two successes of the JM variational method are Seifert's work [189] and Moeckel's work [134].

3.3 What's Known?

3.3.1 Cheating

The direct method works beautifully for the planar N-body problem if we are willing to cheat a bit by replacing Newton's $1/r$ pair potential by a strong force pair potential, meaning one of the form $1/r^\alpha$ with $\alpha \geq 2$. (See Definitions 0.6 and 0.17 and Equation (0.14).) Recall that $q \colon \mathbb{R} \to \mathbb{E} = \mathbb{C}^3$ is called [179] *relative periodic* of period $T > 0$ if $q(t + T) = Rq(t)$ for some rotation $R \in SO(2) = \mathbb{S}^1$.

Theorem 3.12 (Poincaré [179]) *Consider the planar three-body problem with a strong force potential. Fix an angle ω with $0 < \omega < 2\pi$, a positive period $T \in \mathbb{R}$, and two integers n_{13}, n_{23}. Then there is a collision-free relative periodic solution of period T such that edge 12 rotates by ω radians, edge 13 rotates by $\omega + 2\pi n_{13}$ radians, and edge 23 rotates by $\omega + 2\pi n_{23}$ radians in one period. The solution minimizes the action over all relative periodic curves satisfying these constraints.*

Note that the monodromy of Poincaré's solution is $e^{i\omega}$; that is, $q(T + t) = e^{i\omega}q(t)$.

The computation in Lemma 3.13 underlies Poincaré's success.

Lemma 3.13 *Let $L = K + U$ be the Lagrangian for a strong force N-body problem U, and $A = \int L dt$ the corresponding action. Then any curve in $\mathbb{E} = (\mathbb{R}^d)^N$ that suffers a collision has infinite action.*

Proof The key is the observation that for any positive r_1 the integral $\int_0^{r_1} \frac{dr}{r^{\alpha/2}}$ is infinite provided $\alpha \geq 2$. The kinetic energy in center-of-mass frame can be written as $\frac{1}{M}\Sigma_{a<b}m_a m_b |\dot{q}_{ab}|^2$ where $M = \Sigma m_a$ and $q_{ab} = q_a - q_b$. Since $|\dot{q}_{ab}|^2 \geq \dot{r}_{ab}^2$ it follows that $K \geq \frac{m_a m_b}{M}\dot{r}_{ab}^2$ and so $L = K + U \geq m_a m_b(\frac{1}{M}\dot{r}_{ab}^2 + \frac{1}{(r_{ab})^\alpha})$ for any pair a, b. Now suppose that $q(t)$ is a smooth path in $\mathbb{E}$ in which bodies a and b collide at some time $t = t_c$. Write $r = r_{ab}$ so that $L \geq m_a m_b(\frac{1}{M}\dot{r}^2 + \frac{1}{r(t)^\alpha})$. Use $A^2 + B^2 \geq 2|AB|$ to get $L \geq C\frac{\dot{r}}{r^{\alpha/2}}$ or $L dt \geq C\frac{dr}{r^{\alpha/2}}$ where $C = \frac{2m_a m_b}{\sqrt{M}}$ Use the divergence of the integral $\int_0^{r_1} \frac{dr}{r^{\alpha/2}}$ to conclude that $A(q([t_1, t_c])) \geq C|\int_0^{r_1} \frac{dr}{r^{\alpha/2}}| = +\infty$ where $t_1 < t_c$ is any time at which $r(t_1) := r_1 \neq 0$. QED

We give a detailed proof of Poincaré's result (Theorem 3.12) in Section 3.5.1. This proof consists of seven steps. For now, we provide a sketch outline of this proof.

Sketch proof of Theorem 3.12 Poincaré's proof proceeds by the direct method of the calculus of variations. Step 0. Let A_* be the infimum of the action over all relative periodic absolutely continuous curves that satisfy the given winding number conditions on their edges. Consider an action-minimizing sequence $\{q^{(n)}, n = 1, 2, \ldots\}$, meaning a sequence of relative periodic loops having the desired monodromy and winding number constraints and whose action tends to A_*. Step 1. Extract a convergent subsequence from this sequence using the Arzela–Ascoli theorem from real analysis properties of the action and the condition on ω. This condition guarantees *coercivity* of the action, the impossibility of the minimizing sequence having $r_{ab,n} \to \infty$ as $n \to \infty$. The strong force condition on the potential combined with Lemma 3.13 prevents collisions, that is, it guarantees the impossibility of $r_{ab,n} \to 0$ along the minimizing sequence. The limiting curve is thus collision-free. It is our desired minimizer. Steps 2–6. The limiting curve is absolutely continuous and satisfies the other constraints and satisfies Newton's equations. QED

Remark 3.14 Poincaré did not impose the condition $\omega \neq 0, 2\pi$. However, these conditions are needed for his theorem to hold. See Remark 3.30.

Poincaré's method of proof, the direct method of the calculus of variations applied to the action principle, allows us to answer affirmatively versions of both open questions of this chapter for strong force potentials. We begin with the second ($N = 3$) question. Call an eclipse sequence *tied* if all three letters $1, 2, 3$ appear in the sequence.

Theorem 3.15 (Montgomery [145]) *In a strong force planar three-body problem, every tied eclipse sequence is realized by a relative periodic action-minimizing solution. This solution has zero angular momentum.*

Remark 3.16 The untied solutions are *not* realized by action-minimizing solutions. See the chapter notes in Section 3.9.

Theorem 3.17 *Let β be any relative braid type (Definition 3.1) for the planar N-body problem, $R = exp(i\omega) \neq 1$ any non-identity monodromy, $T > 0$ a period, and $U = U_\alpha$ a strong force potential. Then there is a relative periodic solution to the N-body problem with potential U that realizes the relative braid type β, and has period T and monodromy R. This solution minimizes the action over all relative periodic curves satisfying the constraints specified by R, β.*

Theorem 3.12 is the homological $N = 3$ version of Theorem 3.17. Given a periodic curve in $\hat{\mathbb{E}} = \mathbb{C}^3 \setminus \Delta$, we can count the winding number of each edge $q_a - q_b$. This counting defines a surjective homomorphism $P_3 \to \mathbb{Z}^3$, which is the Hurewicz homomorphism $\pi_1(X) \to H_1(X)$ for the case $X = \hat{\mathbb{E}}$. (See Appendix E.) The center Z of the pure braid group P_3 on three strands is generated by rotations. The Hurewicz map sends the center to the diagonal winding numbers, $(n, n, n) \in \mathbb{Z}^3$. It follows that upon forming the quotient of $\hat{\mathbb{E}}$ by the rotation group, we arrive at a surjective homomorphism $P_3/Z \to \mathbb{Z}^3/Z = \mathbb{Z}^2$ realising the Hurewicz homomorphism for the shape sphere (or space) with collisions deleted,

$$\pi_1(S^2 \setminus \Delta) \to H_1(S^2 \setminus \Delta). \tag{3.1}$$

The integers (n_{12}, n_{23}) of Poincaré's theorem parameterize the elements of $H_1(S^2 \setminus \Delta)$. We could call them *relative homology classes*[2]. A conjugacy class in $\pi_1(S^2 \setminus \Delta)$ is a relative braid type on three strands and the Hurewicz map is constant on conjugacy classes. So the Hurewicz map takes Theorem 3.17 to Poincaré's theorem, Theorem 3.12.

For a proof of Theorem 3.8 and the other theorems discussed in this section, see Section 3.5.

3.3.2 $N = 3$, Almost

If we don't want to cheat, that is, if we want to work on Open Question 3 in the context of the $1/r$ gravitational potential problem, then the only significant progress has been made by rejecting variational methods and instead using dynamical methods.

Theorem 3.18 (Moeckel and Montgomery [140]) *Every relative braid type on three strands is realized by some periodic solution to the Newtonian planar three-body problem, provided the masses are equal, or close enough to equal. Those realizing solutions whose existence we guarantee have nonzero angular momentum, many stutters, and come repeatedly close to triple collisions.*

This theorem almost solves the original Open Question 3 for $N = 3$. That question asks us to find, for each free homotopy class of loops in the collision-free configuration space $\hat{\mathbb{E}}(2, 3) = \mathbb{C}^3 \setminus \Delta$, a periodic solution to Newton's three-body problem representing that class. Instead, the theorem guarantees, for each free homotopy class in the collision-free shape sphere $\mathbb{S}^2 \setminus \Delta$, a relative periodic solution that represents that class.

The solutions guaranteed by the theorem repeatedly come very close to triple collision. They have small but nonzero angular momentum. Their eclipse sequences are decidedly not reduced; they can have long stutter runs. (Recall that a stutter within an eclipse sequence is a repeated occurrence of the same symbol.) The lengths of the stutter runs can be arbitrarily long. It follows that the theorem guarantees a countable infinite family of solutions realizing each fixed reduced braid type (see Lemma 3.3).

Variational methods play no role in our proof. See Section 3.7 for the proof.

3.3.3 The Eight, Variationally

The variational method has failed to yield progress in answering the original Open Questions of this chapter concerning motion under the influence of

---

[2] This should not be confused with the standard use of "relative homology" in algebraic topology.

the gravitation $1/r$ potential, except when all masses are equal. When the masses are equal we can combine the method with finite symmetries arising from permuting the masses to great effect. The discovery of the figure eight orbit (Section $-1.2.3$) was one of the earliest successes of this combining of methods. The eight realizes the eclipse sequence 123123 for the equal-mass three-body problem. Being KAM stable, the eight also yields, upon perturbation, nearby relatively periodic solutions for an open set of near-equal masses. We describe here the variational plus symmetry construction of the figure eight solution. The reader may wish to read an abridged version of this discussion that appears in Section C.16 at the end of Appendix C.

In order to obtain the eight, fix attention on one of the Euler configuration points on the shape sphere and one of the two isosceles great circles that do not contain it. Call these EU_1 and $ISOSC_2$. Use these same symbols for the pre-images of these sets in shape space $\mathbb{R}^3$, or in centered configuration space $\mathbb{E}_0 = \mathbb{C}^2$. (See Section 0.4.4.) More generally, let EU_i denote the locus of configurations for which mass i lies at the midpoint of masses j and k, and let $ISOSC_i = \{r_{ij} = r_{ik}\}$ denote the locus of isosceles triangle configurations having mass i as its vertex. Consider the problem of minimizing the action (see Definition 3.9) from EU_1 to $ISOSC_2$. In other words, fix a time $\bar{T}$ and try to minimize the action among all centered paths $q : [0, \bar{T}] \rightarrow \mathbb{E}_0 \cong \mathbb{C}^2$ having $q(0) \in EU_1$ and $q(\bar{T}) \in ISOSC_2$. See Appendix C, Figure C.5.

Assume, for a moment, that such a minimizer $q(t)$ exists and is *collision-free*. Then this minimizer must be a solution arc, must have zero angular momentum, and must intersect the endpoint manifolds EUL_1 and $ISOSC_2$ orthogonally. To see that it has zero angular momentum, take any smooth curve $R : [0, \bar{T}] \rightarrow \mathbb{S}^1$ and replace $q(t)$ by $R(t)q(t)$. Note that the replacement satisfies the boundary conditions. Varying $R(t)$ yields that $\dot{q}$ must be orthogonal to the $\mathbb{S}^1$-fibers of the shape projection in order for $q(t)$ to minimize the action. This orthogonality is equivalent to angular momentum being zero. Varying the curve $q(t)$ subject to the endpoint conditions and computing the resulting first derivative of the action leads to $\dot{q}(0)$ being orthogonal to EU_1 and $\dot{q}(\bar{T})$ being orthogonal to $ISOSC_2$, the claimed orthogonality to the end point manifolds.

Now look at the projection of our minimizing curve to shape space, and to the shape sphere. Because the masses are equal, reflection of shape space about $ISOSC_2$ is an isometry of shape space, one that takes shape projections of zero angular momentum solutions to shape projections of zero angular momenum solutions. Due to the orthogonality of our minimizing arc to $ISOSC_2$, its derivative matches smoothly with that of its reflection. Concatenate the minimizer with its reflection to obtain the projection of a zero angular momentum solution connecting EU_1 to EU_3 in time $2\bar{T}$. We can

(a)   (b)

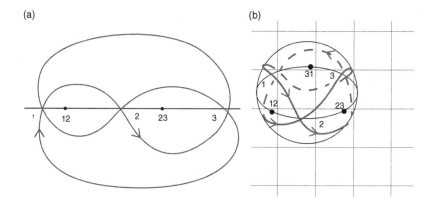

Figure 3.4 Eclipse sequence for the eight. Count crossings of the labeled equato
rial arcs to get sequence 123123. (a) The stereographic projection of (b), with the
31 collision placed at infinity.

proceed to use the reflections about the other Isosceles planes $ISOSC_i$ and the
collinear plane (shape space equator) to continue this curve around the shape
sphere and arrive at a periodic shape curve of period $12\bar{T}$ that solves the zero-
angular momentum Newton's equation. This curve is the shape projection of
the figure eight. See Figure 3.4. Also see [33] or [150] for details.

The reflections about the four great circles generate an action of the 12-
element dihedral group D_6, the symmetry group of a regular hexagon, on the
shape sphere. This same group is the finite symmetry group of the figure eight
and we will explore this symmetry in Section 3.4.2.

Test Curves for the Eight

The figure eight itself is a periodic curve of period $12\bar{T}$ in which the three
masses chase each other around a figure eight. Projected to the shape sphere,
the eight has the same symmetry pattern as the contour $\tilde{U} = c_E$ where $c_E =
\tilde{U}(EU_i)$ is the value of the normalized potential at any one of the three Euler
points. All three Euler points take the same value of $\tilde{U}$ when the masses are
equal. Indeed, the eight, projected to shape space, lies very close to the curve
in shape space defined by the two equations $\tilde{U} = c_E$ and $I = I_m$, where I_m is
obtained by minimizing the action of curves defined by $\tilde{U} = c_E$ and $I = I_m$
and parameterized by arclength. We call this close curve the *test curve*. Again,
see [33].

Shape Sphere Geometry and a Reflection Group

We describe now how to implement reflections about an isosceles great circle $ISOSC_i$ by acting on the configuration space. One might guess that interchanging masses 1 and 3 would induce the reflection about $ISOSC_2$. This interchange does map $ISOSC_2$ to itself, but it interchanges the north and south hemispheres, while mapping the collinear equator $COLL$ to itself, so does not pointwise fix $ISOSC_2$ and so cannot be the desired reflection. Indeed, the northern hemisphere is defined by $w_3 > 0$, where w_3 is the signed area of a triangle times a nonzero constant. Interchange has the effect $w_3 \mapsto -w_3$, and so interchanges the north and south hemispheres of the shape sphere, mapping the collinear equator $COLL$ to itself. To see that this is the effect on w_3, observe that on $\mathbb{E} = \mathbb{C}^3$ the interchange operation is $(q_1, q_2, q_3) \mapsto (q_3, q_2, q_1)$ while $w_3 = C(q_1 - q_2) \wedge (q_3 - q_2)$ for some nonzero constant C. It follows from this description that on the shape sphere the mass interchange $1 \leftrightarrow 3$ implements the spherical isometry of *half-twist* about the axis defined by intersecting $ISOSC_2$ with $COLL$. This axis consists of two antipodal points, namely the binary collision point $r_{13} = 0$ and the Euler point EUL_2 corresponding to the projection of $(a, 0, -a)$, $a \neq 0$. A half-twist about a point (or equivalently, a pair of antipodal points) on the sphere is the rotation by 180 degrees about this point.

So, how would we effect reflection about $ISOSC_2$? Recall a basic fact shared by spherical, hyperbolic, and Euclidean geometry: A rotation by θ radians about a point P is obtained by composing the reflections about two lines passing through P, the angle of these lines being $\theta/2$. A half-twist is a rotation by 180 degrees, and so is achieved when these two lines through P are perpendicular. Thus, the half-twist about our given axis is achieved by composing the reflections about the collinear equator $COLL$ with the reflection about $ISOSC_2$. Any reflection about any line in the plane, applied to the three-body configurations, induces reflection about the collinear equator. In particular, reflection about the x-axis, which is the conjugation $q_a \mapsto \bar{q}_a$, yields the map $(q_1, q_2, q_3) \mapsto (\bar{q}_1, \bar{q}_2, \bar{q}_3)$ of configuration space and induces reflection about the collinear equator. It follows that we induce the reflection about $ISOSC_2$ by composing the interchange operation with this reflection: $(q_1, q_2, q_3) \mapsto (\bar{q}_3, \bar{q}_2, \bar{q}_1)$. Identical considerations apply to generate the reflections about the other isosceles great circles.

Consider, then, the group generated by reflection about the x-axis and all mass interchanges. When acting on the shape sphere, this group coincides with the group generated by reflections about the collinear great circle and the three isosceles great circles. Together, these four great circles cut the shape sphere up into 12 congruent triangles, each a 90–90–60 spherical triangle. Any one triangle is a fundamental domain for the action of this group on the sphere. In particular, the group is a twelve-element group, which Marchal christened P_{12}. A more common name for P_{12} is the dihedral group D_6, which is the name we will be using.

The vertices of any one such fundamental triangle for D_6 consists of an Euler point, a binary collision point, and a Lagrange (equilateral) point. Pick one of these triangles. In our original minimization problem for the eight, we considered the problem of minimizing the action for paths traveling from the Euler point of the triangle to its opposite side, the side joining Lagrange to collision. By a reflection argument, this minimizer, if it exists, must stay within the chosen triangle. Now look at how the 12 arcs fit together by reflection when we propagate this minimizing arc across the sphere by reflection. Our claimed solution has the claimed eclipse sequence 123123, each eclipse being at an Euler configuration.

To understand what goes on with the solution in inertial space, not shape space, we can use a slight generalization of the area law, Equation (0.30), described in Chapter 0. This generalization asserts that if a curve in the shape sphere begins and ends on the collinear equator and is horizontally lifted to $\mathbb{E}_0$ by imposing the condition that its angular momentum is everywhere zero, then the oriented angle between the final and initial lines defined by the lifted curve is half the spherical area bounded by the shape curve and the collinear equator. The figure eight's shape curve repeatedly swings above and below the equator in congruent arcs. The generalized area law asserts that the angles between the lines formed at consecutive collinear instants oscillates, with the rotation associated to an arc above the equator cancelled by the arc below. So if we traverse arc 12 of the eclipse sequence and then arc 23, we return to the same line, with the masses reordered. In particular, by traveling all the way around the shape curve we see that the lifted curve $q(t)$ in $\mathbb{E}_0$ actually closes up, that is, it is periodic and not just relatively periodic.

The hard technical part of all this work is establishing that the minimizer is collision-free. We will address collision elimination in Section 3.6.

3.4 Choreographies and Other Designer Orbits

3.4.1 Symmetries and Their Representations

Combining the direct method of the calculus of variations with discrete symmetries led to the figure eight solution and then to an enormous variety of new periodic orbits. In this section, I take a detour away from Open Question 3 to give a synopsis of this relatively new development in the N-body problem.

When all the masses are equal, permuting them yields yet another symmetry of the N-body problem in addition to the Galilean symmetries. By combining permutations and Galilean symmetries we can form a vast array of finite symmetry types, types that may be realized by solutions.

A good way to describe possible finite symmetries of periodic solutions to the N-body problem is to view periodic curves $q(t) = (q_1(t), \ldots, q_N(t))$ in configuration space as maps,

$$q : \mathbb{S}_T^1 \times [N] \to \mathbb{R}^d, \tag{3.2}$$

by setting $q(t, a) = q_a(t)$. Here, $[N] = \{1, 2, \ldots, N\}$ is the N-element set of mass labels and $\mathbb{S}_T^1 = \mathbb{R}/T\mathbb{Z}$ represents periodic time, so T is the orbit's period: $q_a(t + T) = q_a(T)$, $a \in [N]$. We call $\mathbb{S}_T^1$ the *time circle*. The group of permutations S_N acts on $[N]$ by permuting the mass labels and preserves the action $\int_{\mathbb{S}_T^1} K(\dot{q}) + U(q(t))dt$ of q provided the masses are all equal. We henceforth fix the center of mass by assuming the map satisfies $\Sigma_{\{a \in [N]\}} q(t, a) = 0$ at each instant t of time. The one-dimensional isometry group $O(\mathbb{S}_T^1) = O(2)$ of the time circle acts by time translations and reflections. The group $O(d)$ of orthogonal transformations of $\mathbb{R}^d$ acts on inertial space. All three group actions preserve the action A of a loop when the masses are equal, so altogether we have the group $O(\mathbb{S}_T^1) \times S_N \times O(d)$ acting on our loops, leaving their action unchanged, and mapping solutions to solutions.

We impose a symmetry condition on a loop by choosing a finite group Γ together with a faithful representation

$$\rho : \Gamma \to O(\mathbb{S}_T^1) \times S_N \times O(d),$$

by which we mean an injective homomorphism. Choosing a representation of Γ is the same as choosing a way for Γ to act on $\mathbb{S}_T^1$, on $[N]$, and on $\mathbb{R}^d$. By slight abuse of notation we write this action as $\rho(g)(t, a, x) = (gt, ga, gx)$ for $(t, a, x) \in \mathbb{S}_T^1 \times [N] \times \mathbb{R}^d$ and $g \in \Gamma$. Imposing equivariance defines the symmetry condition on loops.

Definition 3.19 By a Γ-symmetric loop, we mean any loop (Equation (3.2)) that is Γ-equivariant:

$$q(gt, ga) = gq(t, a).$$

Restrict the action A to Γ-symmetric loops and then minimize the action by the direct method. If we can show that the minimum exists and is collision-free, then it is automatically a Γ-symmetric solution to the N-body problem. (This last fact is a special case of the *Principle of symmetric criticality*. See for instance Palais [167].) We call this method the *symmetry-variational method*.

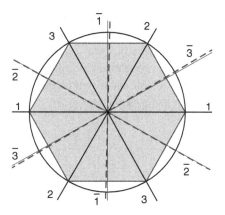

Figure 3.5 The symmetry group of the figure eight is that of a regular hexagon. The reflections about the six indicated lines generate this group.

3.4.2 Representation for the Eight

The figure eight orbit can be obtained by the symmetry-variational method applied to the symmetry group $\Gamma = D_6$, the 12-element group generated by reflections about the symmetry axes of a regular hexagon. See Figure 3.5. We will need to describe the representation of D_6. The $S_3 \times O(2)$ part of the representation, which is to say, D_6's representation on configuration space $\mathbb{E}(2,3)$, induces an action of D_6 on the shape sphere corresponding to the earlier tessellation of the shape sphere into 12 spherical triangles made by the three isosceles great circles and the collinear great circle and used in the eight's construction earlier. The reflections about these four great circles are induced by reflections from D_6 and this is enough to recover the $S_3 \times O(2)$ part of the representation but does not describe D_6's action on the time circle.

We now describe the representation of D_6 yielding the symmetries of the figure eight. The symmetry axes of the regular hexagon come in two flavors, those lines that join opposite vertices and those lines that join midpoints of opposite edges. These lines are indicated in Figure 3.5 as solid and dashed lines, with the three solid lines labeled 1, 2, 3, and the three dashed lines labeled $\bar{1}, \bar{2}, \bar{3}$. We've labeled these lines going around the circle counterclockwise with line 1 the x-axis. We interleave the labels as we go around so that two mass labels (e.g. $1, \bar{1}$) are never are adjacent to each other on the circle. Draw the time circle $\mathbb{S}_T^1$ circumscribing the hexagon. Then each reflectional symmetry of the hexagon acts on the time circle by that same reflection. For example, if we coordinatize the circle so line 1 passes through $t = 0$, then the reflection about this line generates the time transformation $t \mapsto -t$. (The points t and

$t + T$ are the same point of the time circle $\mathbb{S}^1_T$, and so the reflection $t \mapsto -t$ is also represented by $t \mapsto T - t$, showing that $t \to -t$ also fixes the point $t = T/2$ antipodal to $t = 0$.) The reflection about a solid line acts on the physical inertial plane $\mathbb{R}^2 = \mathbb{C}$ by the half-twist, $z \mapsto -z$, while the reflection about any dashed line act on the plane by reflection about the x-axis, $z \mapsto \bar{z}$. Finally, the reflection about a line labeled i or $\bar{i}$ acts on the label space $\{1, 2, 3\}$ as the reflection (jk), which fixes i and transposes the two other mass labels $j, k \neq i$. Note that if we let the reflections of Figure 3.5 act on the labeled lines, then this action on label space is precisely the action on the labeled lines: Reflection about line 1 switches lines 2 and 3 (and lines $\bar{2}$ and $\bar{3}$) so can also be seen from the figure.

A fundamental domain for this action of D_6 on the time circle is one-twelfth of the circle, corresponding to the initial minimizing arc we used in Section 3.3.3 to build the eight. The action that D_6 induces on the shape sphere is identical to the action of Marchal's group P_{12} generated by reflections about $ISOSC_j$ and about $COLL$ as used in our earlier description of the eight. Indeed, D_6 agrees with Marchal's P_{12}.

Theorem 3.20 is a restatement of our existence theorem for the figure eight.

Theorem 3.20 *The loop obtained by minimizing the action over the space of D_6-symmetric loops is collision-free and consequently solves Newton's equations. This solution is the figure eight solution.*

3.4.3 Choreographies

The figure eight solution is a *choreography*: Each of its three masses trace out the same curve, chasing each other around this curve. Any such solution to the N-body problem is a choreography. That the eight must be a choreography can be seen from the way our D_6 representation of the eight's symmetry contains a cyclic three-element group $\mathbb{Z}_3$. Take the product $h \in D_6$ of any four consecutive reflections from Figure 3.5, for example, $1, \bar{3}, 2, \bar{1}$. This h acts by rotation by $1/3$ of a revolution on the time circle, by the identity on the inertial plane and by a cyclic permutation on the labels, and as such generates a $\mathbb{Z}_3$. It follows from equivariance of the eight under h that, upon a relabeling of the masses, we get $q_{a+1}(t) = q_a(t - T/3)$ where a is taken mod 3. This condition is the choreography condition.

Within a few days of announcing the proof and rediscovery of the eight solution, Joseph Gerver found what we called the Gerver super-eight choreographic solution involving four bodies. Within a few months, Carles Simó [193, 194] discovered hundreds of planar three-body choreography solutions and strong indications of the existence of infinite families of N-body choreographic solutions. The dam broke and huge families of choreographic solutions began to spill out. See [26, 35, 61, 162], for some of the early few papers, and [65] and [166] for some appearing about a decade later.

To impose the choreography condition for N equal masses, take $\Gamma = \mathbb{Z}_N$ to be the cyclic group of order N. Take $T = N$ so that the time circle is $\mathbb{S}_N^1 = \mathbb{R}/N\mathbb{Z}$. Define the representation of $\mathbb{Z}_N$ by having the generator of $\mathbb{Z}_N$ act by taking (t, a, x) to $(t + 1, a + 1, x)$, where addition of the mass labels a is taken mod N. We call this representation the *choreographic representation*. Then $\mathbb{Z}_N$-equivariance is the assertion that $q(t + 1, a + 1) = q(t, a)$ or $q_{a+1}(t) = q_a(t - 1), t = 1, 2, \ldots, N$. In other words, all bodies travel the same curve, staggered in phase from each other by $(1/N)$th of a period. If we know how any one of the bodies moves, we know how all the bodies move. Of course, we can use the space-time scaling to change the period from N to any period $T > 0$.

Definition 3.21   A choreographic curve in $\mathbb{R}^d$ is a T-periodic curve $q : \mathbb{S}_T^1 \to \mathbb{E}(d, N)$ for which, after a permutation of the mass indices a, we have $q(t) = (q_1(t), \ldots, q_N(t))$ with the $q_a(t) \in \mathbb{R}^d$ satisfying $q_{a+1}(t) = q_a(t - T/N)$ for $a \in \mathbb{Z}/N\mathbb{Z}$.

If we impose a representation of a finite group Γ that contains $\mathbb{Z}_N$ in such a way that the Γ-representation, restricted to $\mathbb{Z}_N$, is isomorphic to the choreographic representation, then a Γ-equivariant solution is automatically a choreography. Many of the established choreographies are obtained by choosing an appropriate Γ, along with a representation of Γ.

3.4.4 Hip-Hop

Interesting solutions besides choreographies can be found using the direct method applied to symmetry groups that do not contain the choreographic representation. One such solution is the hip-hop solution of Chenciner and Venturelli [34] for the equal-mass spatial four-body problem. See Figure -1.22 back in the Tour, Chapter -1. The symmetry group of the hip-hop is $\Gamma = \mathbb{Z}_2 \times \mathbb{Z}_4$. The generator of $\mathbb{Z}_2$ acts on the time circle by a 180 degree rotation: $t \mapsto t + T/2$, trivially on the mass labels, and by $-Id$ on inertial space $\mathbb{R}^3$. The generator h of $\mathbb{Z}_4$ acts trivially on the time circle, by the cyclic permutation $a \mapsto a + 1$ on the set $\{1, 2, 3, 4\}$ of mass labels, and by the orthogonal transformation $B(x, y, z) = (-y, x, -z)$ on inertial space $\mathbb{R}^3$. Note that B projects onto the plane $\mathbb{R}^2$, and this projected map (which agrees with the restriction of B to $z = 0$) is rotation by 90 degrees. Thus the orbit of any point in $\mathbb{R}^3$ under the $\mathbb{Z}_4$ action projects onto a square in the plane.

As a consequence of being $\mathbb{Z}_4$-symmetric, the motion of any one body determines that of all the others. If $q_1(t)$ is the motion of body 1 then the full four-body loop is $q(t) = (q_1(t), B(q_1(t)), B^2(q_1(t)), B^3(q_1(t)))$. The $\mathbb{Z}_2$ symmetry implies that $q_1(t + T/2) = -q_1(t)$.

Theorem 3.22 *The loop obtained by minimizing the action over the space of $\mathbb{Z}_2 \times \mathbb{Z}_4$-symmetric loops of fixed period is collision-free and consequently solves Newton's equations. This solution is non-planar, with the four bodies oscillating symmetrically between two antipodally related tetrahedra, q_* and $-q_*$, forming a planar square ($z = 0$) at the instant halfway between.*

For q_*, we take the configuration at which the z-coordinate of $q_1(t)$ is maximized. By a time translation, we can assume this happens at $t = 0$. Then $q(T/2) = -q_*$.

3.4.5 Designer Orbits

If we want a solution with a particular pleasing symmetry, we can try our luck by choosing a representation of some finite group Γ that encodes this symmetry. The symmetry-variational method often works, yielding a solution of this type. The hard part of the work is to guarantee that the minimizer is collision-free. We come to that hard part in Section 3.6. Here, we detail a few of the successes of the symmetry-variational method.

An obvious choice of symmetry is that of the Platonic solids for the spatial N-body problem, with N being the number of vertices of that solid, or some other product of the prime factors of the order of the corresponding Platonic symmetry group Γ. See Fusco, Gronchi, and Nagrini [65], Moore and Nauenberg [158], and Figure 3.6. Returning to the plane, one can increase the number N of bodies. For certain families of symmetries containing the $\mathbb{Z}_N$ choreographic symmetry, the symmetry-representation method works beautifully. It yields *chained eights* for each odd number N of bodies. See Ferrario and Terracini [61]. See also [207]. If N is even, one can split the bodies so as to travel on two separate congruent choreographies, half on one and half on the other. See the work of [26] and Figure 3.7.

Ferrario and Terracini obtained a kind of general purpose existence theorem that applies to most of the solutions described in the previous paragraph. It relies on the collision-avoiding Marshal lemma described in Section 3.6. There are many other symmetry types for which minimizers likely exist, but whose existence remains to be rigorously established. Although we can't be sure they

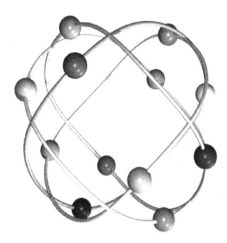

Figure 3.6 A choreography enjoying tetrahedral symmetry. The symmetry groups
of the five Platonic solids have various representations that can be realized by
solutions to the spatial N-body problem by using the direct method. For N you
may take the order of the symmetry group. Courtesy of Davide Ferrario.

exist, we can see them! Numerical methods that implement the variational-
symmetry method and are described in Section 3.4.6 allow us to compute and
visualize approximate solutions that represent these alleged solutions.

Montaldi and Steckles [143] classified all planar ($d = 2$) symmetries Γ that
contain the choreographic representation $\mathbb{Z}_N$. See Figure 3.8 and also [142].
For each N they find two countably infinite families of such Γ and several
exceptional families. One of their infinite families is covered by Ferrario and
Terracini's existence theorem but the other one is not. See [143, section 2.6].
This other family is a family of dihedral-type groups denoted by $D(N,k/\ell)$,
being indexed by positive rational numbers k/ℓ.

The $D(3,k/\ell)$ family of Montaldi–Steckles, assuming existence, yields an
infinite family of planar three-body choreographies. See Figure 3.9. Simó
has found two other infinite families of planar three-body choreographies,
along with many other choreographies. Again these are families we can see
on the screen but their existence is not yet established. Many of Simó's are
free of all symmetries beyond the choreographic one. Simó does not use
variational methods, but rather a shooting method with parameters, shooting
from isosceles to isosceles, with the two isosceles having different vertices
and different symmetry axes. The angle between the symmetry axes is used
as a parameter. See Simó [192], particularly around p. 225. Something like
interval arithmetic could likely guarantee the existence of any finite number of
Simó's solutions. See [90].

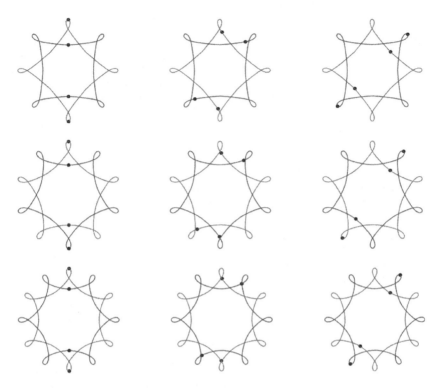

Figure 3.7 Chen established the existence of a countable family of solutions to the $2N$-body problem. Half the masses travel one choreography clockwise. The other half travel a rotation of the same choreography curve counterclockwise. In the figure $2N = 4$. From [26], © Cambridge University Press, reproduced with permission.

An alternative method, in the spirit of Simó's shooting method, has been pursued by Chen, Ouyang, and Xia [27] to establish a number of surprising orbits. See Figure 3.10 for some of these orbits. To motivate their work, observe that in one-sixth of a period, the eight travels from one line to another – one collinear Euler configuration to another, with the order of the masses on the line being permuted. The angle between these two lines is particular to the eight, and the angle bisector of the angle the two lines make is one of the symmetry axes of the eight. Now, as an alternative, lay out N equal masses on the line in some configuration. Rotate the line and permute the masses. In this way we get a shooting problem from one collinear configuration to another. By varying the angle between the lines and the points on the line it may happen that velocities match in such a way that at the next collinearity the same

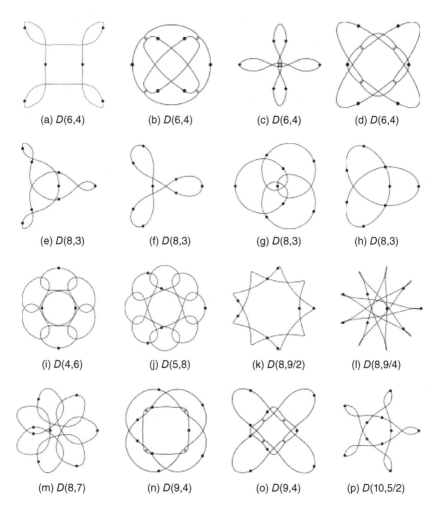

(a) $D(6,4)$ (b) $D(6,4)$ (c) $D(6,4)$ (d) $D(6,4)$

(e) $D(8,3)$ (f) $D(8,3)$ (g) $D(8,3)$ (h) $D(8,3)$

(i) $D(4,6)$ (j) $D(5,8)$ (k) $D(8,9/2)$ (l) $D(8,9/4)$

(m) $D(8,7)$ (n) $D(9,4)$ (o) $D(9,4)$ (p) $D(10,5/2)$

Figure 3.8 Some of the choreographies found by Montaldi and Steckles, with symmetry type indicated. See [175].

angle and set of configurations repeats. In this way, one can make families of quasiperiodic solutions that become periodic when the angle between the two lines is a rational multiple of π.

3.4.6 Numerical Algorithms

Periodic solutions lend themselves to Fourier series. Symmetry conditions imposed on competing orbits can be encoded as symmetry conditions among

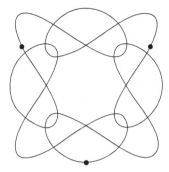

Figure 3.9 Steckles and Montaldi [175] classified possible symmetry types of planar choreographies. All types are realized by strong force potentials. This one they call $D(3,4)$. Numerical evidence suggests it is realized gravitationally. Courtesy of J. Montaldi.

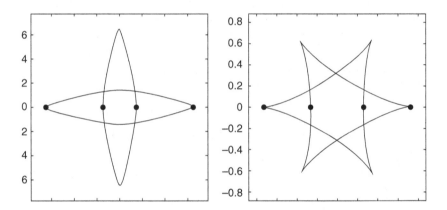

Figure 3.10 Chen, Ouyang, and Xie [36] established the existence of various families of solutions by arranging points on two lines and using the angle between the lines and the arrangements of the points as inputs into the direct method. From [27].

the coefficients of the Fourier series. One can search for numerical approximations to the minimizing solutions realizing a particular symmetry type by truncating the resulting Fourier series to some preordained order and computing the action of the resulting path. This action becomes a function of a finite number of Fourier coefficients. Apply a gradient search to find the Fourier coefficients that minimize. If the result does not change much when the order is increased we call it a victory. See [157, 162, 192]. All of this is beautifully done on the fly in the program written by Gregory Minton [127].

Is Every Braid Realized?

If one wants higher accuracy and more assurances that one has found an actual solution to Newton's equations, one can use the (finite) symmetry to slice up phase space into Poincaré sections associated to the symmetry type. One can then take the approximate minimizer as a seed for numerical integration, and then integrate from one slice to another, or through a series of slices. A fixed point of the resulting map will be the desired periodic orbit. See [192]. All of this can be turned into rigorous existence proofs with bounds using the methods of interval arithmetic. See [90].

3.5 The Direct Method of the Calculus of Variations in Detail

We provide the details of the direct method in the calculus of variations by using it to prove Poincaré's theorem, Theorem 3.12. These details serve two purposes. First, they should be useful to newcomers to the direct method of the calculus of variations. Second, they will serve as a launch point to see what goes wrong with the method in the presence of collisions in the Newtonian case. See the Section 3.5.3.

3.5.1 Proof of Poincaré's Theorem, Theorem (3.12)

We break the proof up into seven steps, the first of which sets up the formalism.

- Step 1: Setup. Let Ω denote the space of all absolutely continuous paths from $[0, T]$ to the configuration space $\mathbb{E} = \mathbb{C}^3$ satisfying the monodromy constraint

$$q(T) = e^{i\omega}q(0), \qquad (3.3)$$

the winding number constraints of the theorem, and having finite action over $[0, T]$. We recall that a path $q : [a, b] \to \mathbb{E}$ is absolutely continuous if and only if it is continuous, is differentiable a.e., and if the path and its derivative are linked through the fundamental theorem of calulus: $q(t) - q(t_0) = \int_{t_0}^{t} \dot{q}(s)ds$ for all t_0, t in the path's domain. See, for example, Royden [178, p. 106, theorem 13].

We can use the monodromy condition (3.3) to extend any path satisfying it to a continuous path and all of $\mathbb{R}$ by imposing the relative periodicity condition

$$q(t + T) = e^{i\omega}q(t). \qquad (3.4)$$

In this way we can identify our path space Ω with a subspace of $C^0(\mathbb{R}, \mathbb{E})$. At a few points in the argument we will use this identification.

Let A_* be the infimum of the action $A(q) = \int_0^T K(\dot{q}(s)) + U(q(s))ds$ as q varies over Ω. By the definition of infimum there exists a sequence $q^{(n)} \in \Omega$, $n = 1, 2, 3, \ldots$, such that $A\left(q^{(n)}\right) \to A_*$ as $n \to \infty$. Such sequence is called a *minimizing sequence*. The remainder of the proof consists of completing the following six steps.

- Step 1. Show that a subsequence of the minimizing sequence converges to some continuous path q_*.
- Step 2. Show that $q_* \in \Omega$.
- Step 3. Show that $A(q_*) = A_*$.
- Step 4. Show that $dA(q_*)(h) = 0$ for all admissible variations h of q_*.
- Step 5. Show that q_* satisfies Newton's equations on $(0, T)$.
- Step 6. Show that $\dot{q}_*(T) = e^{i\omega}\dot{q}_*(0)$.

We proceed to the steps.

Step 1 The initial convergence of the subsequence will be uniform, that is, relative to the C^0-norm, $\|q\|_0 := max_{t \in [0, T]}|q(t)|$. We obtain the subsequence by invoking the *Arzela–Ascoli theorem*, which states that any *bounded, equicontinuous* sequence of paths in $C^0([0, T], \mathbb{E})$ has a convergent subsequence. The sequence $q^{(n)}$ is *bounded* if there exists a positive constant C such that for all n we have $sup_t \|q^{(n)}(t)\| < C$. The sequence is *equicontinuous* if, for every $\epsilon > 0$, there exists a δ *independent of n* such that for all $t_0, s \in [0, T]$ we have $|s - t_0| < \delta$ implies $\|q^{(n)}(s) - q^{(n)}(t_0)\| < \epsilon$. So we must establish the boundedness and equicontiniuity of our minimizing sequence.

The standard way to establish boundedness in the calculus of variations is by verifying that the action with the path constraints is *coercive*.

Definition 3.23 A pair (A, Ω) consisting of a variational principle A and a domain Ω for A is *coercive* with respect to a norm $\|q\|$ on paths if, for $q \in \Omega$, we have $\|q\| \to \infty \implies A(q) \to \infty$.

It is immediate that if (A, Ω) is coercive for the C^0 norm

$$\|q\|_0 = sup_{t \in [0, T]}\|q(t)\|$$

on paths, then any minimizing sequence is bounded.

We have $\omega \neq 0 \implies$ coercivity. We will only use the kinetic energy part of the action. Since $U > 0$ we have

$$A(q) > \int K(\dot{q})dt := \frac{1}{2}\int \|\dot{q}\|^2 dt. \tag{3.5}$$

Cauchy–Schwarz's inequality $\int fg \leq \sqrt{\int f^2}\sqrt{\int g^2}$ applied to $f = \|\dot{q}\|, g = 1$ yields

$$\sqrt{\int \|\dot{q}\|^2 dt}\sqrt{\int dt} \geq \int \|\dot{q}\| dt := \ell(q),$$

where $\ell(q)$ is the Euclidean length of the path q in the Euclidean vector space $\mathbb{E}$. It follows that $\sqrt{2A(q)}\sqrt{T} \geq \ell(q)$ or

$$A(q) \geq \frac{1}{2T}\ell(q)^2.$$

Let t_* be a time such that $\|q(t_*)\| = sup_t\|q(t)\|$. At this juncture we use our identification of Ω with a subspace of $C^0(\mathbb{R}, \mathbb{E})$ satisfying the periodicity condition (3.4). Then $q(t_* + T) = e^{i\omega}q(t_*)$. Also, $\ell(q) > |q(t_*) - q(t_* + T)|$ since Euclidean line segments realize the infimum of distance between any two points, here the points $q(t_*)$ and $q(t_* + T)$ in $\mathbb{E}$. But the angle between the vectors $q(t_*)$ and $q(t_* + T)$ is ω, from which it follows that $\|q(t_*) - q(t_* + T)\| = 2R\sin(\omega/2)$ where $R = \|q\|_0 = |q(t_*)|$. Since the action of our extended path q over $[0, T]$ is the same as its action over $[t_*, t_* + T]$, we get that

$$A(q) \geq \frac{1}{2T}4\|q\|_0^2 \sin^2(\omega/2). \tag{3.6}$$

This inequality implies coercivity as long as $\omega \neq 2\pi k$ for some integer k.

In order to establish the equicontinuity of the minimizing sequence we again only use the kinetic part of the action. We have that $q(s) - q(t_0) = \int_{t_0}^s \dot{q}dt$ because our paths q are absolutely continuous. (Again see Royden [178], for example.) Then

$$\|q(s) - q(t_0)\| \leq \int_{t_0}^s \|\dot{q}(t)\| dt.$$

Using Cauchy–Schwarz again, we have, valid for any $0 \leq t_0 < s \leq T$, that

$$\int_{t_0}^s \|\dot{q}(t)\| dt \leq \sqrt{|s - t_0|}\sqrt{\int_{t_0}^s \|\dot{q}(t)\|^2 dt} \leq \sqrt{|s - t_0|}\sqrt{2A(q)}, \tag{3.7}$$

where in the last inequality we used inequality (3.5) again. Putting these inequalities together yields

$$\|q(s) - q(t_0)\| \leq \sqrt{|s - t_0|2A(q)}, \tag{3.8}$$

which immediately implies equicontinuity for any set of curves such as the $q^{(n)}$ whose action A is bounded by a fixed constant.

Since we have boundedness and equicontinuity we can invoke Arzela–Ascoli and get some subsequence $q^{(n_j)}$ of our original minimizing sequence

$q^{(n)}$ that converges in the C^0-norm to some C^0 curve q_*. As is standard, we relabel the subsequence so as to write it as $q^{(n)}$ simply to not subject ourselves to the pain of writing double, and later on triple, indices. So now we have $q^{(n)} \rightarrow q_*$ in the C^0 topology on curves.

Step 2 Although $q_* \in C^0([0, T], \mathbb{E})$, we do not yet know that q_* is in Ω. In order for q_* to lie in Ω we need to show, among other things, that it is absolutely continuous and has finite action. In order to achieve these things we are going to use the Sobolev space H^1 and the notion of weak convergence.

We define H^1 to be the space of absolutely continuous paths $q : [0, T] \rightarrow \mathbb{E}$ such that $\|\dot{q}\|_{L_2}^2 := \int_0^T |\dot{q}(t)|^2 dt < \infty$. Since any path $q \in H^1$ is absolutely continuous we have $q(t) - q(s) = \int_s^t \dot{q}(u) du$. The same argument that led to inequality (3.8) yields that

$$|q(t) - q(s)| \leq \sqrt{t - s} \|\dot{q}\|_{L_2}, \tag{3.9}$$

which says that the paths in H^1 are Hölder continuous with exponent one-half and, in particular, the collection of paths q with $\|\dot{q}\|_{L_2} \leq C$ is equicontinuous. We can write any $q \in H^1$ as

$$q(t) = q(0) + \int_0^t f(s) ds, f = \dot{q} \in L_2 := L_2([0, T], \mathbb{E}).$$

This representation shows us that H^1 is a Hilbert space isometric to $\mathbb{E} \oplus L_2$. We could use for its squared norm either $\|q(0)\|^2 + \|\dot{q}\|_{L_2}^2$ or the more standard

$$\|q\|_{H^1}^2 = \int |\dot{q}(s)|^2 ds + \int |q(s)|^2 ds = \|\dot{q}\|_{L_2}^2 + \|q\|_{L_2}^2.$$

The inequality $\|q(t) - q(0)\| \leq \sqrt{t} \|\dot{q}\|_{L_2}$ shows us that these two norms are Lipschitz equivalent. (Note that $\|\dot{q}\|_{L_2}$ by itself is not a norm on H^1 since any constant path has $\|\dot{q}\|_{L_2} = 0$.)

We have that $\Omega \subset H^1$. Indeed $\int K(\dot{q}(s) ds = \frac{1}{2} \|\dot{q}\|_{L_2}^2$, which shows that if $A(q) < \infty$ then $q \in H^1$. Moreover, subsets that are action bounded are necessarily H_1-norm bounded. We see this by integrating the C_0 bound of inequality (3.6) from 0 to T to get $\|q\|_{L_2}^2 \leq \frac{T^2}{2 \sin^2(\omega/2)} A(q)$. Adding this to the previous inequality yields

$$\|q\|_{H^1}^2 \leq (2 + C) A(q), q \in \Omega, \tag{3.10}$$

with $C = \frac{T^2}{2 \sin^2(\omega/2)}$.

Let's return to our action-minimizing subsequence $q^{(n)}$. The action of this sequence is bounded, and hence its H^1-norm is bounded according to Equation (3.10). We will use this fact to get $q_* \in \Omega$ by invoking another big theorem from analysis, the Banach–Alaoglu theorem in the context of Hilbert spaces.

We recall the theorem and the context around it. Let H be a Hilbert space. Recall that we say that a sequence $v_n \in H$ *weakly converges to* $v \in H$, written $v_n \rightharpoonup v$, if for any continuous linear functional $\ell : H \to \mathbb{R}$, we have $\ell(v_n) \to \ell(v)$.

Example 3.24   Let $\{e_n\}$ be an orthonormal basis for the separable Hilbert space H. Then $e_n \rightharpoonup 0$, since any linear function ℓ has the form $\ell(v) = \langle u, v \rangle$ for some $u \in H$, and in our case $\ell(e_n) = u_n$ is the nth coordinate of u relative to our basis. Since $\|u\|^2 = \Sigma u_n^2$ we have that $u_n \to 0$. Note that we do not have $e_n \to 0$ in the standard topology of H!

Theorem 3.25 (Banach–Alaoglu) *Any norm-bounded sequence of vectors in a Hilbert space H has a weakly convergent subsequence.*

We showed in Step 1 that H^1 is a Hilbert space. We have also shown that our action-bounded sequence is norm-bounded in H^1. Consequently, the Banach-Alaoglu theorem hands us a weakly convergent subsequence of our sequence. We will now leverage this fact to show that the subsequence converges both in C^0 and in Ω. To this end, we recall what it means for a linear map $F : H \to B$ to be *weakly continuous* when H is a Hilbert space and B is a Banach space. We say that the map F is *weakly continuous* if whenever $v_n \rightharpoonup v$ then $F(v_n) \to F(v)$.

Proposition 3.26 *The inclusion $H^1 \to C^0$ is weakly continuous.*

Remark 3.27   Inequality (3.9) shows that the image of this inclusion is contained in the subspace of C^0 consisting of the functions that are Hölder continuous with exponent one-half.

Proof of Proposition 3.26 Fix $t \in [0, T]$. Then the evaluation map $q \mapsto q(t)$ is a well-defined linear map $H^1 \to \mathbb{E}$. By the definition of weak continuity we must show that each of its components $q_a^i(t)$, $a = 1, 2, 3, i = 1, 2$, are continuous linear functionals with a fixed bound, independent of t. Indeed,

$$|q(t)| = |q(t) - q(0) + q(0)| \le |q(t) - q(0)| + |q(0)| \le \|\dot{q}\|_{L_2} + |q_0|.$$

Now from $a + b \le \sqrt{2(a^2 + b^2)}$, for $a, b > 0$ we have that $|q(t)| \le \sqrt{2} L \|q\|_{H_1}$ where L is the Lipschitz constant relating the norms $\|q\|_{H^1}^2$ and $\|\dot{q}\|_{L_2}^2 + |q(0)|^2$. Thus, for each t we have that $q \mapsto q_a^i(t)$ is a continuous linear functional, uniformly bounded by $L \|q\|_{H^1}$. QED

Equation (3.10) says that our sequence $q^{(n)}$, having its action bounded, is norm bounded. By the Banach–Alaoglu theorem our C^0 convergent subsequence admits a subsequence that converges weakly in H^1 to some $q_{**} \in H^1$. But does $q_{**} = q_*$? Yes. The weak continuity above shows that $q^{(n)}$ also converges to q_{**} strongly in C^0. A strongly convergent sequence cannot have two different limits, so $q_{**} = q_*$.

Since $q_{**} \in H^1$, we have $q_* \in H^1$ and, in particular, q_* is absolutely continuous. To see that $A(q_*) < \infty$ use that $A(q_*) = \frac{1}{2}\|\dot{q}_*\|_{L_2}^2 + \int_0^T U(q_*(s))ds$. The first term is finite since $q_* \in H_1$. The second term is finite since the integrals $I_n = \int_0^T U(q_*(s))ds$ tend to an infimum and the $U(q^n(\cdot))$ converge uniformly to $U(q_*(\cdot))$, which implies that $I_n \to I_* := \int_0^T U(q_*(s))ds$. Note, that by Lemma 3.13, we now also have that q_* is collision-free.

Finally, the conditions fixing the rotation ω and winding numbers n_{ij} used in defining Ω are C^0 closed conditions away from collisions and $q^{(n)} \to q_*$ in C^0, so that q_* must also satisfy these same conditions. We have shown that $q_* \in \Omega$.

Step 3 $A(q_*) = A_*$. Since $q_* \in \Omega$ and since A_* is the infimum of A over Ω, we have that $A(q_*) \geq A_*$. We show that $A(q_*) \leq A_*$. It suffices to show that $\lim_n A\left(q^{(n)}\right) \geq A(q_*)$ since $A_* = \lim_n A(c_n)$.

We showed in the penultimate paragraph of Step 2 that the potential term of the action converges; that is, $I_n = \int_0^T U\left(q^{(n)}(s)\right)ds \to I_* = \int_0^T U(q_*(s))ds$. We must deal with the kinetic term. We will show that $\int_0^T K\left(\dot{q}^{(n)}(s)\right)ds \geq \int_0^T K(\dot{q}_*(s))ds$. Adding I_n to the integral of $K\left(\dot{q}^{(n)}(s)\right)$ then yields the desired $\lim\inf A(q^{(n)} \geq A_*$.

We achieve the kinetic inequality by using the weak H^1 convergence $q^{(n)} \rightharpoonup q_*$. Define the quadratic form

$$\langle\langle v, w \rangle\rangle = \int_0^T \langle \dot{v}(t), \dot{w}(t) \rangle dt, v, w \in H^1, \tag{3.11}$$

so that the integral of the kinetic term is

$$\int K(\dot{c}(t))dt = \frac{1}{2}\langle\langle c, c \rangle\rangle.$$

Now $\langle\langle \cdot, \cdot \rangle\rangle$ is half the derivative part of the H^1 inner product. It follows that any $c \in H^1$ defines a continuous linear functional on H^1 by $h \mapsto \langle\langle c, h \rangle\rangle$. Take $c = q_*$ and use $q^{(n)} \rightharpoonup q_*$ to conclude that

$$\langle\langle q_*, q^{(n)} \rangle\rangle \to \langle\langle q_*, q_* \rangle\rangle. \tag{3.12}$$

Now the quadratic form $\langle\!\langle \cdot, \cdot \rangle\!\rangle$ is non-negative so that

$$0 \le \langle\!\langle q_* - q^{(n)}, q_* - q^{(n)} \rangle\!\rangle = \langle\!\langle q_*, q_* \rangle\!\rangle - 2\langle\!\langle q_*, q^{(n)} \rangle\!\rangle + \langle\!\langle q^{(n)}, q^{(n)} \rangle\!\rangle.$$

Taking *lim inf* and using Equation (3.12), we get that

$$0 \le -\langle\!\langle q_*, q_* \rangle\!\rangle + \langle\!\langle q^{(n)}, q^{(n)} \rangle\!\rangle,$$

or

$$\langle\!\langle q_*, q_* \rangle\!\rangle \le \liminf_n \langle\!\langle q^{(n)}, q^{(n)} \rangle\!\rangle.$$

Dividing by 2 shows us that the kinetic part of the action satisfies the required

$$\int K(\dot{q}_*(t)) \le \liminf_n \int K\big(\dot{q}^{(n)}(t)\big).$$

We have shown that $A(q_*) = A_*$.

Step 4 This step is the standard argument from first-year calculus. The derivative of a function at a minimizer must be zero.

The differential of a function A at a point c_* in the direction h is defined as

$$dA(c_*)(h) = \frac{\partial}{\partial \epsilon} A(c_* + \epsilon h)|_{\epsilon=0}.$$

In our case,

$$A(q_* + \epsilon h) = \frac{1}{2}\langle\!\langle q_* + \epsilon h, c_* + \epsilon h \rangle\!\rangle + \int_0^T U(q_*(t) + \epsilon h(t)) dt.$$

This first kinetic term is a bilinear expression, and so its derivative with respect to ϵ is $\langle\!\langle q_*, h \rangle\!\rangle$. Since the expression $q_* + \epsilon h$ is smooth in ϵ, and since U is C^1 away from collisions, we can use the chain rule to differentiate the second term, arriving at the differential $\int_0^T dU(c_*(t))h(t)dt = \int_0^T \langle \nabla U(c_*(t)), h(t) \rangle dt$. In sum,

$$dA(c_*)(h) = \langle\!\langle c_*, h \rangle\!\rangle + \int_0^T \langle \nabla U(c_*(t)), h(t) \rangle dt,$$

showing that A is indeed differentiable at any path $q_* \in H^1$ *that is collision-free* and that this derivative is continuous as a function from H^1 to its dual space (which we may identify with H^1 if we use the weak topology).

Now set

$$\Omega_\omega = \{h \in H^1 : h(T) = e^{i\omega}h(0)\}.$$

This is a closed linear subspace of H^1 representing the tangent space to $\Omega \subset H^1$. Indeed, our inital path space Ω is an open subset of Ω_ω. Notice that since the paths $h \in \Omega_\omega$ are continuous, absolutely continuous, and in H^1 we have that $q_\epsilon := q_* + \epsilon h \in \Omega$ for any fixed element $h \in \Omega_0$ and all ϵ small enough. Small ϵ may be needed to ensure that the deformed curve remains collision-free, and hence has $A(q_\epsilon) < \infty$ and, for the same reason, so that the winding conditions defining Ω are well defined and constant for q_ϵ.

We now argue by contradiction that $dA(q_*)(h) = 0$ for all $h \in \Omega_0$. Suppose not. Then there is an $h \in \Omega_\omega$ such that $dA(q_*)(h) \neq 0$. Upon multiplying h by a nonzero scalar we can assume $dA(q_*)(h) = -1$. Taylor expanding $A(q_* + \epsilon h)$ with respect to ϵ yields $A(q_* + \epsilon h) = A(q_*) - \epsilon + O(\epsilon^2) = A_* - \epsilon + O(\epsilon^2)$, which shows that we can lower the value of $A(q_*)$ to below A_* while staying in Ω. This contradicts the fact that $A(q_*) = A_*$ is the infimum of A over Ω. We have shown that $dA(q_*)(h) = 0$ for all admissible h.

Step 5 If $q_* \in \Omega$ and $dA(q_*)(h) = 0$ for all $h \in \Omega_0$, then q_* satisfies Newton's equation.

Remark 3.28   The Euler–Lagrange equations for L are precisely Newton's equations. They are derived by an integration by parts, *assuming the extremal is C^2*. The main point of step 5 is that we do not yet know that our minimizer q_* is C^2 , so the integration by parts step is illegal.

Step 5 is based on the fundamental lemma of the calculus of variations, which we set out as Theorem 3.28.

Theorem 3.29 (Fundamental lemma of the calculus of variations) *Let $\langle\langle v, w \rangle\rangle$ be the symmetric bilinear form of Equation (3.11). Suppose $c_* \in H^1$ and $\langle\langle c_*, h \rangle\rangle = 0$ for all $h \in \Omega_0$ where $\Omega_0 \subset \Omega_\omega$ consists of all paths in H^1 for which $h(0) = 0 = h(T)$. Then $\dot{c}_*$ is a constant path.*

For alternate proofs and a beautiful discussion around the fundamental lemma, see L. C. Young [218, p. 18].

Proof of the fundamental lemma, Theorem 3.29   We use Fourier series. Since $\dot{c}_*$ is square integrable, we can expand it as an L_2 convergent cosine series,

$$\dot{c}_* = A_0 + \Sigma_{k=1}^\infty a_k \cos(2\pi k / T); A_0, a_k \in \mathbb{E},$$

with

$$\langle\langle c_*, c_* \rangle\rangle = T|A_0|^2 + (T/2)\Sigma|a_k|^2.$$

The functions $\sin(2\pi k/T)$, $k = 1, 2, \ldots$, all vanish at 0 and T, so, for any fixed $e \in \mathbb{E}$, we have that $h(t) = \sin(2\pi kt/T)e$ is in $h \in \Omega_0$. Now $\dot{h} = 2\pi k/T \cos(2\pi kt/T)e$, and so, by the orthogonality of the cosines, we have

$$\langle\langle c_*, h\rangle\rangle = \pi\langle a_k, e\rangle.$$

But by assumption, this latter is zero. Since it is zero for all $k = 1, 2, \ldots$, and all $e \in \mathbb{E}$, it follows that all the nonzero Fourier series a_k of $\dot{c}_*$ are zero and hence $\dot{c}_* = A_0 = const.$, as claimed.

We proceed to step 5, working out the logic backwards. If we integrate Newton's equations, $\ddot{q} = \nabla U(q)$, we get $\dot{q}(t) = \int_0^t \nabla U(q(s))ds + v_0$, where v_0 is a constant vector. With this in mind,

$$G(t) = -\int_0^t \nabla U(q_*(s))ds,$$

so that $dG/dt = -\nabla U(q_*(t))$. Now take $h \in \Omega_0$. Use $\frac{d}{dt}\langle G(t), h(t)\rangle = \langle(-\nabla U(q_*(t))), h(t)\rangle + \langle G(t), \dot{h}(t)\rangle$ where the equality holds a.e. and both sides are integrable. Integrate by parts to obtain that

$$\int \langle G(t), \dot{h}\rangle(t)dt = +\int \langle \nabla U(q(t)), h(t)\rangle dt,$$

where the boundary term vanishes upon integration by parts because $h(0) = 0 = h(T)$. Set

$$\dot{\beta}(t) = \dot{q}_*(t) + G(t),$$

and observe that $\beta \in H_1$. (This uses that q_* is collision-free. If q_* had collisions, then the integral defining $G(t)$ could blow up fast enough at collision times so that $\dot{\beta}$ would not be square-integrable.) The condition $dA(q_*)(h) = 0$ for all $h \in \Omega_0$ now reads $\langle\langle \beta, h\rangle\rangle = 0$ for all $h \in \Omega_0$. By the fundamental lemma, we have that $\dot{\beta} = const. = v_0$, which means that $\dot{q}_* = \int_0^t \nabla U(q_*(s))ds + v_0$ where v_0 is a constant vector. Now q_* is absolutely continuous, and hence so is $t \mapsto \nabla U(q_*(t))$. Thus the integral defining $G(t)$ is differentiable with continuous derivative. This shows that $\dot{q}_*$ is continuously differentiable, that is, q_* is C^2, and that $\ddot{q}_*$ is equal to $\nabla U(q_*(s))$, concluding the proof of the fundamental lemma. QED

Step 6 $\dot{q}_*(T) = e^{i\omega}\dot{q}_*(0)$.

The traditional proof of this fact relies on keeping track of the boundary term when we integrate by parts to derive the Euler–Lagrange equations. Note that integration by parts is now allowed since we know that q_* satisfies Newton's equations on $(0, T)$ and is absolutely continuous and collision-free on $[0, T]$. See almost any text on the calculus of variations in one dimension.

We give an alternative proof based on the identification of Ω with a subspace of $C^0(\mathbb{R}, \mathbb{E})$ satisfying the periodicity condition (3.4). Let q_* continue to denote the path defined on all of $\mathbb{R}$ so extended. Take any number a with $0 < a < T$. The action of q_* over $[a, T + a]$ and over $[-a, T - a]$ equals its action over $[0, T]$ and so must realize the minimum of the action over this interval, with the corresponding constraints on winding numbers over these intervals imposed. So all the above steps hold with $[0, T]$ replaced by either interval. In particular, q_* satisfies Newton's equations across $t = 0$ and $t = T$ and so is differentiable. Differentiating the monodromy condition $q(T + h) = e^{i\omega} q(h)$ with respect to h now yields $\dot{q}_*(T) = e^{i\omega} \dot{q}_*(0)$. QED

Remark 3.30 (Non-coercive failures of Poincaré's theorem) Theorem 3.12 fails to hold when $\omega = 0 = n_{12} = n_{23} = 0$. Then none of the edges wind. The action infimum is $A_* = 0$. To obtain an action-minimizing sequence, place the bodies farther and farther away, say at the vertices of an equilateral triangle of radius R, and insist that they do not move. Now let $R \to \infty$. So, one could say the minimizing sequence *converges* to an ideal *solution* in which all bodies lie infinitely far away from each other and do not move. The theorem also fails if $\omega = 0$ and $n_{23} = 0$. In this case, the minimizing sequence converges to an ideal minimizer in which masses 1 and 2 perform a period T orbit with winding number $n_{12} \neq 0$ for our $1/r^\alpha$ potential's central-force law, while mass 3 is infinitely far away from them. However, the theorem seems to hold for $\omega = 0$ as long as $n_{12} n_{23} \neq 0$. Alternatively, we could introduce an n_{23}, in which case we get a minimizer as long as two of the $n_{ij} \neq 0$. In particular, the analogue of "n-times around Lagrange," which is to say $n_{12} = n_{23} = n_{31} = n \neq 0$ with $\omega = 0$, allows for a minimizing solution.

3.5.2 Other Applications

The proof of Theorem 3.8 goes through almost verbatim for any coercive class when the action is that of a strong force potential. Theorems 3.15 and 3.17 immediately follow.

3.5.3 Breakdown of the Argument for the Standard *N*-Body Problem

Steps 2 and 4 are not valid when paths to collision have finite action. Step 2 breaks down because the action no longer acts like an infinite wall separating distinct braid types, and the limit q_* could lie in the closure of the fixed braid class Ω and hence have collisions. Step 4 breaks down because the action is not differentiable at classes with collision.

The limit q_* still exists and is in H^1 and hence C^0. Its collision times form a closed set of measure zero, closed since the path is continuous, measure zero since its action is finite. Think about the Cantor set. The complement of such a set is a countable collection of open intervals. Upon being restricted to any one of these intervals, q_* satisfies Newton's equations. These solution segments are concatenated continuously along the Cantor set and suffer an uncountable number of derivative jumps as they enter and leave the collision set. Such a q_* is not a very satisfying solution.

3.6 Eliminating Collisions

We just described how the presence of collisions arising in the limit of an action-minimizing sequence effectively destroys the utility of the direct method. Gordon's theorem (Theorem 3.10) asserts that within the Newtonian two-body problem such collisions *do* occur for a large class of problems. We prove Gordon's theorem in Section 3.6.1.

We will need methods for showing that collisions don't arise if we are to succeed with the direct method. In Section 3.6.2 we describe the test path method for getting rid of collisions. In Sections 3.6.3 to 3.6.6 we describe and give a partial proof of Marchal's lemma, the most powerful general tool we have for getting rid of collisions.

3.6.1 Gordon's Work

The key to Gordon's work is the realization that the action of a periodic solution to Kepler's problem only depends on its period T and that this dependence is sublinear: $A(kT) < kA(T)$ for $k > 0$.

Proposition 3.31 *The action of one period of any negative energy Keplerian solution depends only on the period T and equals $CT^{1/3}$, where $C = \frac{3}{2}(4\pi^2\mu^2)^{1/3}$ and where the constant μ is the Kepler constant arising in the definition of Kepler's problem: $\ddot{q} = -\mu q/|q|^3$.*

See [30, 70, 146] for more details and discussion.

Proof The Lagrangian and Hamiltonian are related by $L(q, \dot{q}) = p\dot{q} - H(q, p)$ provided p and $\dot{q}$ are related to each other by the Legendre transformation. (See Section A.0.) The relation inducing the Legendre transformation can be understood as an equality of one-forms,

$$Ldt = pdq - Hdt,$$

in an extended phase space, pulled back to solution curves. The form pdq denotes the canonical one-form on phase space. (Again, see Appendix A.) Integrate this relation over one period, using the fact that the energy and period are related by Kepler's third law and, in particular, that fixing the energy fixes the period. We get

$$\int_0^T L dt = \int_c pdq - HT.$$

The integral of Ldt is the action. Now let c_1, c_2 be two solutions having the same energy, and hence the same period T. We can subtract the two integrals to get

$$A(c_2) - A(c_1) = \int_{c_2} pdq - \int_{c_1} pdq.$$

Now the energy level set is path connected for Kepler's problem. Parameterize the two curves and connect their initial points $c_1(0), c_2(0)$ by a smooth curve $\gamma(s), 0 \leq s \leq 1$, which lies within the fixed energy level. Because $\omega(X_H, v) = dH(v)$ we have that $\omega(X_H, \dot{\gamma}) = 0$. But $\omega = dp \wedge dq$. Applying the Hamiltonian flow to all the points of the arc γ for one full period, we sweep out an annulus whose boundary is $c_2 - c_1$. We have just shown that this annulus is ω-isotropic: The pull-back of $dp \wedge dq$ to this annulus is zero. It follows by Stoke's theorem that $\int_{c_2} pdq - \int_{c_1} pdq = 0$ or $A(c_2) = A(c_1)$.

To get the formula for the action $A = \int_0^T \frac{1}{2}(|\dot{q}|^2 + \mu/|q|)dt$ as a function of the period T, evaluate the action for one period of the circular Keplerian solution $q(t) = a\exp(i\omega t)$, using the defining equation of Kepler's problem to get Kepler's 3rd law in the form $\omega^2 = \mu/a^3$. QED

Proof of Gordon's Theorem, Theorem 3.10

Fix the class of going twice around the Sun (the origin) counterclockwise. We show any action-minimizing sequence in this class tends to a single period of the collision-ejection solution. Let's begin by going twice around a Keplerian orbit. If the period of once around this orbit is P then to go twice around takes time $T = 2P$. Additionally, the action of this twice-circled orbit is $2A(P) > A(2P)$, where $A(2P) = A(T)$ is the action of the collision-ejection solution of period T. The inequality $2A(P) > A(2P)$ follows from the formula for the action in Proposition 3.31.

This fact regarding actions implies that going twice around the Sun with the classical solution cannot be an action minimizer. In order to see this, take the collision-ejection solution and just back up a bit from collision, pause, and concatenate with a tiny quick circle traversed twice. For example, if collision occurs when $t = 0$, then it occurs again at $t = T$. Using $r(t) = r(-t) = r(T-t)$, travel along the collision-ejection solution from $t = \epsilon/2$ to $t = T - \epsilon/2$. At that stopping time we are at a point a distance $a = c\epsilon^{2/3}$ from the origin. Now take the Keplerian circle of this radius. Its action, twice around, is $4\pi\mu^{1/2}a^{1/2}$, which tends to zero with ϵ. Now, in the concatenation process we've messed up the period but only by $O(a^{3/2}) = O(\epsilon)$, and this can be renormalized back to T by a time rescaling at the risk of a vanishingly small change to the action.

Without much difficulty one can show the scenario just sketched is optimal: Any minimizing sequence converges to a collision-ejection orbit. Moreover, we can show that the theorem holds for any winding number k as long as k is not $0, 1$, or -1.

3.6.2 The Test Path Method

In the test path method we need to be able to estimate a lower bound a_{coll} for the actions of all *collision paths* lying in the *closure* (either H^1 or C^0) of our initial path space Ω. We must also find some test path γ_{test} lying in the interior of Ω for which $A(\gamma_{test}) < a_{coll}$. Finding the test path is a matter of art and luck. Since $A_* := inf_{q \in \Omega} A(q) \leq A(\gamma_{test})$ we are then guaranteed that our minimizing sequence converges to a non-collision path, one in the interior of Ω.

The test path method was the original method that yielded existence for the figure eight solution. The associated test path is one-twelfth of the test path described in Section 3.3.3.

3.6.3 The Workhorse Lemma of Christian Marchal

Soon after the rediscovery of the eight, Christian Marchal built a lemma that has become the main tool for eliminating collisions in action-minimizers for N-body problems.

Lemma 3.32 (Marchal) *Consider the action for the standard gravitational N-body problem in d dimensions, $d > 1$. Fix endpoints $q_0, q_1 \in \mathbb{E} = (\mathbb{R}^d)^N$ and a travel time $T > 0$. Then every action minimizer $q: [0, T] \to \mathbb{E}$ connecting q_0 to q_1 in time T has no interior collision points. In other words, if $q: [0, T] \to \mathbb{E}$ is an action minimizer, then $q(t) \notin \Delta$ for all $t \in (0, T)$.*

Remark 3.33

- Either endpoint q_0 or q_1 of the lemma may be a collision point.
- The minimizers of the lemma are solutions to Newton's equations up to either endpoint.

- From the proof of the lemma, one sees that the fixed endpoint conditions can be replaced by fixed submanifold conditions and the lemma continues to hold.
- The lemma has been generalized by [61] so as to hold for any power law potential $-U_\alpha$ in place of the standard Newtonian potential $V = U_1$.

Let $C = q^{-1}(\Delta) \subset [0, T]$ be the set of collision times of an action-minimizing curve. Before Marchal's lemma we could only guarantee that C was a closed set of measure zero; for example, C could be a Cantor set. Thus the complement of C consisted of a possibly countable set of open intervals of total measure T. Upon restriction to any of these intervals, q must solve Newton's equations. However, a priori, the energy and angular momentum constants for different intervals may be different. These various solution arcs glue together along C to form the continuous solution curve q. This affords us with a very strange notion of solution: a countable set of solutions defined on various vanishingly small open intervals, concatenated continuously along a Cantor set.

Conversely, armed with Marchal's result, we get that $C \subset \{0, T\}$ and that the minimizer q is a bona fide solution on the entire open interval $(0, T)$.

3.6.4 Proving Marchal's Lemma

For the original proof and insights see Marchal [117]. For full proofs that eliminate all types of subcluster collisions and work for all power-law potentials we recommend [30], [61], or [210]. We will not provide a full proof. Instead, we will give a proof under the restrictive assumption that only binary collisions occur for the alleged minimizers.

We proceed in two steps:

- Step 1. Get rid of isolated binary collisions. This step contains Marchal's averaging idea.
- Step 2. Show that, indeed, the collision times are isolated.

Step 1: Marchal's Averaging Method

We give the proof for the standard N-body problem under the simplifying assumption that the collisions arising in the interior $(0, T)$ are isolated binary collisions. The proof is by contradiction. Assume our minimizer has an isolated binary collision at time $t_c \in (0, T)$. We show that we can decrease its action while maintaining the endpoint conditions, thereby arriving at a contradiction and showing that the collision could not have happened in the first place.

Let δ_1 be a time such that no other binary collision arises within the interval $(t_c - \delta_1, t_c + \delta_1)$. Shift time by t_c so as to center the collision time to $t = 0$ and so that our minimizer starts at q_0 at time $-t_c$ and ends at q_1 at time $q(T - t_c)$. For each point $s \in \mathbb{S}^2$ we will construct a perturbation $q_s(t)$ of $q(t)$ that agrees with $q(t)$ outside the critical interval $[-\delta_1, \delta_1]$ and has no collisions within the interval. We will not be able to compare the action $A(q_s)$ to $A(q)$ for any particular s to any advantage. Marchal's brilliance was to instead *average* $A(q_s)$ over the whole sphere and show that the average action decreases:

$$\Delta A := \fint_{s \in \mathbb{S}^2} A(q_s) d\sigma(s) - A(q) < 0, \qquad (3.13)$$

where the symbol $\fint_{s \in \mathbb{S}^2}$ means to take the average over the sphere with respect to the usual area measure $d\sigma(s)$ of the sphere. If the average action of the perturbations is less than that of $A(q)$, then for most values $s \in \mathbb{S}^2$ the perturbed action $A(q_s)$ is less than $A(q)$ and, in particular, there has to be *some* $s_* \in \mathbb{S}^2$ for which $A(q_{s_*}) < A(q)$, contradicting minimality.

The key to Marchal's proof of inequality (3.13) is potential theory. He identified the leading term in the averaged perturbed action with the potential due to a homogeneous gravitating spherical shell and used the relation between harmonic functions and averaging.

The particular family of perturbations of Marchal is elegant. Suppose that it is bodies 1 and 2 that collide at time $t = 0$. Let $f(t; \epsilon)$ be the continuous even non-negative piecewise linear scalar function equal to ϵ for $t = 0$ and having support on

$$I_c := [-\delta, \delta],$$

where $\delta \leq \delta_1$. (A particular choice of δ will be made later, one guaranteed so that within I_c our alleged minimizer has well-described asympotics.) See Figure 3.11. Thus

$$f(t; \epsilon) = \epsilon(1 - t/\delta), 0 \leq t \leq \delta.$$

Let μ_1, μ_2 be the usual "reduced mass" coefficients so that $m_1 \mu_1 - m_2 \mu_2 = 0$, $\mu_1 + \mu_2 = 1$. Define the perturbation $q_s^\epsilon(t) = (q_{1,s}^\epsilon(t), q_{2,s}^\epsilon(t), \ldots, q_{N,s}^\epsilon(t))$ by

$$q_{1,s}^\epsilon(t) = q_1(t) + \mu_1 f(t; \epsilon)s,$$

$$q_{2,s}^\epsilon(t) = q_2 - \mu_2 f(t; \epsilon)s,$$

while

$$q_{a,s}^\epsilon(t) = q_a(t), a > 2.$$

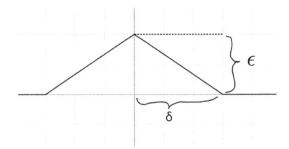

Figure 3.11 The piecewise linear function $f(t, \epsilon)$ defining the perturbation used in our proof of Marchal's lemma.

Note that

$$m_1 q_{1,s}^\epsilon + m_2 q_{2,s}^\epsilon = m_1 q_1 + m_2 q_2,$$

so the perturbation does not change the 1-2 center of mass, nor the overall center of mass. Also note that

$$q_{1,s}^\epsilon - q_{2,s}^\epsilon = q_{12}(t) + f(t; \epsilon)s,$$

where $q_{12}(t) = q_1(t) - q_2(t)$. The perturbation has destroyed the collision at $t = 0$, shifting it by the vector s.

Since q and q_s agree outside of $I_c = [-\delta, \delta]$, we have

$$A(q_s^\epsilon) - A(q) = \int_{I_c} (K(q_s(t)) - K(q(t))) + U(q_s(t)) - U(q(t)) dt$$

for each s. Remember, the proof is achieved by averaging this action difference over $s \in \mathbb{S}^2$ and showing that it becomes negative for appropriate $\delta > 0$ and all ϵ sufficiently small. To this end we will show that as $\epsilon \to 0$, the dominant term of the action difference comes from the 1-2 potential term,

$$\Delta A_{12} := \fint_{s \in \mathbb{S}^2} \left(\int_{I_c} \frac{m_1 m_2}{|q_{12}(t) + f(t, \epsilon)s|} - \frac{m_1 m_2}{|q_{12}(t)|} dt \right) ds.$$

We estimate ΔA_{12} now. Switch the order of integration, so that

$$\Delta A_{12} := \int_{I_c} \left(\fint_{\mathbb{S}^2} \frac{m_1 m_2}{|q_{12}(t) + f(t, \epsilon)s|} - \frac{m_1 m_2}{|q_{12}(t)|} ds \right) dt.$$

At this point the promised potential theory enters, potential theory going back to Newton. Newton established that a uniform spherical shell of mass exerts no force inside the shell, while outside the shell it exerts the same force as would be exerted if all of its mass were concentrated at a point at its center.

This reasoning implies that if $f > 0$ denotes a constant radius and if we define a function $W(Q)$ on $\mathbb{R}^3$ by

$$W(Q;f) = \fint_{s \in \mathbb{S}^2} \frac{ds}{|Q - fs|},$$

then

$$W(Q) = \begin{cases} \frac{1}{|Q|}, & |Q| > f, \\ \frac{1}{f}, & |Q| \leq f. \end{cases} \tag{3.14}$$

Note also that $W(Q, f) = \fint_{s \in \mathbb{S}^2} \frac{ds}{|Q+fs|} ds$ since the integral is invariant under $s \mapsto -s$. Applying this realization regarding $W(Q)$ to the integrand inside our last expression for ΔA_{12}, setting $Q = q_{12}(t)$, $f = f(t)$, multiplying by $m_1 m_2$, and using $|Q| = r_{12}$ we see that the two terms of this integrand cancel unless $r_{12} < f$, in which case we get the integrand to be $\frac{m_1 m_2}{f} - \frac{m_1 m_2}{r_{12}}$. Thus

$$\Delta A_{12}^\epsilon = m_1 m_2 \int_{t \in I_c : r_{12}(t) \leq f(t;\epsilon)} \left(\frac{1}{f(t;\epsilon)} - \frac{1}{r_{12}(t)} \right) dt.$$

We are done with our use of potential theory.

Since $1/r_{12} > 1/f$ on the region of integration, we have that $\Delta A_{12} < 0$. We will need the leading asymptotics in ϵ of ΔA_{12} to ensure it dominates the rest of the average action, keeping it negative. To obtain these asymptotics we use that

$$r_{12}(t) \sim \gamma t^{2/3} + O(t), \qquad \gamma = \left(\frac{9}{2}(m_1 + m_2) \right)^{1/3}, \text{ as } t \to 0. \tag{3.15}$$

This estimate can be found in Wintner [215, p. 269, section 350, equation (20)]. (See the chapter notes, Section 3.9, regarding other routes to the estimate and its generalizations to more collisions and other power laws.) We choose δ independent of ϵ so that these asymptotics are valid on the interval $[-\delta, \delta]$ for both collision branches with fixed constant bounds.

I claim that estimate (3.15) implies that

$$\Delta A_{12}^\epsilon = -\frac{4m_1 m_2}{\gamma^{3/2}} \sqrt{\epsilon} + O(\epsilon). \tag{3.16}$$

Focus on the branch leaving collision: $0 \leq t$. We have that $f(t)$ decreases monotonically from ϵ to 0 while r_{12} increases monotonically from 0. It follows that there is a unique first time $t_0 = t_0(\epsilon)$, close to zero, such that $r_{12}(t_0) = f(t_0)$. See Figure 3.12. We can estimate that time by replacing $r_{12}(t)$ with $\gamma t^{2/3}$ and solving $f(t) = \gamma t^{2/3}$ to leading order, using $f(t) = \epsilon(1 - t/\delta)$, which yields

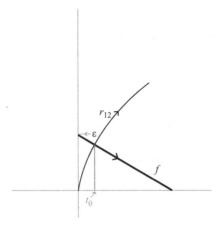

Figure 3.12 The point t_0 is where $r_{12}(t)$ first exceeds $f(t)$.

$$t_0 = \left(\frac{\epsilon}{\gamma}\right)^{3/2} + O(\epsilon^2).$$

The errors made by replacing r_{12} by $\gamma t^{2/3}$ in the integrand of ΔA_{12} are $O(\epsilon)$. The integral is now estimated by

$$m_1 m_2 \left(\int_0^{t_0(\epsilon)} \frac{dt}{f(t;\epsilon)} - \int_0^{t_0(\epsilon)} \frac{dt}{\gamma t^{2/3}} \right).$$

The second integral we do immediately to get $-\frac{3}{\gamma} t_0^{1/3}$. The first integral could be done, but instead we estimate it by $\frac{t_0^{1/3}}{\gamma} \geq \int_0^{t_0} \frac{dt}{f(t)}$ as follows. The function f is monotonically decreasing on $[0, t_0]$ so $1/f(t_0) \geq 1/f(t)$ on that interval. Thus $\frac{t_0}{f(t_0)} = \int_0^{t_0} \frac{1}{f(t_0)} dt \geq \int_0^{t_0} \frac{dt}{f(t)}$. Now use $f(t_0) = \gamma t_0^{2/3} + O(t_0)$. Summing the two integrals, we get $(1-3) m_1 m_2 \frac{t_0^{1/3}}{\gamma} + $ lower order terms $\geq \Delta A_{12}|_{t>0}$ or

$$-2 \frac{m_1 m_2}{\gamma^{3/2}} \epsilon^{1/2} \geq \Delta A_{12}|_{t>0},$$

where $\Delta A_{12}|_{t>0}$ denotes the contribution to the averaged action change obtained by integrating over the interval $[0, t_0]$. The estimate for $\Delta A_{12}|_{t<0}$ (the averaged 12 potential part of the action for the interval $[-\delta, 0]$ before collision) proceeds identically and yields the same upper bound. Adding the two upper bounds yields the claimed result, inequality (3.16).

It remains to show that ΔA_{12} dominates the other terms in the averaged change in action. Expand out the total averaged change in action:

$$\Delta A = \Delta A_K + \Sigma_{j>2}(\Delta A_{1j} + \Delta A_{2j}) + \Delta A_{12},$$

where ΔA_K denotes the averaged change in the kinetic part of the action, and where ΔA_{ij} denote the differences in perturbed and unperturbed averaged action arising from the $1/r_{ij}$ term of the potential. There are no terms ΔA_{bc} for $b, c > 2$ since the motions of all the bodies $b, c > 2$ are unchanged by the perturbation. We will show that

$$\Delta A_K = O(\epsilon^2/\delta)$$

and

$$\Delta A_{1j} = O(\epsilon\delta), \Delta A_{2j} = O(\epsilon\delta).$$

(Recall that δ is independent of ϵ and is chosen so that the collision asymptotics of both branches hold.) These estimates show that the negativity of ΔA_{12} dominates the rest of the action as claimed, and will complete the proof.

The integrand for the difference in kinetic parts of the action is

$$\frac{1}{2}\Sigma m_a(|\dot{q}^\epsilon_{a,s}(t)|^2 - |\dot{q}_a(t)|^2).$$

Since we've only changed the trajectories for $a = 1, 2$ these are the only ones that contribute. Since

$$\dot{q}^\epsilon_{1,s} = \dot{q}_1 + \mu_1\dot{f}(t;\epsilon)s,$$

the kinetic difference term in the integrand for mass 1 is

$$\frac{1}{2}m_1(2\mu_1\dot{f}_1\dot{q}_1 \cdot s + \mu_1^2(\dot{f})^2).$$

Now

$$\dot{f}(t;\epsilon) = \epsilon\frac{\Theta(t;\delta)}{\delta},$$

where

$$\Theta(t,\delta) = \begin{cases} -1, & 0 \leq t \leq \delta, \\ 1, & -\delta \leq t < 0, \\ 0, & |t| > \delta. \end{cases}$$

From this expansion of the integrand it looks like the averaged kinetic is $O(\epsilon)$ right away. However, the velocity $\dot{q}_1$ blows up as $t \to 0$ so we would need its asymptotics to estimate this term which is apparently linear in ϵ. The integral arising from this term is at least finite since the rate of blow up of the velocity is $t^{-1/3}$. But viewed as a function of s, this first term is linear in s and linear functions on the sphere average to zero! It follows that the only term that contributes to the averaged kinetic action from mass 1 is the term $m_1 \mu_1^2 (\dot{f})^2$. The same argument holds for the kinetic term arising from mass 2. Adding the two, using $\dot{f}^2 = \epsilon^2 / \delta^2$ on I_c and doing the integral, we get

$$\Delta A_K = \frac{\epsilon^2}{\delta} \left(m_1 \mu_1^2 + m_2 \mu_2^2 \right)$$

exactly.

In order to estimate ΔA_{1j} and $\Delta A_{2,j}$ we use the fact that the collision is isolated. Thus, there is a positive constant C such that $r_{1j}, r_{2j} > C$ as long as $t \in I_c$ and $j > 2$. Write $q_{ij} = q_i - q_j$. One expands out $\frac{1}{r_{1j}^\epsilon} - \frac{1}{r_{1j}}$ using

$$(r_{1j}^\epsilon)^2 = r_{1j}^2 \left(1 + \frac{2}{r_{1j}^2} \mu_1 f(t, \epsilon) q_{1j} \cdot s + \frac{1}{r_{1j}^2} f(t, \epsilon)^2 \right)$$

and the binomial expansion to find that

$$\left| \frac{1}{r_{1j}^\epsilon} - \frac{1}{r_{1j}} \right| \le \frac{\epsilon}{2} \frac{\mu_1}{C^2} + O(\epsilon^2),$$

where we used that $|f| \le \epsilon$ on I_c. An identical estimate holds for $\frac{1}{r_{2j}^\epsilon} - \frac{1}{r_{2j}}$. Since the integrands arising in $\Delta A_{1j}, A_{2,j}$ are bounded by ϵ and their interval of integration is of size δ, we get that $\Delta A_{1j}, \Delta A_{2,j} = O(\epsilon \delta)$ as claimed.

This completes the proof of step 1.

3.6.5 Step 2: Isolating Collisions

Marchal's averaging argument required that the collision times be isolated points within the time interval. If, instead, the collision times gang together to form an impenetrable continuum like the Cantor set, the perturbation described does not work. Having no idea how to apply perturbation arguments to an impenetrable continuum, we instead show this continuum cannot arise in the first place. Collision times must be discrete.

Lemma 3.28 provides formal statement of the required result.

Lemma 3.34 *If the collisions occuring among a minimizer for fixed endpoint problem (as per Marchal's lemma) are all binary collisions of type 12 then these collision times form a discrete, and hence finite, set.*

The proof of the lemma rests on energy conservation. See property (iii) of the proof.

Proof of Lemma 3.34   We will need the following a priori properties of any minimizer γ. Note that, as dscussed above, the complement of the collision times is a countable union of disjoint open intervals.

(i) Restricted to any collision-free interval, the minimizer satisfies the Euler–Lagrange equations.

(ii) In the neighborhood of a binary collison time the motion of all the masses *not* participating in the collision, and of the 12 center of mass, is twice continuously differentiable in a neighborhood of any one of the collision points.

(iii) The energy $H = \frac{1}{2}|\dot{\gamma}|^2 - U(\gamma)$ is constant almost everywhere.

Proof of property (i)   The minimizer is continous. Let $I = (a, b)$ be a collision-free open solution interval. If the Euler–Lagrange equations are not satisfied, then $dA(\gamma)(\delta\gamma) < 0$ for some variation $\delta\gamma$ whose support is contained within I. We can follow this variation with an actual perturbation $\gamma^\epsilon = \gamma + \epsilon\delta\gamma + O(\epsilon^2)$, thus decreasing the action while leaving γ's endpoints fixed. This contradicts minimality.

Proof of property (ii)   Although U, and consequently A, are not differentiable near binary collision, they are differentiable in the direction of perturbations in which only the nonparticipating mass is varied. Suppose that m_1 and m_2 collide at time t_c, while $r_{ab} > \delta > 0$ for all $(a, b) \neq (1, 2)$ and all t in our interval. Recall that $U = \Sigma m_i m_j / r_{ij}$. Observe that U is differentiable with respect to the q_a, $a \neq 1, 2$. Consider variations of the action in which $q_1(t), q_2(t)$ are fixed, while $q_a(t)$ varies to $q_a(t) + \epsilon\delta q_3(t)$. The action functional is differentiable, even at 12 collision points, for such a perturbation. Consequently, the q_a-Euler–Lagrange equations must be satisfied all along the interval, for the same reasons as in (i). These q_a Euler–Lagrange equations are simply the ath Newton equation $m_a\ddot{q}_a = \Sigma_{b \neq a}F_{ab}$. An inspection of the two-body forces shows the forces are continuous functions, bounded away from zero. Consequently, the q_a have continuous second derivatives. The proof that the center of mass vector $Q_{12} = \frac{1}{m_1+m_2}(m_1q_1 + m_2q_2)$ also has continuous second derivative is the same. Simply change variables from q_1, q_2, q_a to $q_{12} = q_1 - q_2, Q_{12}, q_a$, and rewrite the Lagrangian in these new variables. Note that the result is differentiable in Q_{12} provided 1 and 2 are the only masses colliding.

Proof of property (iii)   Vary the action with respect to changes of parameterization: $q_i^\tau(t) = q_i(\tau(t))$ where $\tau: [0, T] \to [0, T]$ is a smooth invertible time change. Write $t = t(\tau)$ for the inverse function, and $\lambda = dt/d\tau$, evaluated at τ. Write $V = -U$ for the potential, so that the Lagrangian is $K - V$, and the energy is $H = K + V$. We compute $L(q^\tau, \dot{q}^\tau) dt = [\frac{1}{\lambda} K - \lambda V] d\tau$. Consequently, the differential of the action with respect to parameterization change at $\lambda = 1$ (corresponding to $\tau = t$) is $\int \delta\lambda(t) H(t) dt$. Here $\delta\lambda(t)$ denotes the change in the differential of parameterization. This energy $H(t)$ is defined almost everywhere since the solution intervals of (i) have full measure. Remember the constraint $\int_0^T \lambda d\tau = \int dt = T$, and use the method of Lagrange multipliers to conclude that $\int (-H(t) + c)\delta\lambda(t) dt = 0$ for all possible parameterization changes. We conclude that $H(t) = c$ a.e.

We now proceed to the proof of the lemma. It is a proof by contradiction. Suppose that γ is such a minimizer, and has only 12 binary collisions. Suppose that there is a collision time t_c with a sequence of binary collision times t_n converging to t_c. Because the complement of the collision set is open and of full measure we may further assume that the t_n are ordered so as to form open collision-free intervals $I_n = (t_{2n}, t_{2n+1})$. Now r_{12} is continuous on the closure of I_n and zero at its boundary points t_{2n}, t_{2n+1}, from which it follows that $t \mapsto r_{12}(t)$ achieves its maximum at some point $t_{*,n} \in I_n$. The function r_{12}^2 achieves its maximum over I_n at this same time. Being a local maximum of a C^2 function on I_n, we must have that

$$\frac{d^2}{dt^2} r_{12}^2(t_{*,n}) \le 0. \tag{3.17}$$

However, the a.e. constancy of energy H (property (ii)) implies that

$$\frac{d^2}{dt^2} r_{12}^2 = \frac{2\mu}{r_{12}} + O(1), \quad \text{with } \mu = m_1 m_2/(m_1 + m_2), \tag{3.18}$$

holds (off of C) in a neighborhood of t_c. Equation (3.18) is a version of the Lagrange–Jacobi identity $\ddot{I} = 4H + 2U$ with I replaced by r_{12}^2, using the fact that H is a shared constant away from C. (See Venturelli's thesis [210, p. 45] for a derivation of Equation (3.18) and its generalization to collisions involving more bodies. Alternatively, see Levi-Civita [108, bottom of p. 117].) Now since $r_{12}(t_n) \to 0$ and $t_{2n} < t_{*,n} < t_{2n+1}$ we have, by continuity, that $r_{12}(t_{*,n}) \to 0$. In particular, the right-hand side of inequality (3.18) is eventually, for n large enough, greater than any given positive constant K, which is to say that $\frac{d^2}{dt^2} r_{12}^2(t_{*,n}) > K > 0$, contradicting inequality (3.17). The only way out of the contradiction is that the sequence t_n does not exist in the first place. Our collision point t_c must be isolated.

3.6.6 The General Argument and Blow-Up

In our proof of Lemma 3.34, it was important that the 12 binary collision time t_c was not also the limit of, say, a set of 13 binary collision times. If it were, by continuity, t_c would then be a triple or quadruple collision instant, and not a binary collision instant. More generally, if two separate clusters of bodies have the same collision time t_c as an accumulation point, then t_c is a collision time for the union of the two clusters. This observation allows us to set up an inductive argument to isolate all collisions, the induction being on the number of bodies p that suffer collisions. We have just proved the $p = 2$ case. For the full argument see [30].

A crucial ingredient in the full proof is a method called blow-up, borrowed from PDEs, and which we have not yet discussed. In blow-up one imagines an alleged collisional minimizer in which some cluster k of the N bodies has collided. Using the space–time scaling symmetry of the N-body problem, we replace the alleged minimizer by a family of rescaled solutions for which the origin is the common collision point and the energy of the cluster tends to zero. One shows that the rescaled family converges to a new action minimizer associated to the k-body problem. This new minimizer must be of *dropped* central configuration type – it evolves by scaling. Its motion is of collision-ejection type: radially inward from one endpoint to k-fold collision, then radially outward to another endpoint. One then applies the Marchal averaging trick to the limit minimizer and shows that its action can be decreased by perturbing in some random direction, thereby obtaining a contradiction. This use of blow-up replaces and simplifies our explicit elimination of isolated binary collisions so as to apply to the case of isolated k-collisions. See [30, section 3.2] for details.

Variants of Marchal

One cannot directly apply Marchal to eliminate collisions for the minimizers obtained by applying the direct method to paths with Γ-symmetry. This is because Marchal had to consider perturbing away from the collision in random directions, some of which could violate the symmetry condition. Ferrario and Terracini [61] constructed an equivariant version of Marchal's lemma in which only a circle's worth of perturbed directions were required. Their method does not apply to any symmetry representation, but instead only to those satisfying their "rotating circle" condition. See [143, section 2.6] and [61, section 9] for the statement of this condition.

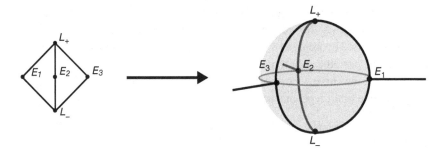

Figure 3.13 The abstract graph embedded in the shape sphere. Adapted from [140].

3.7 A Dynamical Method

Theorem 3.18 was proven using dynamical methods developed by Moeckel in the 1980s [129, 131, 133, 135]. The theorem itself is a corollary of Theorem 3.35, the statement of which uses the notion of a *stutter block*. Use exponential notation to denote stutters. For example, $1^2 2^3 = 11222$. We say that an eclipse sequence consists of stutter blocks in the range $[N, N'] \subset \mathbb{Z}$ if it has the form $\ldots a_0^{n_0} a_1^{n_1} \ldots a_i^{n_i}$, where the integers n_i lie in the interval $[N, N']$ and the a_i come from our alphabet $\{1, 2, 3\}$.

Theorem 3.35 *Consider the planar three-body problem for fixed negative energy, small but non-vanishing angular momenta J, and mass ratios equal or sufficiently close to equal. There is a positive integer N such that for any $N' \geq N$ every bi-infinite stutter sequence with range $[N, N']$ is realized by a collision-free solution. The angular momentum of these solutions satisfies $0 < |J| < \mu(N')$ with $\mu(N') \to 0$ as $N' \to \infty$. If the stutter sequence is periodic then it is realized by a reduced periodic solution.*

Properties of These Solutions
The projection to the shape sphere of the solutions guaranteed by Theorem 3.35 stay in a small neighborhood of the embedded graph shown on the right-hand side of Figure 3.13.

When transitioning from one stutter block to the next, the solutions pass close to triple collision in near-Lagrange configurations.

Theorem 3.35 $\implies$ **Theorem 3.18** Given a relative braid type, there is a *reduced* (no stutters) periodic eclipse sequence $w = a_1 a_2 \ldots a_{2k}$ representing it. (Lemma 3.3.) Replace each a_i in w with the stutter block $a_i^{n_i}$ with n_i odd and in the range $[N, N']$ to get a new word. Repeat the new word to make it periodic. By Theorem 3.18 this new periodic eclipse sequence is realized by a relative periodic solution. Stutters cancel in pairs (Section 3.1.1), so this relative periodic curve, when projected to the shape sphere, is freely homotopic to the relative braid type encoded by w.

Figure 3.13 depicts the main idea driving Theorems 3.35 and 3.18. It shows a graph embedded in the shape sphere. The graph's vertices are the five central configurations. When the masses are equal the edges of the graph map to geodesic arcs – quarter great circles. These quarter circles lie on corresponding isosceles great circles. The shape sphere minus binary collisions deformation retracts onto the embedded graph, and so the graph faithfully carries all the topology; that is, all reduced free homotopy classes. Periodic walks on the graph thus represent any desired reduced free homotopy classes.

A version of this graph with its dynamical meaning described can be found in the papers by Moeckel mentioned at the beginning of this section. The main tool used in these papers is the McGehee blow-up as described in Chapter 2. We showed there that to each central configuration $R \in \{L_+, L_-, E_1, E_2, E_3\}$ we have a pair R, R^* of equilibrium points on the collision manifold, together with a dropped solution RR^* which is heteroclinic between the pair. (See the last sentence of Proposition 1.18.) For each vertex R of the graph, consider the closure of this dropped solution: so the dropped central configuration solution along with its pair of equilibria. Edges $E_i L_\pm$ in the graph indicate the existence of a heteroclinic set whose closure connects the two associated dropped solutions. The spikes in Figure 3.13 coming out of the E_j represent an excursion taking us far from the collision, shadowing $E_j E_j^*$. The main point of the proofs is that by turning on a tiny bit of angular momentum, any walk on the abstract graph is realized by a solution whose projection to the shape sphere stays within a small neighborhood of the embedded graph. Consequently, this solution has an eclipse sequence that faithfully records the walk on the graph with its consequent free homotopy class.

Qualitative Description of These Solutions

The solutions guaranteed by Theorem 3.35 suffer repeated close approaches to triple collision, one such approach per stutter block. Far from collision, the solution looks like one of the three Eulerian homothetic solutions E_i, perturbed a bit and given a small amount of angular momentum so that the three bodies move along very eccentric nearly collinear ellipses. As the solution collapses to near triple collision, the shape of the triangle oscillates for a while around the collinear shape E_i, producing the syzygy block i^n and then it shoots off into one or the other of the two Lagrange equilateral triangle shapes $L_\pm$, all the while staying close to triple collision. The shape during the transition from E_i to L_+ or E_i to L_- remains near-isosceles, and thus stays near the corresponding edge of the embedded graph. Once the shape is nearly equilateral, the triangle spins a half turn and morphs again, following close to another isosceles edge of the embedded graph to an E_j. Arriving near E_j, the shape oscillates for a while about collinearity and then makes an excursion away from triple collision, shadowing close to the dropped E_j solution and the process repeats. During each close approach to collision, we can choose whichever Lagrange configuration L_+ or L_- to shoot off to, and which E_j to go to next, forcing evenness or oddness in the stutter block, and following any given sequence as per the theorem.

3.7.1 Why We Need Equal or Near Equal Masses: Spiralling

The hypothesis that the masses are equal or near-equal in Theorem 3.35 implies two properties of the collision flow essential to our proof. The first is that the eigenvalues of the linearization of the collision flow at the Euler configuration equilibria E_j and E_j^* are not real. The set of mass ratios for which this property holds is indicated in Figure 3.14. The second property is a kind of transversality between some stable and unstable manifolds on the collision manifold. See propositions 1 and 2 of [135].

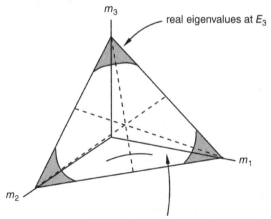

Figure 3.14 The spiralling condition holds for masses lying in the *non-shaded* region indicated on the mass simplex $m_i \geq 0, m_1 + m_2 + m_3 = 1$. From [135], figure 10.

Figure 3.15 When the abstract graph of Figure 3.13 is doubled and placed on the collision manifold then its vertices account for all the equilibria on the collision manifold.

The first property is referred to as the *spiralling condition*. This spiralling condition implies that $W^u(L)$ and $W^s(L^*)$ (for $L = L_\pm$) wind around the $E_j E_j^*$ dropped central solution, forming something like an infinite scroll. One of these manifolds, say $W^u(L)$, winds around the dropped solution counterclockwise while the other winds around clockwise, resulting in infinitely many intersections between the two manifolds, the intersections occuring very close to the dropped central solution $E_j E_j^*$. Each intersection point represents an connection $L \rightarrow L^*$ that repeatedly comes very close to the Euler E_j configuration during its travel. One can think of these solutions as forming drawn bows or violin strings during their time near E_j, so that the three masses form a nearly collinear string with mass j in the middle, oscillating about the midpoint of the line formed by the other two masses for a long time. On either side of this vibrational behavior the configuration collapses to total collision in a near-equilateral way as $t \rightarrow \pm\infty$. The number of oscillations of mass j is the number of stutters.

3.7.2 Why We Need Some Angular Momentum

We require a bit of angular momentum in order to get solutions that connect small neighborhoods of different dropped Eulerian central configuration solutions. We explain here how this tiny bit of angular momentum enters, paraphrasing the discussion of the last paragraph of section 3 of [140]. See also [151].

Each central configuration R corresponds to a pair $\{R, R^*\}$ of equilibria on the collision manifold; R corresponds to explosion out from total collision while R^* represents collapse to collision. Double the abstract graph, with one copy for the explosions and another for the collapses, as shown in Figure 3.15.

Now join two vertices v, w of the doubled graph by a directed edge $v \rightarrow w$ if there is an orbit in the blown-up flow whose α limit set contains v and whose ω limit set contains w. The dropped solutions associated to each R yield a directed edge $R \rightarrow R^*$. Let L denote either of $L_\pm$ and E any of

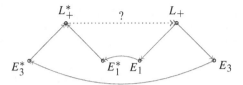

Figure 3.16 After angular momentum is turned on, we can implement the arrow marked by a question mark. All the other arrows are implemented by stable-unstable connections at angular momentum zero.

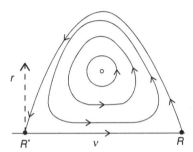

Figure 3.17 The projection of a central configuration family having fixed energy onto the v, r plane. The arch and "floor" $r = 0$ comprise the rest cycle.

the E_j. Devaney [41] showed that both $L \to E$ and $E^* \to L^*$ happen by looking into the isosceles three-body problem. His results require the spiraling eigenvalue condition just discussed. Concatenating Devaney solutions with dropped solutions, we arrive at concatenated solutions representing $L \to E_i \to E_i^*$ and $E_j \to E_j^* \to L^*$. In order to complete the circuit to get $E_i \to E_j$ we need a connection $L^* \to L$. This connection requires some angular momentum. Figure 3.16 shows an indication of scenarios for connecting E_1 and E_3, with the needed last connection indicated with a question mark.

The dropped solution $R \to R^*$ forms one end of the one-parameter family of homographic solutions corresponding to the central configuration R as described in Lemma 1.10. Figure 3.17 depicts this family at fixed negative energy. The family is parameterized by angular momentum J with the dropped solution having $J = 0$. All the solutions except the dropped one are periodic. As the angular momentum tends to zero, the trajectories of the family do not limit to just the dropped solution, but rather they limit to a rest cycle consisting of the dropped solution and a return heteroclinic path $R^* \to R$ that lies on the collision manifold. Shadowing this return part of the cycle with real solutions requires an ϵ of angular momentum.

The return part of the cycle and Devaney's connecting paths all lie on the collision manifold $r = 0$. The energy H and angular momentum J are not defined on the collision manifold; instead their renormalized versions $\tilde{H} = rH$ and $\tilde{J} = r^{-1/2}J$ are well defined with finite analytic limits as $r \to 0$. The original researchers working on the collision manifold consistently set $\tilde{H} = 0$ *and* $\tilde{J} = 0$ in addition to setting $r = 0$. The rationale for imposing $\tilde{H} = 0$ is that any real solution has a constant finite energy H so that $\tilde{H} = rH \to 0$ when $r = 0$. The rationale for imposing $\tilde{J} = 0$ is Sundman's theorem, which asserts that any solution tending to triple collision has angular momentum J equal to zero. The Devaney connecting paths live on this traditional $r = \tilde{J} = \tilde{H} = 0$ collision manifold. By relaxing the condition $\tilde{J} = 0$ and investigating the flow in parts of the collision manifold with $\tilde{J} \neq 0$, Moeckel was able to include the return part of the rest cycle as an object to be shadowed. But this shadowing requires some angular momentum.

We have the ingredients in place to follow arbitrary walks in our abstract graph with concatenated orbits. We want a real orbit, not a concatenated one, for each such walk. Moeckel establishes these real orbits using the technique of "correctly aligned windows" initiated by Easton. See, in particular, figures 10 and 8 of [140]. To get the technique to work, Moeckel needed to turn on a tiny bit $\epsilon > 0$ of angular momentum in order to ensure solutions traveling through his windows access the part of Figure 3.17 having $r > 0, J = \epsilon \neq 0$ close to the return part of the rest cycle from $L^* \to L$.

If we restrict ourselved to $J = 0$, then the closure of the energy momentum level set $\{H = const., J = 0\}$ within the blown-up phase space does not contain the return part of the rest cycle. We cannot find a return path $L^* \to L$ and are unable to find correctly aligned windows to execute shadowings of paths in the abstract graph. However, the closure of $\{H = 0\}$ does contain the return leg and is reached by letting $J \to 0$. As $J \to 0$ for fixed H the projections of the guaranteed orbits suffer longer and longer stutter blocks and stay within a smaller and smaller neighborhood of the embedded graph.

3.7.3 The Variational Excess of Stuttering

None of the solutions guaranteed by Theorem 3.18 are minimizers in their free homotopy classes.

Lemma 3.36 *The action of any relative periodic solution that has a stutter can be decreased while remaining relative periodic and in the same reduced free homotopy class.*

Proof Our proof is by orbit surgery. Suppose q is the solution and the stutter is realized by two successive times t_0, t_1. Look at the image curve $s(t)$ in the shape sphere. The operation of reflecting the arc $s([t_0, t_1])$ about the equator while leaving the rest of $s(t)$ unchanged keeps its eclipse sequence the same. Some thought shows that this *surgery operation* on the shape curve can be lifted. To do so, let ℓ_0 be the line through the three masses at time t_0 and ℓ_1 the line at t_1. Reflect the arc $q([t_0, t_1])$ about ℓ_0 while keeping $q([0, t_0])$ unchanged to arrive at a new arc whose shape projection on the interval $[t_0, t_1]$ is the reflection of $s([t_0, t_1])$. This new arc ends in a collinear configuration lying on the line $\tilde{\ell}_1$ obtained by applying the reflection to ℓ_1. Let g be the rotation taking ℓ_1 to $\tilde{\ell}_1$. Apply g to the final segment $q([t_1, T])$ of q and concatenate the result with the previous part of the curve. By this process we have made a continuous relative periodic curve $\tilde{q}$ with the same eclipse sequence as q and the same action.

Now, by way of contradiction, suppose q cannot be shortened in its homotopy class. Then q must satisfy the Euler–Lagrange equations and, in particular, be analytic. But $\tilde{q}$ has the same action and represents the same homotopy class as q so it, also, cannot be shortened and so must be a solution. But our surgery process has destroyed the analyticity of $\tilde{q}$ at t_0 (and t_1). This contradicts minimality. QED

3.8 Closed and Open Questions

The following questions concern the planar three-body problem.

1. (Closed) Does the zero angular momentum three-body problem admit periodic solutions whose projection to the shape sphere minus collisions is contractible?

Yes. Periodic collision-free brake orbits are contractible. Such orbits exist for the equal-mass three-body problem according to [110, 111]. See also the commentary in [155].

2. (Closed) Does the zero angular momentum three-body problem admit periodic solutions whose projection realizes the tight binary loop (syzygy sequence ij) around one of the three collisions?

Yes. See the orbits whose existence is sketched by Carles Simó in his featured review of my "Infinitely Many Syzygies" [148] in the AMS's MathSci-Net.[3]

---

[3] This review can be found at Mathematical Reviews. (2002). Review of Journal Article 'Infinitely Many Syzygies'. Retrieved from https://mathscinet.ams.org/mathscinet/article?mr=1933631.

3. (Open) Relax the constraint that three-body solutions be periodic or relative-periodic.
Is every finite eclipse sequence realized by some solution?

4. (Open) I proved [148, 149] that any solutions asymptotic in both time directions to a Lagrange equilateral triangle collision solution has finitely many syzygies. Moeckel proved that there are infinitely many such solutions.
Can any finite syzygy sequence be achieved by such a bi-asymptotic collision solution?

5. (Open) Is there a solution whose projection to the shape sphere has dense image?
Wu-yi Hsiang asked this question during a conversation in 1997.

3.9 Chapter Notes

On Section 3.1

Wu-yi Hsiang [82, 83, 84] posed a version of Open Question 3 to me in a coffee shop on Euclid Avenue in Berkeley, California in 1997. Hsiang is a force of nature. His enthusiasm, confidence, and questions dragged me into the world of the three-body problem and had me believing that I could actually say something new about it. Hsiang was insistent about making progress by using the Jacobi–Maupertuis (JM) metric reformulation of the three-body problem. So far, the JM method has been of limited use. However, see [134, 189].

On Section 3.2

The variational method has a long history in mechanics, going back to Fermat, Maupertuis, the Bernoulli brothers, Euler, and Lagrange. The method is central to Feynman's path integral reformulation of quantum mechanics.

Gordon [69, 70] coined the phrase "tied to collisions" for a coercive free homotopy class of loops. The image is that a loop that is *not* tied to collisions can be slid out to infinity while keeping its action finite. However, the action of the loops in a tied class will tend to infinity if you try to slide the loop out to infinity. Gordon also coined the term "strong force," which has stuck among mathematical celestial mechanicians but is somewhat unfortunate given quarks and the name for the force binding them together in the standard model of elementary particle physics. His strong force condition is rather more general than the one we wrote down, but the end result is the same: With his strong force condition imposed, collision paths have infinite action. Gordon brought a functional analyst's precision to bear down on the essential issue – collisions – when applying variational methods to the standard N-body problem. Whatever happened to W. B. Gordon?

On Section 3.3

I was proud of Theorems 3.15 and 3.17. I did not know of Poincaré's work [169]. Phil Holmes told Alain Chenciner and I about Poincaré's paper. From there I slowly began to realize that Poincaré probably already knew my theorem. After all, he essentially invented the fundamental group a few years before he published his theorem, when he was unearthing the difference between homology and homotopy. It seems likely that Poincaré knew my result and just found it easier to write his short paper using winding numbers instead of the largely unknown, brand-new, and still evolving language of homotopy.

Holmes also told Chenciner and I about the work of his student, Moore [157]. You can find a sketch of the eight in Moore's paper. Moore used the same variational thinking I was using, combined with the writing of programs to implement numerical minimization of the action. If I had known how far Poincaré and Moore had gone along the lines I was beginning to explore I quite likely would have been demoralized to the point of quitting my work in this new-to-me N-body variational world.

Before the figure eight and hip-hop were established, several papers used variational methods to establish the existence of periodic *generalized solutions* to Newton's equations $\ddot{q} = -\nabla V(q)$ under various conditions on the potential V. These conditions on V sometimes include Newton's $1/r$ potential. See, for example, Bahri and Rabinowitz [14] and the references to this paper found in Math Reviews.[4] A generalized solution consists of the concatenation of a possibly countably infinite set of solution arcs to Newton's equations, glued together at collisions (points where $V = -\infty$) so as to form a continuous finite-action curve. The set of collision times along which solution arcs are glued, one to the next, is a closed set of measure zero. For example, the Cantor set could be the set of collision times. I have a hard time accepting generalized solutions as solutions. The generalized solutions of these papers arise as limits of some variational construction, perhaps a min-max or mountain pass family construction.

On Section 3.6

I met Christian Marchal in December of 1999 at the Don Saari birthday/Northwestern retirement party. Marchal gave an astounding talk that had nothing to do with celestial mechanics. He talked about how the United Nations had all their models wrong for human population growth. They

---

[4] See https://mathscinet-ams-org.us1.proxy.openathens.net/mathscinet/article?mr=1145561.

universally over-predicted world population. Data showed that birth rates were actually declining precipitously around the world, in contradiction to all models. (He justified giving this talk in a celestial mechanics conference by relating the history of human population dynamics study to the misguided epicycles of Ptolemy before Kepler.) For me, this was the best news I had heard in a decade! I was brought up in Northern California in the 1970s and had an oceanography teacher, Mr. Sikora, in Junior High for whom the upcoming population bomb, predicted by the Ehrlich couple, was the worst disaster the Earth was facing.[5] For Christian, this birth rate data was horrible news! Who was going to pay for his retirement? He himself had seven children. What would Europe be coming to? Later, my wife and I were blessed with a few dinners with him and his wife. He lived in the apartment building he had grown up in, and could remember the bombs dropping and Nazi soldiers occupying Paris during the war.

Venturelli came up with the blow-up method in his thesis under Chenciner's directions [210]. I met Venturelli at the Bureau des Longitudes in Paris in 1998. We talked a lot about getting rid of collisions. A number of years later, my wife and I visited him for a week in Avignon. She still talks about the mean pot-au-feu that Andrea showed her how to cook. Terracini simultaneously came up with the use of the blow-up method in the service of eliminating collisions. See the discussion in Venturelli's thesis [210]. The method had been well known in the PDE world. See the references in the survey paper by Musso [161].

There is a hoary history of deriving collision estimates such as that of Equation (3.15). See [108, 202, 215]. One obtains the correct asymptotics for a cluster of k masses colliding under an α-power law potential as follows. Ignore the non-colliding masses. Make the parabolic central configuration ansatz $q(t) = t^\beta s_*$ for some fixed k-body configuration s_*. Now plug into the α-Newton's equations and solve for β. If you do this for standard gravity ($\alpha = 1$) you will get the claimed $\beta = 2/3$ of Equation (3.15). See Ferrario–Terracini [61] for full derivations in the k-clusters α-power law case. For the case of the planar or spatial N-body problem with isolated binary collisions, one can derive the standard Newtonian estimate of equation (3.15) by Levi-Civita or Kuustanheimo–Stiefel regularization, viewing the full N-body problem as a perturbed Kepler problem with the non-colliding bodies supplying the perturbation.

---

[5] See https://en.wikipedia.org/wiki/The_Population_Bomb.

On Section 3.7

Giving up on variational methods and moving over to dynamical methods was the result of a visit to Carles Simó in Barcelona. In [152] I wrote up the story of how a single remark of Simó made me change tack and give up on variational methods to solve Open Question 3 (as far as we were able to). Luckily, I had befriended Rick Moeckel, who had done almost all of the dynamical systems work that was needed in the 1980s.

4

Does a Scattered Beam Have a Dense Image?

Open Question 4 Is the scattered image of a beam full of N-body solutions open and dense within the sphere of outgoing directions in configuration space?

Unlike Open Questions 2 and 3, this question concerns the positive energy N-body problem. Positive energy solutions are unbounded. By a beam of solutions we mean a family of positive energy solutions whose velocities all have the same limiting nonzero value as $t \to -\infty$. Recording their limiting velocities as $t \to +\infty$ defines the scattering map. A fair amout of this chapter is taken up with defining the map more precisely. The range of the map is the sphere of outgoing directions (or velocities) in $\mathbb{E}$. The question asks "Is the image of this map open and dense in the outgoing sphere?".

4.1 Motivation: Rutherford and the Discovery of Nuclei

The discovery of the nucleus relied on Rutherford's analysis [179] of classical two-body scattering, which includes defining the scattering map for the two-body problem. See Figure 4.1. Rutherford, Geiger, and Marsden had been aiming beams of alpha particles (helium nuclei) at gold foil and measuring the directions of the outgoing particles scattered by the foil. They were surprised to find a measurable fraction (1/20,000) suffered over a ninety degree change in direction with a few suffering nearly complete recoil. To explain their results, Rutherford posited replacing Thompson's "plum pudding" model of the atom by one in which nearly all of its mass is concentrated in a point-like positive charge at the center (the nucleus), this center being surrounded by a diffuse cloud of very light negative charges (electrons). He assumed the alpha particles to also be positively charged and that in order to suffer their strong deviations

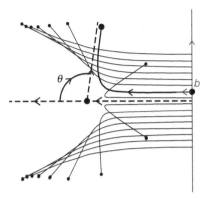

Figure 4.1 Rutherford Scattering. The variable b is the impact parameter and θ coordinatizes the direction of outgoing rays.

they must have penetrated through the electron cloud so as to be strongly interacting with the repulsive Coulomb force of the gold nuclei. Ignoring the electrons, Rutherford was faced with a well-known problem, the Coulomb problem, which is just the Kepler problem (Equation (0.32)) with the sign of the Kepler constant μ reversed so as to yield a repulsive instead of an attractive center.

Figure 4.1 indicates a beam of parallel lines coming in from infinity, interacting with the Coulomb force at the origin, getting deflected, and retreating to infinity in a splay of directions. We parameterize the rays of the incoming beam by a real parameter b, a coordinate on a line orthogonal to the beam's direction. We take $b = 0$ to correspond to the ray heading directly to the origin. This b is called the *impact parameter* as it measures how close the corresponding trajectory would come to impact with the nucleus if we were to turn off the Coulomb force. We label the outgoing trajectories by the angle θ made by their asymptotic direction and the original incoming beam direction. Coulomb dynamics then defines the scattering map $b \mapsto \theta = f(b)$. Rutherford computed

$$\theta = f(b) = 2\arctan\left(\frac{-\mu}{2Eb}\right). \tag{4.1}$$

Here E is the energy of the incoming beam and μ is the Kepler–Coulomb constant of $\ddot{q} = -\mu q/|q|^3$, so that $\mu < 0$ for Coulomb and $\mu > 0$ for Kepler. See [96], particularly p. 286, equation (12.3.3), or [101] for this computation. *It is this map, extended and made sense of in the context of the N-body problem, that Open Question 4 refers to.*

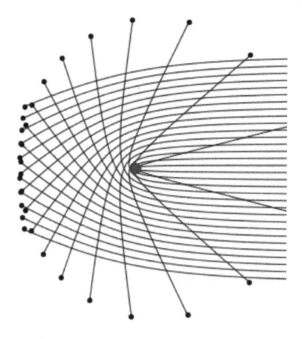

Figure 4.2 Kepler scattering.

We pause to reiterate that for the attractive Kepler problem, expression (4.1) for the scattering map continues to hold. See Figure 4.2 for a picture of the attractive scattering map. (It is remarkable that the differential cross sections as described in Section 4.1.1 for the attractive and repulsive cases are identical, with μ entering only through its absolute value.)

The map of Equation (4.1) is a smooth invertible map

$$f : \mathbb{R} \setminus \{0\} \to S^1 \setminus \{0, \pi\}.$$

In particular, its image is open and dense, missing exactly two angles, $\theta = 0$ and $\theta = \pi$, and so, in the two-body case the answer to our question is "yes." The Coulomb scattering map extends smoothly to the closure $\mathbb{R} \cup \{\infty\} = \mathbb{S}^1$ of its domain. This extension sends $b = \infty$ to $\theta = 0$, corresponding to no deflection at all for infinitely distant trajectories, and sends $b = 0$ to $\theta = \pi$, corresponding to the ray reflecting off the nucleus.

Scattering as Stereographic Projection

Rutherford and crew built their experiment in three-space, not in the plane. By rotational symmetry, Coulomb scattering in three dimensions can be obtained from two-dimensional Coulomb scattering by spinning the picture (Figure 4.2)

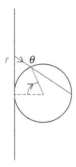

Figure 4.3 Stereographic projection.

for two-dimensional scattering about the beam axis. Take the beam axis to be $-e_3 = -(0,0,1)$. Then the incoming beam is parameterized by the xy plane and outgoing directions are parameterized by the unit sphere $\mathbb{S}^2 \subset \mathbb{R}^3$ so the relevant map is a map $\mathbb{R}^2 \to \mathbb{S}^2$. Use polar coordinates (b, θ) on the xy plane and spherical coordinates (φ, θ) on the sphere. Fixing θ defines the plane of motion for the trajectory of the beam and this does not change during the motion. Then, by rotational symmetry, the three-dimensional scattering map is $(b, \theta) \mapsto (f(b), \theta)$. But $(r, \theta) \mapsto (2 \arctan(1/r), \theta)$ is the expression for the inverse to stereographic projection, $\mathbb{R}^2 \to \mathbb{S}^2$. Upon setting $r = \pm 2Eb/\mu$ we find that Coulomb scattering and stereographic projection are the same map! See Figure 4.3. It is remarkable that the scattering map analytically extends so as to yield a diffeomorphism $\mathbb{R}^2 \cup \{\infty\} \to \mathbb{S}^2$ covering the entire sphere of outgoing directions. This fact is one reason to suspect the answer to Open Question 4 is "yes."

4.1.1 Scattering Cross Section

Scattering experiments are inherently probabalistic. One does not have a single nucleus, but rather an enormous number of gold nuclei whose precise location are unknown and essentially unknowable. One thinks of the result of the experiment as a sampling of impact parameters through the scattering map.

Assume that the impact parameters b are uniformly distributed according to Lebesgue measure db. Then the experiment is sampling from the push-forward of Lebesgue measure db by the scattering map

$$d\sigma := f_* db. \tag{4.2}$$

This measure, or its Radon–Nikodym derivative, is known as the differential cross section, and written $d\sigma$, It is a measure on the sphere of outgoing directions (with total measure infinite). In the case of planar scattering this sphere is the circle $\mathbb{S}^1$ of directions whose standard spherical measure is $d\theta$. To compute $d\sigma$, invert the scattering map (4.1) to get

$$b = b(\theta) = \frac{-\mu}{2E}\cot(\theta/2),$$

and compute the derivative $db = |\frac{db}{d\theta}|d\theta$. In other words, $d\sigma = |\frac{db}{d\theta}|d\theta$. Traditionally, one writes $\frac{d\sigma}{d\theta}$ instead of $\frac{db}{d\theta}$ for this function, the Radon–Nikodym derivative relating the two measures. This function is usually what is called the differential cross section. One computes

$$\frac{d\sigma}{d\theta} = \frac{|\mu|}{4E\sin^2(\theta/2)} \tag{4.3}$$

for planar Rutherford scattering.

More generally, for Coulomb–Rutherford scattering in any dimension d, the Radon–Nikodym derivative relating the standard spherical measure and $d\sigma$ is still written $\frac{d\sigma}{d\theta}$. By rotational symmetry it depends only on the angle θ between the incoming beam and the outgoing scattered ray. One computes

$$\frac{d\sigma}{d\theta} = \left(\frac{|\mu|}{4E\sin^2(\theta/2)}\right)^{d-1}$$

for Rutherford scattering in d dimensions. See [96, p. 295–297] for a derivation and clear descriptions.

4.2 Scattering in the N-Body Problem

Classical scattering is based on the expectation that if the forces between bodies decay with distance and the bodies are far apart and getting further, then each body ought to be moving asymptotically along a straight line, so N straight lines in $\mathbb{R}^d$ yield a single line in the N-body configuration space $\mathbb{E}$. Conversely, a single line in $\mathbb{E} = (\mathbb{R}^d)^N$ corresponds to N individual particle straight lines in $\mathbb{R}^d$. A line $\ell(t)$ in $\mathbb{E}$ can be written as

$$\ell(t) = At + B, \qquad A, B \in \mathbb{E}, \quad \|A\| = 1, \quad B \perp A. \tag{4.4}$$

Here A represents the direction of the line and B is its impact parameter, the unique point on the line closest to the origin.

Additionally, A is a point in the unit sphere $\mathbb{S}^{M-1} \subset \mathbb{E}$ of dimension $M-1$ where $M = dN = dim(\mathbb{E})$. Together, $(A, B) \in T\mathbb{S}^{M-1}$, the tangent bundle of the sphere, so that we identify the space of lines in $\mathbb{E}$ with this tangent bundle. We may want to center the solutions by going to the center-of-mass frame, in which case we replace $\mathbb{E}$ by $\mathbb{E}_0$, the centered configuration space, and this dimension M becomes $M = d(N - 1)$.

As a zeroth approximation to the scattering map, which we will eventually define, we form a map, from the space of oriented lines to itself. This map sends an incoming line representing the asymptotics of a solution at $t = -\infty$ to the corresponding outgoing line representing the same solution at $t = +\infty$. Identify $T\mathbb{S}^{M-1}$ with $T^*\mathbb{S}^{M-1}$ using the mass metric, so as to bring in ideas from symplectic geometry. The scattering map, SC, should be a symplectic map of the form

$$SC \colon T^*\mathbb{S}^{M-1} - \to T^*\mathbb{S}^{M-1} \tag{4.5}$$

defined by the dynamics. The broken arrow notation for the map indicates that its domain will not be all of $T^*\mathbb{S}^{M-1}$ due to the incompleteness of the flow. Some incoming solutions will end in collision (or a non-collision singularity) and cannot be continued into the future. We will call the eventual map inspired by Equation (4.5) the *full scattering map*. The scattering map of Open Question 4 is built from the full scattering map by restricting this map to a single fiber representing a single incoming beam direction, and then projecting the resulting image to the base sphere of directions.

A *beam* is the space of all oriented lines having a fixed direction A in Equation (4.4) so that a beam is identified with a single fiber $T_A^*\mathbb{S}^{M-1} = A^\perp$. The beam is parameterized by fixing A and varying B in Equation (4.4). The map whose image is the subject of Open Question 4 is the composition

$$T_A^*\mathbb{S}^{M-1} \xrightarrow{i_A} T^*\mathbb{S}^{M-1} \xrightarrow{SC} T^*\mathbb{S}^{M-1} \xrightarrow{\pi} \mathbb{S}^{M-1} \, , \tag{4.6}$$

where i_A is the inclusion $T_A^*\mathbb{S}^{M-1} \to T^*\mathbb{S}^{M-1}$ and π is the projection. This map takes all of the solutions that fill out a fixed incoming beam, integrates the flow along them from time $t = -\infty$ to $t = +\infty$, and then projects out the final limiting direction for each resulting outgoing solution. We call this composition the *beam-restricted scattering map* when we need to distinguish it from the full scattering map of Equation (4.5). The beam-restricted scattering map is a map $\mathbb{R}^{M-1} - \to \mathbb{S}^{M-1}$. Open Question 4 asks if its image, like that of Coulomb scattering's image, is open and dense.

What makes this zeroth approximation inaccurate is that N-body solutions are not asymptotic to lines in a traditional sense. A $\log(|t|)$ occurs in the $t \to \pm\infty$ asymptotics of hyperbolic solutions (see Equation (4.7)) and this logarithmic term wrecks the closeness of the solution to any particular line, making it more complicated to rigorously define the scattering map than the strategy we have just outlined.

The Coulomb case   We return to Rutherford's scattering in the plane (Figure 4.1) to illustrate the preceding formalism. Then $M = 2$ so that

$$SC : T^* \mathbb{S}^1 \to T^* \mathbb{S}^1,$$

where $T^* \mathbb{S}^1 = \mathbb{S}^1 \times \mathbb{R}$ is a cylinder parameterized by $(\theta, b) \in \mathbb{S}^1 \times \mathbb{R}$. The angle $\theta \in S^1$ defines the direction of the oriented line and the impact parameter $b \in \mathbb{R}$ (see Figure 4.1) is the signed distance of the line from the origin. We have $A = (\cos \theta, \sin \theta)$ and $B = b A^{\perp} = b(-\sin \theta, \cos \theta)$ in Equation (4.4). The impact parameter equals the angular momentum times a constant depending on mass and energy, and hence is unchanged in scattering. Using its constancy and rotational symmetry, it follows that SC must have the form $SC : (\theta, b) \mapsto (\theta + f(b), b)$ for some function $f(b)$. Scattering for any decaying central force problem will take this same form. The function $f(b)$ for Kepler is the one given in Equation (4.1).

4.2.1 Defining the Image of the Scattering Map

In order to start work on scattering, we need a family of unbounded solutions to scatter, hence our requirement that the energy be positive in Open Question 4 for the N-body problem. See Corollary 0.16. Despite being unbounded, positive energy N-body solutions are *not* asymptotic to lines in the standard sense. Equation (4.7) describes their actual asymptotics. The presence of the unbounded $\log(|t|)$ term in that equation, combined with the fact that typically A and $\nabla U(A)$ are linearly independent, implies that the solution asymptotically diverges from any single line. This lack of a standard asymptotics means that the above description of N-body scattering given in Section 4.2, although conceptually correct, is not accurate. For this reason it will be easier to define the *image* of the beam-restricted scattering map directly rather than to attempt to define the full scattering map SC. (We define the full scattering map in Section 4.4.)

Definition 4.1   A solution $q(t)$ is called *backward hyperbolic* if it has positive energy and if the limit of $s(t) = q(t)/\|q(t)\|$ as $t \to -\infty$ exists and is collision-free. Let $s_- \in \mathbb{S}^{M-1} \setminus \Delta$ denote this limit and refer to it as the backward asymptotic shape of $q(t)$.

We trust the reader to formulate definitions for *forward hyperbolic* and *forward asymptotic shape*, s_+.

Definition 4.2   A solution is *hyperbolic* if it is both forward and backward hyperbolic.

A hyperbolic solution $q(t)$ has backward and forward asymptotic shapes, s_- and s_+. We say that the solution connects s_- to s_+.

Definition 4.3 The *incoming beam* with direction s_- and energy E is the set of all backward hyperbolic solutions having fixed backward asymptotic shape $s_- \in \mathbb{S}^{M-1} \setminus \Delta$ and fixed energy $E > 0$.

Definition 4.4 The *scattered image* of the incoming beam with direction s_- is the set of all outgoing directions s_+ connected to hyperbolic solutions lying in this beam.

With these definitions in place, the terms appearing in Open Question 4 are now defined.

4.3 What's Known?

In this section we summarize what we know about the scattering map.

(i) The scattered image has nonempty interior. See [46] for the proof. The proof relies on analysis at infinity and a perturbation argument.

(ii) The "one-sided" scattering map is onto. What I mean is that a given beam hits every point of configuration space before receding back to infinity as $t \to +\infty$. Thus, if $q(t)$ are the solutions in the given beam, by a time translation we can ensure that the solutions are defined on $-\infty < t < 0$ and that the points $q(0)$ sweep out all of configuration space. Maderna and Venturelli establish this result in [113] through a deep analysis of the associated Hamilton–Jacobi equation $H(q, dS(q)) = E$ using weak KAM ideas. Their rays $q(-\infty, 0]$ are global JM minimizers connecting the asymptotic beam direction at $t = -\infty$ to the desired $q_0 \in \mathbb{E}$.

(iii) Rays in the beam that encounter collisions cannot continue to infinity. But (ii) immediately above we are guaranteed that some beams hit collision. Consequently the beam-restricted scattering map *never* has domain all of $T^*_{s_-} \mathbb{S}^{M-1}$.

(iv) For power law potentials $U = 1/r^\alpha$ with $0 < \alpha < 1$, the scattering map is *not* onto for the two-body problem. Indeed, in the planar version of the problem, in center-of-mass coordinates, the image of the map forms a circular arc of arclength $2\pi \frac{\alpha}{2-\alpha} < 2\pi$ centered on the beam direction. See Figures 4.4 and 4.5. Conversely, when $1 < \alpha < 2$ we find that the map is onto, with some outgoing directions covered more than once. As $\alpha \to 2$, the scattered image of the beam covers the circle more and more times, tending to cover each outgoing direction infinitely many times as

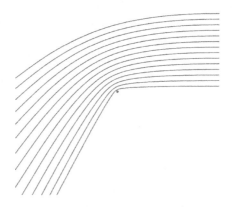

Figure 4.4 Scattering a half beam with power law $r^{-1/2}$.

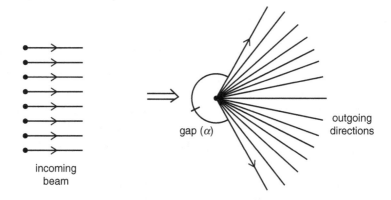

Figure 4.5 Gap in the scattering image for $\alpha = 1/2$.

$\alpha \to 2$. These results can be found in the paper [98]. We reprove these results in Appendix H. Figure 4.4 describes the scattering of a half-beam for $\alpha = 1/2$.

(v) Numerical experiments done by Moeckel for an incoming Lagrange beam indicate that the scattering map has an open dense image when $d = 2$ and $N = 3$, in keeping with a "yes" answer to Open Question 4. See Figure 4.6. In these dimensions, the scattered image of a centered beam is a subset of the three-sphere $\mathbb{S}^3 \subset \mathbb{E}_0(2, 3) \cong \mathbb{C}^2$. What is depicted in the figure is the further projection of the scattered image from $\mathbb{S}^3$ to the shape sphere $\mathbb{S}^2$ and, from there, onto the shape disc $\mathbb{D}^2$. The shape disc is the quotient of the shape sphere by reflections so its points represent (unoriented) similarity classes of triangles. The reflection about

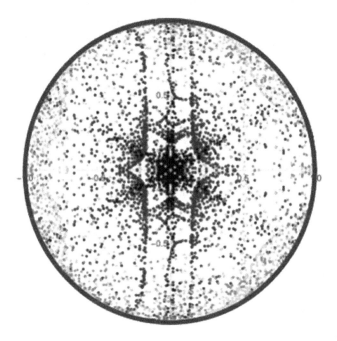

Figure 4.6 The scattered image of a Lagrange beam, projected to the shape disc. Courtesy of Richard Moeckel.

any line in the plane generates $\mathbb{Z}_2 = O(2)/SO(2)$, which acts on shape space by $(w_1, w_2, w_3) \mapsto (w_1, w_2, -w_3)$. It follows that we can identify the space of congruence classes of triangles with the closed upper half space, $w_3 \geq 0$, and the space of similarity classes of triangles with the closed upper hemisphere of the shape sphere. Project out w_3 to get a convenient picture of the space of similarity classes of triangles as the *shape disc*: the unit disc in the $w_1 - w_2$ plane. The boundary of the disc represents collinear triangles and has the same six marked points as before, alternating between binary and Euler points. In the disc's interior we have a single marked point representing the equilateral triangle. When the masses are equal, this point is at the center and the three isosceles circles become three equispaced diameters. Moeckel made the figure by taking a sampling of initial conditions in the unstable manifold at infinity of the Lagrange shape s (see the end of Section 4.4.3), so a Lagrange beam, numerically evoling the trajectories of the beam into the distant future, and plotting their resulting shapes on the shape disc.

4.4 Building the Full Scattering Map

We proceed to construct the full scattering map, starting with Chazy's 1922 paper on asymptotics for hyperbolic solutions [25]. His asymptotics allow us to define the full scattering map by Equation (4.10). In our own reading of Chazy's paper it became problematic as to whether or not he had established the *existence* of solutions with given Chazy scattering parameters (A, B). (See Equation (4.7) and the chapter notes in Section 4.6.) Partly in order to assuage this doubt, we wrote paper [46] where we established the needed existence as Theorem 4.5 (see Section 4.4.2). This paper yielded an alternative formulation of scattering in terms of stable and unstable manifolds and we will follow that formulation.

4.4.1 Chazy Asymptotics

Chazy [25] established the asymptotic expansion

$$q_+(t) = At - (\nabla U(A))\log(t) + B + O\left(\frac{\log t}{t}\right), \ t \to \infty, \qquad (4.7)$$

valid for any forward hyperbolic solutions $q_+(t)$. The dominant linear coefficient is related to the solution's energy $E > 0$ and asymptotic shape s_+ by

$$A := A_+ = \sqrt{2E}s_+.$$

The secondary $\log t$ term is needed for the expansion to solve Newton's equations to leading order $1/t^2$. The coefficient of this term is $\nabla U(A) = \mathbb{M}^{-1}F(A)$. Compare with Equation (0.38). There is no reasonable sense in which our solution is asymptotic to a line since its distance from any fixed line as $t \to \infty$ is typically unbounded.[1] Despite this lack of an asymptotic line we simply drop the $\log(t)$ term and let $B = B_+$ play the role of the impact parameter of a line and so declare that the line associated to q_+ is the one whose parameters are $a = A_+, b = B_+ (\text{modulo } A_+)$ in Equation (4.4). (Now this line has speed $\sqrt{2E}$ instead of speed 1. No matter.) We call (A, B) the *Chazy parameters* of $q_+(t)$.

Similarly, if the solution $q = q_-(t)$ is backward hyperbolic with energy $E > 0$ and asymptotic shape s_-, then Chazy's asymptotics in the backward time direction are

---

[1] If A and $\nabla U(A)$ are linearly dependent, then q_+ and the line with parameters $a = A_+, b = B$ are asymptotic as unparameterized curves. But this special case does not help us develop a general scattering map.

$$q_-(t) = A_-t - (\nabla U(A_-)) \log(|t|) + B_- + O\left(\frac{\log |t|}{t}\right), \; t \to -\infty, \quad (4.8)$$

with

$$A_- = -\sqrt{2E}s_-. \quad (4.9)$$

To justify the choice of minus sign here, note that Equation (4.7) implies that

$$\lim_{t \to +\infty} \dot{q}_+(t) = A_+.$$

The minus sign relating A_- to s_- guarantees that

$$\lim_{t \to -\infty} \dot{q}_-(t) = A_-.$$

We call (A_-, B_-) the Chazy parameters of q_-.

If $q(t)$ is hyperbolic, then both its forwards and backwards ends have Chazy parameters. The full scattering map takes one to the other:

$$SC: (A_-, B_-^\perp) \mapsto (A_+, B_+^\perp), \quad (4.10)$$

where $\perp$ in each case denotes orthogonal projection onto the space orthogonal to the beam direction A so that $B_\pm^\perp = B_\pm - \langle s_\pm, B_\pm \rangle s_\pm$.

4.4.2 Scattering Parameters: Existence and Uniqueness

In order for our formulation (4.10) of the scattering map to be well defined we must know that a backward hyperbolic solution exists for any given scattering parameters $(A_-, B_-^\perp)$ and that this solution is unique modulo time translations. Once we know this, the domain of the scattering map is the set of (A_-, B_-) for which its solution extends indefinitely into the future, and the map is defined by taking this extension and working out its future asymptotics. Theorem 4.5 establishes the needed existence and uniqueness.

Theorem 4.5 *(A) Two trajectories that have the same Chazy parameters (A_-, B_-) are the same up to a time translation.*

(B) The incoming beam with direction s_- and energy E modulo time translation forms an affine space parameterized by the Chazy impact parameter $B_- \in s_-^\perp$. Any two trajectories in the beam remain a bounded distance apart as $t \to -\infty$.

Theorem 4.5 says that we can think of the Chazy parameters as initial conditions at infinity.

We prove (A) in Section 4.4.3.

Proof of Theorem 4.5, part (B) Take two solutions $q, \tilde{q}$ in this incoming beam and subtract one from the other. Because they have the same leading Chazy parameter $A = -\sqrt{2E}s_-$, their difference has asymptotic expansion $B_- - \tilde{B}_- + O(\log|t|/|t|)$ whose limit as $t \to -\infty$ is $B_- - \tilde{B}_-$, showing that the two trajectories remain a bounded distance apart and that we can use the impact parameter B_- as an affine parameter.

The affine parameter B_- is defined modulo A_-. To see this, consider a solution $q(t)$ in the incoming beam and its time translate $q(t - t_0)$, which is also in the incoming beam and has the same impact parameter B_-. We have $q(t) - q(t - t_0) = At_0 - \nabla U(A)\log(|t|) + \nabla U(A)\log(|t - t_0|) = At_0 + O(1/t)$, where we have used that $\log|t| - \log(|t - t_0|) = \log(|t|/|t - t_0|) = \log(1/(1 - t_0/t)) = O(1/|t|)$, $t \ll -1$. It follows that the "B_-" for the time translate is $B_- - At_0$, so that the impact parameter is only defined modulo A_-.

4.4.3 Blowing Up Infinity: Proof of Theorem 4.5, Part (A)

We are to show that for given Chazy parameters A_-, B_- there is *exactly one* backward hyperbolic solution, modulo time translation, having these Chazy parameters. There are several proofs available. We will follow [46]. Other approaches are Chazy's original approach [25] and one based on the Dollard–Möller transformation that can be found in [60].

In the same spirit as the McGehee blow-up, we add a manifold at infinity as a place for scattering solutions to go. The flow, extended to infinity, has a manifold of fixed points that is normally hyperbolic and that splits in two parts according to incoming and outgoing beams. The unstable manifold of the incoming fixed point will correspond to an incoming beam.

To begin, introduce spherical coordinates $(r, s) \in (0, +\infty] \times \mathbb{S}$ on $\mathbb{E} \cong \mathbb{R}^{dN}$, as in the McGehee transformation.

$$q = rs, \quad \text{where } r = \|q\|, \; s = q/r \in \mathbb{S}.$$

Decompose the velocity vector $v := \dot{q}$ into radial and tangential or *spherical* components,

$$v = vs + w, \quad \text{where } v = \langle s, v \rangle \text{ and } \langle s, w \rangle = 0.$$

Change the independent time variable according to,

$$dt = r \, d\tau, \tag{4.11}$$

writing $' = \frac{d}{d\tau}$. Finally, introduce

$$\rho = 1/r,$$

so that $\rho = 0$ corresponds to spatial infinity. Under these changes of variables, Newton's equations transform into

$$
\begin{aligned}
\rho' &= -\nu\rho, \\
s' &= w, \\
\nu' &= \|w\|^2 - \rho U(s), \\
w' &= \rho\tilde{\nabla}U(s) - \nu w - \|w\|^2 s,
\end{aligned}
\tag{4.12}
$$

where

$$\tilde{\nabla}U(s) = \nabla U(s) + U(s)s$$

is the component of $\nabla U(s)$ orthogonal to s since $\langle s, \nabla U(s)\rangle = -U(s)$ due to U's homogeneity. In these new variables the energy H is given by

$$\frac{1}{2}\nu^2 + \frac{1}{2}\|w\|^2 - \rho U(s) = H.$$

The transformed equations extend analytically to $\rho = 0$. Further setting $w = 0$ yields all equilibria e of the extended flow. The equilibria e form a manifold $\mathcal{E}$ coordinatized by $\nu, s \in \mathbb{R} \times \mathbb{S}$. Concretely, write elements of our extended phase space as quadruples (ρ, ν, s, w) and set $e = (0, \nu, s, 0)$ to coordinatize the equilibria. The dimension of $\mathcal{E}$ is $M = dim(\mathbb{E})$, which is half the dimension of the extended phase space $\mathcal{P}$.

The energy at infinity is $\frac{1}{2}\nu^2 + \frac{1}{2}\|w\|^2 = H$, so at equilibrium $H = \frac{1}{2}\nu^2$. We are only interested in the case $H > 0$ of positive energy solutions in which case the equilibrium manifold $\mathcal{E}$ falls into two (diffeomorphic) components $\mathcal{E}_-$ and $\mathcal{E}_+$ according to whether $\nu > 0$ or $\nu < 0$, where $\nu = \pm\sqrt{2H}$.

This equilibrium manifold $\mathcal{E}$ is normally hyperbolic. To be normally hyperbolic means that when we linearize the vector field at any equilibrium $e \in \mathcal{E}$, the resulting operator on the tangent space $T_e\mathcal{P}$ of phase space $\mathcal{P}$ splits into a two-by-two block diagonal form where one block is zero and corresponds to the tangent space to $\mathcal{E}$ and the other block corresponds to the generalized eigenspaces for the nonzero generalized eigenvalues. Moreover, and this is the hyperbolic part, all these nonzero generalized eigenvalues have nonzero real part. In our case there is only one nonzero generalized eigenvalue and it is $-\nu$ when $e = (0, \nu, s, 0) \in \mathcal{E}$.

A nontrivial simultaneous linearization theorem, theorem 3.1 of [46], now comes in to play, and implies that the N-body flow near infinity is locally conjugate in a neighborhood of $\mathcal{E}_-$ to its linearization, with this conjugation varying analytically with e. In particular, this theorem yields, for each $e = (0, \nu, s, 0) \in \mathcal{E}_-$ an M-dimensional unstable manifold $W^u(e)$. The part of $W^u(e)$ lying in the interior $\rho > 0$ of $\mathcal{P}$ is the incoming beam with direction s and energy $\frac{1}{2}\nu^2$. When the parameters of the linearized data are compared with the Chazy parameters, the existence and uniqueness theorem follows. See [46, theorem 4.1]. QED

Remark 4.6   There is a single generalized eigenvector associated to the eigenvalue $-\nu$. The corresponding nontrivial Jordan block in the linearization propagates to twist the relations between the linearizing variables and the Chazy variables (see the equations in theorem 4.1 of [46]) and also yields the $\log(|t|)$ term in the Chazy asymptotics.

Summary: Reimagining beams   The proof of Theorem 4.5, part (I), yields the following picture of what a beam is in our Open Question 4. The incoming beam with direction s and energy E is equal to $W^u(e) \cap \{\rho > 0\}$, where $e = (0, \nu, s, 0)$ as in the proof.

4.4.4 Systems of Rays and Scattering as Lagrangian Intersection Theory

Arnol'd [10] defines a system of rays on a manifold to be essentially any Lagrangian submanifold of the cotangent bundle of that manifold. He uses this idea and ideas from optics and singularity theory to formulate a number of interesting theorems and conjectures. Trusting his perspectives, we believe that the following facts may be of eventual use.

Theorem 4.7   *The set of phase points* $(q(t), p(t)) \in T^*\mathbb{E}_0$ *swept out by the incoming beam with fixed direction* s_- *and energy* E *forms a Lagrangian submanifold* $\mathcal{L}_-(s_-; E)$ *lying within the given energy level set.*

The obvious analogues of Theorems 4.5 and 4.7 hold for the *outgoing beam* for an $s_+ \in \mathbb{S} \setminus \Delta$. We write $\mathcal{L}_+(s_+; E)$ for the corresponding outgoing Lagrangian submanifold. If $\mathcal{L}_-(s_-; E)$ and $\mathcal{L}_+(s_+; E)$ intersect, then, since they are invariant under the backward and forward flows respectively, they must intersect along a union of energy E hyperbolic orbits. Thus s_- is connected to s_+ if and only if $\mathcal{L}_-(s_-; E) \cap \mathcal{L}_+(s_+; E) \neq \emptyset$ for some $E > 0$. (By scaling symmetry, whether they intersect or not does not depend on the energy E.) We have reformulated Open Question 4 in terms of the popular subject of "Lagrangian intersection theory." Namely: given s_-, is it true that for an open dense set of s_+'s we have that $\mathcal{L}_-(s_-; E) \cap \mathcal{L}_+(s_+; E) \neq \emptyset$.

4.5 More Open Questions

In this section we ask three more questions regarding the scattering map.

(i) What is the differential cross section f for the incoming Lagrange beam of Figure 4.5?

Commentary. In more detail, the scattering map for a beam of solutions for the planar three-body problem is a map $SC: \mathbb{R}^3 \to \mathbb{S}^3$. Let $d^3 b$ and $d^3 \omega$ be the standard measures on the domain and range of the map. The differential cross section is the Radon–Nikodym derivative f in the expression $SC_* d^3 b = f(\omega) d^3 \omega$. This question asks that we compute the function f on the sphere. Open Question 4, the subject of this chapter, asks if the support of f is an open dense set.

(ii) How tangled can the trajectories of a hyperbolic-hyperbolic solution be?

Commentary. This question is a marriage of ideas from Chapters 3 and 4. The scattering of Chapter 4 concerns hyperbolic-hyperbolic solutions. In the planar three-body problem, such a solution asymptotes to one fixed triple of lines in the distant past, and another in the distant future. The set of all collision-free paths having the past and future asymptotics specified by such a pair of three lines has topology encoded in the braid group as described in Chapter 3. Indeed, the set of free homotopy classes of such paths is in bijection with conjugacy classes of the pure braid group, which, in turn, is encoded by a reduced eclipse sequence. Are all finite reduced eclipse sequences realized in this way through scattering? If not, what restrictions are there? Is there a bound on the length of these hyperbolic eclipse sequences say, in the equal mass case? For example, must a hyperbolic-hyperbolic planar three-body solution have less than 100 eclipses in the region within which the three masses interact?

(iii) Does broken geodesic flow govern the limiting scattering map at infinity?

Commentary. The flow at infinity, that is, on $\rho = 0$ in terms of our blown-up McGehee variables as described in Equations (4.12), is free flow (straight lines in $\mathbb{E}_0$) projected onto the sphere $\mathbb{S}$ representing the sphere of directions in $\mathbb{E}_0$. A picture shows you that the trajectories are halves of great circles on the sphere, these semicircles connecting antipodal points so that $s_+ = -(s_-)$. However, this picture ignores the potential, which tends to infinity somewhat like a δ-function at the binary collisions. Knauf has argued that if collisions are not avoided then the scattering map at infinity yields what he called "train tracks," what Vasy calls "broken geodesic flow," and what in [59] we called "linear point billiards." The idea is that even at infinity we still have to take into account the potential when two of the particles get very close while the others remain far away. The suggestion, and the numerical evidence, is that binary collisions at infinity act as *billiard walls* for the collision flow at infinity. Thus an arc of a great circle on the sphere at infinity that encounters a binary collision (at infinity) can bounce off of it in any direction and then continue along another great circle. The fact that we can go in any direction is connected to the fact that the microscopic dynamics near the binary collisions at infinity are governed by Rutherford two-body scattering. The fact that when we come into a binary collision at infinity we can choose to leave in any direction at all tangent to the sphere at infinity is equivalent to the fact that Rutherford scattering maps onto the sphere of outgoing directions.

This question has two associated versions.

(A) Prove that these broken geodesics really happen. In other words, establish that families of true hyperbolic solutions shadow any of these broken geodesic trajectories.

(B) Modify the statements and proofs to get a broken geodesic flow at infinity appropriate for the other homogeneous potentials $1/r^\alpha$ when $0 < \alpha < 2$.

The point of part (B) is that the analysis of the central force problem in Appendix H shows that the scattering map for the two-body problem is not onto when $0 < \alpha < 1$, and that the set of accessible angles $\Delta\theta(\alpha)$ goes to zero with α. See Figures 4.4 and 4.5. We must take this accessibility problem into account when working out the correct "billiards at infinity" model. Presumably, we must modify the "Rutherford" broken geodesic flow appropriate for $\alpha = 1$ in the obvious way so that instead of all outgoing angles being allowed, only angles that lie within an outgoing sector of size $\Delta\theta(\alpha)$ are allowed.

4.6 Chapter Notes

Scattering theory forms a large subfield of mathematical physics. See, for example, Reed and Simon [174]. The theory is primarily concerned with linear PDEs such as Schrödinger's equation (quantum scattering) or the wave equation. The book by Derezínski and Gérard [40] is a kind of bible of scattering in classical mechanics.

Andreas Knauf introduced me to classical scattering problems at a conference in Banff in 2012, and we began working together. Knauf has numerous papers on scattering. See, for example, [95] and [96].

On Section 4.4.2

Chazy's paper [25] is a joy to read but is not easy to read. It is unclear to me whether or not Chazy proved the existence result stated as part (A) of Theorem 4.5. It is possible that he did using ideas from convergent power series expansions.

On Section 4.4.3

Variants of the McGehee transformation such as the one described here have been used a number of times to analyze solutions near infinity. For example, see [120, 159, 176].

On Section 4.4.4 – Lagrangian Relations

The graph of a symplectic map is a Lagrangian relation but not all Lagrangian relations are graphs of symplectic maps. Sometimes scattering processes cannot be described by a symplectic map and instead need to be represented by a Lagrangian relation. See [60].

In 1981, Weinstein asserted the symplectic creed: "Everything is a Lagrangian submanifold" [212]. This creed has been important in the development of symplectic topology. Various theorems assert that in various circumstances two Lagrangian submanifolds must intersect and that the number of their intersections can be bound from below by the number of critical points of a Morse function on either one. This idea, a version of the Arnol'd conjectures, has driven developments in symplectic topology.

APPENDICES

Appendix A Geometric Mechanics

A.1 The Lagrangian and Hamiltonian Formalisms

This appendix introduces geometric mechanics to someone who knows some differential geometry. For more complete treatments we recommend [2], [11], or [96]. Begin with a manifold Q, called configuration space, whose points encode the positions and orientations of the parts comprising the mechanical system of interest. In this book $q \in Q$ lists the positions of the N bodies of the N-body problem.

Classical mechanics can be expressed in terms of the Lagrangian formalism, the Hamiltonian formalism, and the relations between them. The Hamiltonian formalism lives on the cotangent bundle T^*Q of the configuration space. Points of the fiber of T^*Q are thought of as momenta. The Lagrangian formalism lives on the tangent bundle TQ of the configuration space. Points of the fiber of TQ are thought of as velocities. The Legendre transformation is a bundle map that intertwines the two formalisms. The Hamiltonian formalism plays a central role in Chapter 2 of this book, while the Lagrangian formalism plays a central role in Chapter 3.

Definition A.1   Let Q be a manifold representing the configuration space of a mechanical system. A Hamiltonian is a smooth function $H \colon T^*Q \to \mathbb{R}$ on the cotangent bundle of configuration space. A Lagrangian is a smooth function $L \colon TQ \to \mathbb{R}$ on the tangent bundle of configuration space. Either TQ or T^*Q are referred to as the phase space of the system and will sometimes be denoted by $\mathcal{P}$.

Coordinates q^i on Q induce fiber-linear coordinates v^i on TQ and p_i on T^*Q by writing a vector $v \in T_qQ$ as $v = \Sigma v^i \frac{\partial}{\partial q^i}$ and a covector $p \in T_q^*Q$ as $p = \Sigma p_i dq^i$. Then $p(v) = \Sigma p_i v^i$. (Here $q \in Q$ is any point covered by our coordinate neighborhood.) In terms of these coordinates, the Hamiltonian and Lagrangian are functions of the forms $H = H(q^1, \ldots, q^n, p_1, \ldots, p_n)$ and $L = L(q^1, \ldots, q^n, v^1, \ldots, v^n)$ where $n = dim(Q)$. The coordinates q^i, p_i on T^*Q are called canonical coordinates. Hamilton's equations, written in canonical coordinates, is the system of equations

$$\dot{q}^i = \frac{\partial H}{\partial p_i},$$

$$\dot{p}_i = -\frac{\partial H}{\partial q^i}. \tag{A.1}$$

The Euler–Lagrange equations for L, written in the above coordinates, are

$$\frac{d}{dt}\left(\frac{\partial L}{\partial v^i}\right) = \frac{\partial L}{\partial q^i}. \tag{A.2}$$

When performing the time derivative $\frac{d}{dt}$ on the left-hand side of the Euler–Lagrange equations in order to write them out as a system of second-order ODEs, we use the chain rule, and when we are finished we impose the equation $\dot{q}^i = v^i$, which says that v is indeed the velocity associated to a curve $q^i(t)$ in M.

To relate the Lagrangian and Hamiltonian systems, set $p^i = \frac{\partial L}{\partial v^i}$ within the left-hand side of the Euler–Lagrange equations, thus defining a map $(q,v) \mapsto (q,p) = (q, p(q,v))$ called the Legendre transform. The Legendre transformation is an intrinsically defined bundle map (generically not linear),

$$FL: TQ \to T^*Q,$$

with the following coordinate-free definition as a fiber derivative: If $v \in T_q Q$ then $p = FL(q,v) \in T_q^*Q$ is the linear functional $p: T_q Q \to \mathbb{R}$ given by $p(w) = \frac{d}{dt}\big|_{t=0}L(q, v + tw)$.

When the Legendre transformation is invertible, we can write its inverse as $(q,p) \mapsto (q, v(q,p))$ and using this inverse we can view the function

$$H = p(v) - L(q,v)$$

as a function of q and p. The resulting function $H: T^*Q \to \mathbb{R}$ is called the Legendre transform of L. The Legendre transform, when invertible, turns the Euler–Lagrange equations for L into Hamilton's equations for H.

Exercise A.2   1. Take $Q = \mathbb{R}$ and write $x \in \mathbb{R}$ for the coordinate. The standard Lagrangian of one-dimensional mechanics is $L(q,v) = \frac{1}{2}mv^2 - V(x)$, where the constant $m > 0$ is the particle's mass. Verify that its Euler–Lagrange equations are Newton's equations: $m\ddot{x} = -V'(x)$. Verify that its Legendre transform is $(x,v) \mapsto (x,p) = (x,mv)$. Verify that its Hamiltonian is $H = \frac{1}{2}\frac{1}{m}p^2 + V(x)$ and that Hamilton's equations are equivalent to Newton's equations.

2. Take $Q = \mathbb{R}^n$ be Euclidean space with its standard inner product and let $V: \mathbb{R}^n \to \mathbb{R}$ be a smooth function called the potential. Let M be a positive definite symmetric matrix on $\mathbb{R}^n$ introduced to play the role of masses. Then Newton's equations are $M\ddot{q} = -\nabla V(q)$. Verify that Newton's equations are the Euler–Lagrange equations of the Lagrangian $L(q,v) = \frac{1}{2}\langle v, Mv\rangle - V(q)$. Verify that the Legendre transformation is given by $p = Mv$ if we use the standard inner product to identify $(\mathbb{R}^n)^*$ with $\mathbb{R}^n$, and so identify both TQ and T^*Q with $\mathbb{R}^n \times \mathbb{R}^n$. Verify that the Hamiltonian is $H(q,p) = \frac{1}{2}\langle p, M^{-1}p\rangle + V(q)$. Verify that Hamilton's equations are equivalent to Newton's equations.

3. Modify part 2 of this Exercise so as to work for the N-body problem, by restricting to the non-collision configurations so derivatives of the potential are defined. You

have rewritten the N-body problem as Euler–Lagrange equations and as Hamilton's equations.

Exercise A.3 Let X_H denote the vector field defined by Hamilton's equations (A.1). Show that $div(X_H) = 0$, where the divergence is taken relative to the canonical coordinates. Conclude that the flow defined by Hamilton's equations preserves phase space volumes $vol(U) = \int_U dq^1 dq^2 \cdots dq^N dp_1 dp_2 \cdots dp_N$. This volume measure is called the Liouville volume and this fact is called "Liouville's theorem" in mechanics.

A.2 Natural Mechanical Systems

When our configuration space Q comes to us along with a Riemannian metric ds_Q^2 and a function $V : Q \to \mathbb{R}$ then we call it a *natural mechanical system*. If a Lie group G acts on Q by isometries of the metric that leave the function invariant, then we say we have a "natural mechanical system with symmetry." The N-body problem is such a system.

Write $\langle \cdot, \cdot \rangle_q$ for the fiber inner product on $T_q Q$ defined by the metric on Q and $\langle \cdot, \cdot \rangle_q^*$ for the associated dual metric on $T_q^* Q$. Write $K(q, v) = \frac{1}{2} \langle v, v \rangle_q$ and $K^*(q, p) = \frac{1}{2} \langle p, p \rangle_q^*$ for the kinetic energies. We take for the Lagrangian $L(q, v) = K(q, v) - V(q)$, while the Hamiltonian is

$$H(q, p) = \frac{1}{2} K^*(q, p) + V(q). \tag{A.3}$$

Coordinate expressions for K and K^* may be useful. If $ds_Q^2 = \Sigma g_{ij}(q) dq^i dq^j$ expresses the metric in coordinates, then $K = \frac{1}{2} \Sigma g_{ij}(q) v^i v^j$ and $K^* = \frac{1}{2} \Sigma g^{ij}(q) p_i p_j$, where g^{ij} is the inverse matrix to g_{ij}.

Exercise A.4 Verify that the Legendre transform for the L just described is the operation of "lowering indices," which takes $v = T_q Q$ to the covector $p = FL(q, v)$ with $p(w) = \langle v, w \rangle_q$. In coordinates: $p_i = \Sigma g_{ij}(q) v^j$. Verify that the Legendre transform of L is the Hamiltonian H given in Equation (A.3).

Exercise A.5 Verify that both Hamilton's and Lagrange's equations of motion for a natural mechanical system are equivalent to their Newtonian form $\nabla_{\dot{q}} \dot{q} = -\nabla V(q)$, where ∇ is the Levi-Civita metric for the Riemannian metric on Q.

A.3 Symplectic Structure: The Hamiltonian Side

The Hamiltonian side of mechanics enjoys the benefits of symplectic geometry, a geometry associated to a particular type of two-form on a manifold known as a *symplectic form*. The symplectic form for mechanics is called the *canonical two-form* on $T^* Q$ and is given in our local coordinates by

$$\omega = \Sigma dq^i \wedge dp_i. \tag{A.4}$$

Exercise A.6 Verify that ω is well defined, independent of the choice coordinates q^i on Q used to induce the coordinates q^i, p_i for T^*Q.
Verify that ω is closed; that is, $d\omega = 0$.
Verify that ω is nondegenerate.

To say that ω is nondegenerate means that the fiber-linear *index lowering* map $v \mapsto i_v\omega := \omega(v, \cdot)$ from the tangent bundle (of T^*Q) to the cotangent bundle (so to $T^*(T^*Q)$) is invertible. Equivalently, any two-form ω has the local coordinate expression $\omega = \Sigma\omega_{ij}dx^i \wedge dx^j$ relative to coordinates x^i on the manifold. Nondegeneracy of ω is equivalent to the skew-symmetric matrix ω_{ij} being invertible.

Definition A.7 A symplectic form is a closed nondegenerate two-form on a manifold. A manifold with such a form is called a symplectic manifold.

Symplectic manifolds must be of even dimensional since skew-symmetric matrices in odd dimensions always have kernels.
Two symplectic manifolds of the same dimension are locally diffeomorphic as symplectic manifolds by Lemma A.8.

Lemma A.8 (Darboux) *About any point of a symplectic manifold there exist coordinates, called "Darboux" or "canonical" coordinates, such that the expression for the two-form in these coordinates is given by Equation (A.4).*

The matrix of our ω in terms of the canonical coordinates $(q^1, \ldots, q^n, p_1, \ldots, p_n)$ is

$$\mathbb{J} = \begin{pmatrix} 0 & I_n \\ -I_n & 0 \end{pmatrix},$$

where I_n is the $n \times n$ identity matrix and $n = dim(Q)$. One has $\mathbb{J}^2 = -I_{2n}$, and hence the matrix of ω is invertible and the canonical two-form is indeed nondegenerate.

Exercise A.9 Write Θ for the one-form on T^*Q given in local coordinates as $\Theta = \Sigma p_i dq^i$. Then $\omega = -d\Theta$. Verify that Θ and hence ω are globally defined forms on T^*Q, independent of the choice of coordinates q^i used to induce the system of canonical coordinates (q^i, p_i) on T^*Q.

See [2] for a coordinate-free definition of Θ.
Given a function H on such a symplectic manifold P, we define the associated Hamiltonian vector field X_H on P by solving

$$\omega(X_H, \cdot) = dH \tag{A.5}$$

for the vector field X_H. The nondegeneracy of ω implies the solvability of this linear equation. We also write $i_{X_H}\omega$ for the one-form $\omega(X_H, \cdot)$.

Definition A.10 The vector field X_H defined in Equation (A.5) is called the Hamiltonian vector field of the Hamiltonian H.

Exercise A.11 Verify that Hamilton's equations (A.1) are the ODEs associated to the Hamiltonian vector field X_H (as defined in Equation (A.5)) on $P = T^*Q$, relative to the canonical symplectic form.

Cartan's "magic formula"[1] asserts that $L_X\alpha = di_X\alpha + i_Xd\alpha$, where L_X is the Lie derivative with respect to a vector field X and where α is any k-form. Applied to $X = X_H$ and $\alpha = \omega$, Cartan's formula yields $L_{X_H}\omega = 0$. Thus the flow Φ_t^H of any Hamiltonian vector field X_H is a flow through symplectomorphisms: diffeomorphisms that preserve ω. Such transformations are also refered to as *canonical transformations*.

Definition A.12 A symplectic transformation, also known as a canonical transformation, is any diffeomorphism of P preserving the symplectic form. We write $Symp(P)$ for the (infinite-dimensional) group of symplectic transformations.

The huge and varied nature of canonical transformations give the symplectic formalism much of its power. We can use canonical transformations to put the Hamiltonian into a normal form or an approximate normal form in order to better understand aspects of its flow.

We have that $Diff(Q) \subset Symp(T^*Q)$. The diffeomorphism group of Q embeds in the symplectomorphism group of T^*Q as follows. Given $\psi\colon Q \to Q$ a diffeomorphism, define $\psi_*\colon T^*Q \to T^*Q$ by $\psi_*(q,p) = (\psi(q),(d\psi_q)^{-1,T}(p))$. This induced map is called the *cotangent lift* of ψ and is easily verified to be a canonical transformation. The process of cotangent lift defines an injective homomorphism $Diff(Q) \to Symp(T^*Q)$.

At the level of Lie algebras, cotangent lift induces an inclusion of the Lie algebra of smooth vector fields on Q into the Lie algebra of Hamiltonian vector fields on T^*Q as demonstrated in Exercise A.13.

Exercise A.13 If Y is a vector field on Q, define $P_Y\colon T^*Q \to \mathbb{R}$ by $P_Y(q,p) = p(Y(q))$. Show that the Hamiltonian vector field X_H of $H = P_Y$ projects to Y by the projection $T^*Q \to Q$. Write $exp(tY)\colon Q \to Q$ for the flow of Y. Show that the Hamiltonian flow defined by P_Y is the cotangent lift of the flow of Y.

In canonical coordinates, $P_Y = \Sigma p_i Y^i(q)$ where $Y = \Sigma Y^i(q)\frac{\partial}{\partial q^i}$. In other words, Hamiltonians linear in the momentum variable correspond to the vector fields on configuration space.

Definition A.14 Given a vector field Y on Q we call the associated function P_Y on T^*Q defined in Exercise A.13 its *momentum function*.

Momentum functions let us quantify the vast gulf between diffeomorphisms of Q and symplectomorphisms of T^*Q. At a formal level, we can view the space $C^\infty(T^*Q)$ of all Hamiltonians H as the Lie algebra of the group of symplectomorphisms $G = Symp(T^*Q)$, with the exponential map $Lie(G) \to G$ taking H to the time one flow of X_H. Since the fibers of the cotangent bundle are vector spaces, we can Taylor expand any given H in terms of momenta,

$$H(q,p) = H_0(q) + H_1(q,p) + H_2(q,p) + \cdots + H_k(q,p) + \cdots, \qquad (A.6)$$

where $p \mapsto H_k(q,p)$ is a homogeneous degree k polynomial in p for each fixed q. We have seen that this H corresponds to a diffeomorphism of the configuration space if and only if $H = H_1$, which is to say, if and only if $H_0 = H_2 = \cdots = H_k = \cdots = 0$.

---

[1] See https://en.wikipedia.org/wiki/Cartan_formula or [2].

This characterizes the image of the embedding of $Diff(Q)$ in $Symp(T^*Q)$ as having infinite codimension, requiring as it does infinitely many equalities to hold.

Exercise A.15   Show that $\frac{1}{n!}\omega^n = dq^1 \wedge dq^2 \cdots dq^n \wedge dp^n \wedge dp^n$. We call this form the Liouville form and denote it by $dvol$. Show that $L_{X_H} dvol = 0$, giving an alternative proof of Liouville's theorem (see Exercise A.3).

A.4   Variational Structure: The Lagrangian Side

The Lagrangian defines a function on paths in configuration space called the action.

Definition A.16   The action A associated to the Lagrangian $L: TQ \to \mathbb{R}$ is the function

$$c \mapsto A(c) := \int_I L(c(t), \dot{c}(t)) dt$$

defined on the space of absolutely continuous paths $c: I \to Q$. Here, $I \subset \mathbb{R}$ is any closed bounded interval.

Recall that a path is absolutely continuous if it is differentiable almost everywhere (a.e.) and if the fundamental theorem of calculus holds locally: If $c(t) = (q^1(t), \ldots, q^n(t))$ is the coordinate expression for a path $c: I \to Q$, valid in some coordinate neighborhood, then $q^i(t) = q^i(a) + \int_a^t \dot{q}^i(s) ds$ for $a, t \in I$. (See [178].) Being absolutely continuous does not depend on the choice of local charts.

One starts off the calculus of variations by differentiating the action with respect to sufficiently smooth paths c. If we restrict the action to the space of all smooth paths joining two fixed points of Q in some fixed interval I of time, then the Euler–Lagrange equations are necessary conditions for a path to be a critical point of the restricted action [66, 114]. We now verify this assertion by hand for a natural mechanical system on a Euclidean space $Q = \mathbb{E}$, as is relevant for the N-body problem. To do so we make explicit the differentiation process.

Definition A.17   By a compact perturbation of a path $q: (a,b) \to \mathbb{E}$ we will mean a family $q_\epsilon(t) = q(t) + \epsilon h(t)$ of curves where the support of h is some compact subinterval J of (a,b).

Theorem A.18   *Consider the case of the action $A = \int L dt$ for a natural mechanical system $L = K - V$ on the Euclidean vector space $\mathbb{E}$, where K corresponds to half the inner product. Suppose that the path q is twice differentiable and that the derivative of A is zero at q for all compact twice differentiable perturbations $q_\epsilon = q + \epsilon h$ of q. Then q solves Newton's equations.*

Proof Differentiate $A(q + \epsilon h)$ with respect to ϵ. Use $\dot{q}_\epsilon = \dot{q}(t) + \epsilon \dot{h}(t)$ to expand and find

$$K(\dot{q}_\epsilon(t)) = K(\dot{q}(t)) + \epsilon \langle \dot{q}(t), \dot{h}(t) \rangle + \epsilon^2 K(\dot{h}(t)).$$

Also,

$$V(q_\epsilon(t)) = V(q(t)) + \epsilon \langle \nabla V(q(t)), h(t) \rangle + O(\epsilon^2),$$

where we have assumed that V is C^2 along the path $q(t)$. It follows that

$$A(q_\epsilon) = A(q) + \epsilon \left(\int \langle \dot{q}, \dot{h} \rangle - \langle \nabla V(q), h \rangle \right) dt + O(\epsilon^2),$$

so that

$$\frac{d}{d\epsilon}|_{\epsilon=0} A = \int_J (\langle \dot{q}, \dot{h} \rangle - \langle \nabla V(q), h \rangle) dt,$$

where J is the closed interval supporting h. We note that $\frac{d}{dt} \langle \dot{q}, h \rangle = \langle \ddot{q}, h \rangle + \langle \dot{q}, \dot{h} \rangle$, or

$$\langle \dot{q}, \dot{h} \rangle = \langle -\ddot{q}, h \rangle + \frac{d}{dt} \langle \dot{q}, h \rangle.$$

It follows that

$$\frac{d}{d\epsilon}|_{\epsilon=0} A = \int_J \langle -\ddot{q} - \nabla V(q), h \rangle dt + (\langle \dot{q}, h \rangle)|_c^d,$$

where $J = [c, d]$. The final boundary term is zero since h has compact support within the open interval J, and so $h = 0$ at the endpoints c, d of J. So, we have shown that

$$\frac{d}{d\epsilon} A = \int_J \langle -\ddot{q} - \nabla V(q), h \rangle dt.$$

Now assume the hypothesis of the theorem, that this derivative is zero for all twice differentiable compactly supported h. Since the set of such h is dense in $L_2 = L_2(J, \mathbb{E})$ for this interval J, where the L_2 pairing is $(w, h) \mapsto \int_J \langle w(s), h(t) \rangle ds$, we must have $w = -\ddot{q} - \nabla V(q)$ is zero a.e. But $q(t)$ is assumed C^2, so we have that Newton's equations $\ddot{q} = -\nabla V(q)$ are satisfied on J. Since $J \subset I$ is arbitrary, we have q satisfying Newton's equations on all of I. QED

We may not know, a priori, that an action extremal q is twice-continuously differentiable. For this reason the following weakening of the previous Theorem A.18, is essential.

Theorem A.19 *Suppose that the path $q : I = [a, b] \to \mathbb{E}$ is absolutely continuous and V is C^2 in a neighborhood of $q((a, b))$. Then A is Frechet-differentiable at q and the derivative is zero at q if and only if $q(t)$ is twice-differentiable on the interior of I and satisfies Newton's equations there.*

This theorem is proved in a manner almost identical to our proof in Chapter 3, Section 3.5, where we proved a theorem of Poincaré.

An extremal path need not exist. For example, take $Q = \mathbb{R}$ with standard coordinate q and $L = \dot{q} + q$. Then the Euler–Lagrange equations for L read $\frac{d}{dt}(1) = 1$, or $0 = 1$. There is no solution. Correspondingly, the action of a path $q : [0, T] \to \mathbb{R}$ is $\int_0^T \dot{q}(t) + q(t) dt = q(T) - q(0) + \int_0^T q(s) ds$. This function has no extremals q for fixed values of T. Indeed, this action functional is linear so has no extremal and, in particular, no minimum or maximum.

For the Lagrangians of classical mechanics, however, minimizers exist, at least locally, as shown in Proposition A.20.

Proposition A.20 *For the action for the Lagrangian of a natural mechanical system (see Section A.2), minimizers exist between any two sufficiently close points provided*

the flight time T is short enough. Specifically, restrict the action to absolutely continuous paths joining two fixed points of Q in a time T. If the points are distinct but sufficiently close and if the time sufficiently short, then the minimum exists, is smooth, and solves the Euler–Lagrange equations. Conversely, any sufficiently short subarc of a solution to the Euler–Lagrange equations minimizes the action among all paths connecting its endpoints.

For a proof of a somewhat more general version of this proposition we refer to what Mañé calls "Weierstrass's theorem" [114, p. 24–26]. See also section 3.4 and theorem 3.4.1 of Fathi's book [56].

Can we join any two distinct points of Q by a smooth minimizer in any given time $T > 0$? In the case of Riemannian geometry (a natural mechanical systems with potential $V = 0$) the answer is inextricably linked with completeness of the flow. The Hopf–Rinow theorem asserts that the answer is "yes" if and only if the associated metric is complete, which is in turn true if and only if the geodesic flow is complete.

The answer to this question is also "yes" for the power-law N-body problems (in $d > 1$ dimensions). This existence result for minimizers follows directly from Marchal's lemma (Chapter 3, Lemma 3.32). And it is true despite the fact that the N-body flow is incomplete.

A.5 Noether, Symmetries, and the Momentum Map

If our Lagrangian L is independent of one of the position coordinates, say q^1, then $\frac{\partial L}{\partial q^1} = 0$ and the first Euler–Lagrange equation, Equation (A.2) reads

$$\frac{d}{dt}\left(\frac{\partial L}{\partial v^1}\right) = 0.$$

Thus, the symmetry of L with respect to translation of the q^1 coordinate yields conservation of the corresponding dual momentum coordinate $p_1 = \frac{\partial L}{\partial v^1}$. This is the simplest instance of Noether's theorem, relating symmetries to conservation laws. Recall that under the Legendre transform, $\frac{\partial L}{\partial v^1} = p_1$ is in fact the momentum variable dual to q^1. A quick computation shows that our Hamiltonian, $H = H(q, p)$, is also independent of q_1. Hamilton's equations (A.1) yield this same conservation law $\dot{p}_1 = 0$. The classical language for this situation is to say that q^1 is a *cyclic variable* for our system. The variable p_1 plays another role here by way of its Hamiltonian vector field. Recall that $\{f, p_1\} = \frac{\partial f}{\partial q^1}$, so that the Hamiltonian vector field of the conserved coordinate yields translation of q_1, providing a basic link between the conserved quantity and the symmetry it is associated to, namely, translation of q^1.

This situation arises when the Lagrangian is invariant under the action of a circle $\mathbb{S}^1$ acting on configuration space. Take coordinates adapted to the circle action so that $q^1 = \theta \in \mathbb{S}^1$ labels points along an individual circle orbit and the additional coordinates $q^2, q^3, \ldots, q^n$ label the orbits. In these coordinates, the circle action becomes $(\theta_0, q^2, \ldots, q^n) \mapsto (\theta_0 + \theta, q^2, \ldots, q^n)$. The corresponding conservation law for circular invariance of L is the momentum $p_1 = p_\theta$ and is called the *angular momentum* or *momentum map* associated to the action. We write this function J and will also refer to it as the momentum map for the circle action.

Exercise A.21 Verify that $J = p_\theta$ agrees with the angular momentum for the planar N-body problem as defined in Chapter 0.

This discussion generalizes from the circle to a Lie group G acting on our configuration space Q so as to leave the Lagrangian L invariant. Thus, we assume that for all $g \in G$, $q \in Q$, and $v \in T_q Q$, we have

$$L(q, v) = L(gq, dg_q v).$$

Here, the action of g on Q is written $q \mapsto gq$ and the differential of the action at q is $dg_q : T_q Q \to T_{gq} Q$. Under the Legendre transformation, the Hamiltonian will be invariant under the corresponding cotangent lifted action $g(q, p) = (gq, (dg)_q^{-1T} p)$.

Noether's theorem gives us $r = dim(G)$ conservation laws. We package these conservation laws together into a single vector-valued map

$$J : T^* Q \to \mathfrak{g}^*, \tag{A.7}$$

where $\mathfrak{g}^*$ is the dual space to the Lie algebra $\mathfrak{g}$ of G. The map J is constructed from the infinitesimal generator σ of the G action on Q,

$$\sigma : Q \times \mathfrak{g} \to TQ; \qquad \sigma(q, \xi) := \frac{d}{d\epsilon}|_{\epsilon=0} exp(\epsilon\xi) q,$$

which is a vector bundle map over Q. In this expression, $\epsilon \mapsto exp(\epsilon\xi) \in G$ is the one-parameter subgroup corresponding to $\xi \in \mathfrak{g}$. The dual of the infinitesimal generator map yields

$$\sigma^* : T^* Q \to Q \times \mathfrak{g}^*,$$

which we project on to the second factor to get the desired map,

$$J = pr \circ \sigma^*. \tag{A.8}$$

(Here, $pr(q, \mu) = \mu \in \mathfrak{g}^*$ is the projection on to the second factor.)

Definition A.22 The map J defined in Equation (A.8) is called the momentum map for the action of G on $T^* Q$ arising from the cotangent lift of G's action on Q.

For fixed $\xi \in \mathfrak{g}$, write $J^\xi = \langle J, \xi \rangle$ where the pairing is by duality so that J^ξ is a scalar function on $T^* Q$. Also, let $\xi_Q(q) = \sigma_q(\xi)$ be the vector field on Q corresponding to the action of the one-parameter group in the direction ξ. Recall the momentum function of Exercise A.13 associated to any vector field X on Q. Then

$$J^\xi = P_{\xi_Q} \tag{A.9}$$

since $J^\xi(q, p) = p(\xi_Q(q))$. Equation (A.9) provides an equivalent definition of the momentum map.

Remark A.23 Earlier, with $G = \mathbb{S}^1$ we wrote $J = p_\theta$ – the momentum map for a circle action has just a single component. This corresponds to the fact that $Lie(\mathbb{S}^1) = \mathbb{R} = Lie(\mathbb{S}^1)^*$ is one-dimensional.

Proposition A.24 (Noether) *Suppose, as above, that the Lagrangian is G-invariant and that its Legendre map is invertible so that the associated Hamiltonian H is defined on all of $T^* Q$. Then the J^ξ, $\xi \in \mathfrak{g}$ are all conserved: $\{J^\xi, H\} = 0$.*

Proof Return to Exercise A.13. We showed there that if ψ_t is the cotangent lift of the flow of the vector field Y then $\frac{d}{dt}\psi_t^* F = \{F, P_Y\}$. Now, from $J^\xi = P_{\xi_Q}$ we see that $\frac{d}{dt} g_t^* F = \{F, J^\xi\}$ where $g_t \colon T^*Q \to T^*Q$ is the cotangent lift of the action of $exp(t\xi)$ on Q. It follows that if F is G-invariant then $\{F, J^\xi\} = 0$ for all $\xi \in \mathfrak{g}$. But H is G-invariant since L is, and $\{H, J^\xi\} = -\{J^\xi, H\}$. QED

The result of Exercise A.25 is important when we get to symplectic reduction in Appendix B.

Exercise A.25 The group G has a represententation on its dual Lie algebra given by $g \cdot \mu = Ad^*_{g^{-1}}\mu$, where Ad_g is the usual adjoint action (conjugation for matrix groups). Upon using this action, show that the momentum map as defined in Equation (A.9) is G-equivariant, $J(gp) = g \cdot J(p)$.

Momentum Maps for Natural Mechanical Systems

For a natural mechanical system (Q, ds_Q^2, V) with symmetry group G, we suppose that the Lie group G acts on Q by isometries that preserve the potential V. The lifted G-actions preserve both L and H. In this case, the Legendre transformation sends $v \in T_qQ$ to the linear functional $\langle v, \cdot \rangle_q \in T_q^*Q$ and is G-equivariant. We can use it to transfer the momentum map to a function on the tangent bundle. We compute that when viewed this way we get

$$J^\xi(q, v) = \langle v, \xi_Q(q) \rangle_q.$$

Exercise A.26 The N-body problem is a natural mechanical system with symmetry group $G = SE(d) = \mathbb{R}^d \times SO(d)$, the group of rigid motions of d-space acting on its configuration space $Q = \mathbb{E}(d, N) \cong (\mathbb{R}^d)^N$. Identify $\mathfrak{g}$ and $\mathfrak{g}^*$ with $\mathbb{R}^d \oplus \wedge^2\mathbb{R}^d$. (See Appendix D.) Verify that the tangent version of the momentum map after this identification is the map $(P, J) \colon \mathbb{E} \times \mathbb{E} \to \mathbb{R}^d \oplus \wedge^2\mathbb{R}^d$, where $P = \Sigma m_a v_a$ is the linear momentum from the introductory chapter (see Equation (0.22)) and where $J = \Sigma m_a q_a \wedge v_a$ (from Equation (0.23)).

A.6 Notes

1. See Arnol'd [11], Abraham – Marsden [2], or Knauf [96] for more on the geometric foundations of classical mechanics. See Young [218] for another perspective on the Legendre transformation and its use in the calculus of variations. See Dirac [44] for a way to obtain a Hamiltonian and for Hamiltonian formalism when the Legendre transformation is not invertible, which is the situation that arises in special and general relativity.

2. Birkhoff [19] takes the perspective of Equation (A.6) in classifying and understanding Hamiltonians.

   Natural mechanical systems correspond precisely to those Hamiltonians of the form $H = H_0 + H_2$ with H_2 positive-definite in the fiber. If we add on a first-order piece H_1 to such an H, then there is a one-form dual with respect to H_2 to the vector field X of $H_1 = P_X$. The resulting Hamiltonian dynamics is that of a particle in a static electromagnetic field whose magnetic part is given by H_1 and whose electric potential is given by H_0.

3. In physics texts one reads the adage "physical trajectories minimize the action" not "extremize the action." The difference between minimizing and extremizing is quantified by the Hessian and its relation to Jacobi fields. See for example [126], and specifically the chapter on Riemannian geometry.

The effort to minimize the action was essential to the development of the calculus of variations and yielded one of the big tools of Chapter 3, the direct method in the calculus of variations.

Feynman dragged the action principle and Lagrangian formalism back to center stage in physics through his realization that the Schwartz kernel $K(t, x, y)$ for solving the Schrödinger equation could be expressed as a path integral whose integrand was $exp(i \hbar A(q))$ and whose domain of integration was "all paths" joining x to y in time t.

Appendix B   Reduction and Poisson Brackets

B.1 Poisson Bracket Formalism and Reduction

At various places in this book we use reduced dynamics, meaning the dynamics for the N-body problem after being modded out by the Galilean group. The idea of reduced dynamics can be found in Section 0.4. This appendix implements reduction through the Poisson bracket formalism. We will use that formalism to derive the reduced planar three-body equations (0.71) in Section B.5. Then we relate the Poisson bracket reduction to the more traditional symplectic reduction.

The phase space of the N-body problem is a symplectic manifold $\mathcal{P}$ on which the Lie group G of rigid motions acts. There are two primary methods for forming the quotient of a symplectic manifold $\mathcal{P}$ by the action of a group G. The most direct and computationally effective method is to simply do it: to form the quotient space $\mathcal{P}/G$. We call this method "Poisson reduction" because the quotient space is no longer a symplectic manifold, but rather something more general, a *Poisson manifold*. Hence this section.

The other method of forming the quotient, called symplectic reduction, yields a symplectic manifold, or rather a family of such manifolds parameterized by a function called the "momentum map," generalizing the idea of angular momentum. These symplectic reduced spaces all sit within the Poisson reduced space, as its "symplectic leaves."

To begin Poisson geometry, return to symplectic geometry: a manifold $\mathcal{P}$ endowed with a symplectic form ω. We define the Poisson bracket on $\mathcal{P}$ by

$$\{F, H\} = \omega(X_F, X_H),$$

where X_F, X_H are the Hamiltonian functions of the smooth functions F, H on $\mathcal{P}$. We could also write $\{F, H\} = dF(X_H) = -dH(X_F)$. This bracket defines a bilinear skew-symmetric operation on functions on $\mathcal{P}$. In terms of our canonical coordinates q^i, p_i we have

$$\{F, H\} = \Sigma \frac{\partial F}{\partial q^i} \frac{\partial H}{\partial p_i} - \frac{\partial F}{\partial p_i} \frac{\partial H}{\partial q^i}.$$

Exercise B.1   Verify that when ω is the canonical symplectic form as above, we have the bracket relations, also known as the canonical commutation relations,

$$\{q^i, q^j\} = 0, \{q^i, p_j\} = \delta^i_j, \{p_i, p_j\} = 0. \tag{B.1}$$

Exercise B.2   If X is a vector field on Q, write $P_X : T^*Q \to \mathbb{R}$ for the momentum function as defined in Exercise A.13. If $f : Q \to \mathbb{R}$ is any smooth function, write $\pi^* f : T^*Q \to \mathbb{R}$ for its pull-back to T^*Q so that $\pi^* f(q, p) = f(q)$. Let X, Y be any two vector fields on Q, with $[X, Y]$ their Lie bracket, and let f, g be any two functions on Q and write $X[f] = df(X)$ for the directional derivative of f along X. Verify the Poisson bracket relations on $\mathcal{P} = T^*Q$:

- $\{P_X, P_Y\} = -P_{[X,Y]}$,
- $\{\pi^* f, P_X\} = \pi^*(X[f])$,
- $\{\pi^* f, \pi^* g\} = 0$.

Definition B.3 formalizes Poisson brackets.

Definition B.3   A Poisson bracket on a manifold $\mathcal{P}$ is a Lie algebra structure $F, H \mapsto \{F, H\}$ on the $\mathbb{R}$-vector space $C^\infty(\mathcal{P})$ of its smooth functions that, in addition, satisfies the Liebnitz identity

$$\{F, GH\} = G\{F, H\} + \{F, G\}H.$$

(There is a second Liebnitz identity relative to the first slot "F" of the Poisson bracket that follows automatically from the skew-symmetry of the bracket.) The reader may wish to verify that the bracket previously defined on a symplectic manifold is actually a Poisson bracket, the main issue being the Jacobi identity.

A Poisson manifold is, by definition, a manifold with a Poisson bracket operation on its smooth functions. The Liebnitz identity allows us to associate a tensor $B \in \Gamma(\wedge^2 T\mathcal{P})$ to a Poisson manifold, called the "Poisson tensor." This B is a contravariant skew-symmetric two-tensor, defined pointwise by $B_p(\alpha, \beta) = \{f, g\}(p)$ where $\alpha = df(p), \beta = dg(p) \in T_p^*\mathcal{P}$. If x^i are local coordinates on $\mathcal{P}$ then set $B^{ij}(x) = \{x^i, x^j\}(x)$. One verifies that $B = \Sigma_{i<j} B^{ij}(x)\frac{\partial}{\partial x^i} \wedge \frac{\partial}{\partial x^j}$ is well defined, independent of the choice of coordinates and that $\{F, H\} = B(dF, dH) = \Sigma_{ij} B^{ij}(x)\frac{\partial F}{\partial x^i}\frac{\partial H}{\partial x^j}$ where we use $B^{ji} = -B^{ij}$.

Exercise B.4   If $\omega = \Sigma\omega_{ij}dx^i \wedge dx^j$ is the coordinate expression for a symplectic form ω on a symplectic manifold, verify that the Poisson tensor corresponding to its Poisson bracket is given by $B = -\Sigma\omega^{ij}\frac{\partial}{\partial x^i} \wedge \frac{\partial}{\partial x^j}$, where ω^{ij} is the inverse matrix to ω_{ij}.

B.2 Hamilton's Equations via Brackets

We can associate a Hamiltonian vector field X_H to each smooth function H on a Poisson manifold by using the Poisson bracket to define X_H in terms of its action as a derivation on functions $f : X_H[f] = \{f, H\}$. This definition agrees with our previous definition of X_H in the symplectic case (see Appendix A). Write Φ_t^H for the flow of X_H. The Jacobi identity implies that this flow preserves the Poisson bracket. If we set $f_t = f \circ \Phi_t^H$ for $f \in C^\infty(\mathcal{P})$ then we have that

$$\frac{d}{dt} f = \{f, H\},$$

which provides us with another equivalent way to express Hamilton's equations when $\mathcal{P} = T^*Q$. Just let f vary over q^i, p_i and write out this equation to recover our original version of Hamilton's equations (A.1).

Symplectic Leaves

A Poisson manifold can be viewed as a disjoint union of symplectic manifolds called symplectic leaves. The symplectic leaves are the integral manifolds of a singular foliation. To form the symplectic leaf through the point $p \in \mathcal{P}$, take the collection of all Hamiltonian vector fields $X_H = \{\cdot, H\}$ and integrate them, starting from p. The endpoints of the resulting integral curves form the leaf through p. It is an immersed symplectic manifold. The Jacobi identity shows us that this collection of vector fields is involutive, so, by an extension of the Frobenius integrability theorem due to Sussman [203], the collection of all endpoints of such vector fields is a locally embedded submanifold of $\mathcal{P}$ – the leaf through p. The tangent space to the leaf through p at the point p is the image of the Poisson tensor B_p viewed as a linear map $T_p^*\mathcal{P} \to T_p\mathcal{P}$. The symplectic form ω on the leaf can be obtained by writing $\omega(X_F, X_G) = \{F, G\}$. All of these facts are described in detail in Weinstein [213].

B.3 New Poisson Manifolds by Quotient

If $\mathcal{P}$ is a Poisson manifold on which a compact Lie group acts freely on $\mathcal{P}$ by Poisson automorphisms, then $\mathcal{P}/G$ inherits the structure of a Poisson manifold. (The "compact" and "freely" assumptions are only used to ensure the quotient is a manifold.) Write the action of pulling a function back by the action of g as $F \mapsto g^*F$. Then to say that G acts by Poisson automorphisms is to say that for all $g \in G$ and all $F, H \in C^\infty(\mathcal{P})$ we have that $\{g^*F, g^*H\} = g^*\{F, H\}$. Since G is compact and acts freely, the quotient space $\mathcal{P}/G$ is automatically a smooth manifold. We can identify the ring of smooth functions on $\mathcal{P}/G$ with $C^\infty(\mathcal{P})^G$ – the ring of G-invariant smooth functions on $\mathcal{P}$. The fact that G acts by Poisson automorphisms implies that the Poisson bracket of two G-invariant functions is also G-invariant. Thus, we have defined a Poisson structure on the quotient space $\mathcal{P}/G$.

In the case where $\mathcal{P} = T^*Q$ and the compact Lie group G acts freely on Q, then its cotangent lift acts freely on T^*Q by symplectic, and hence Poisson, automorphisms. It follows that $(T^*Q)/G$ is a Poisson manifold. This class of examples arises in the N-body problem and turns out to be quite simple in the case of the planar N-body problem.

B.4 Reduction by the Circle

Suppose that $G = \mathbb{S}^1$ acts by symmetries on a configuration space Q. Write the action as $(\theta, q) \mapsto e^{i\theta}q$ and its infinitesimal generator as $\frac{\partial}{\partial \theta}$. (The infinitesimal generator is the vector field $q \mapsto \frac{\partial}{\partial \theta}(q) := \frac{d}{d\theta}|_{\theta=0} e^{i\theta}(q)$ on Q.) Here, $\mathbb{S}^1$ also acts on T^*Q by cotangent lift and this action preserves the Poisson brackets. We then have the corresponding momentum map as described in Exercise A.13,

$$J = P_{\frac{\partial}{\partial\theta}} : T^*Q \to \mathbb{R}.$$

See also Section (A.5). We will call J the "angular momentum" since it agrees with the standard angular momentum for the case of the planar N-body. In terms of the coordinates adapted to the circle action as in Exercise A.21, we have that $J(\theta, q^2, \ldots, p_\theta, p_2, \ldots) = p_\theta$.

It follows from Exercise A.13 that the flow of J generates the lifted circle action. It then follows that a function f on T^*Q is $\mathbb{S}^1$ invariant if and only if $\{f, J\} = 0$. We repeat,

$$f \in C^\infty(T^*Q)^{\mathbb{S}^1} \iff \{f, J\} = 0. \tag{B.2}$$

In particular, since $\{J, J\} = 0$ we have that J is $\mathbb{S}^1$ invariant and so can be viewed as a function on the quotient space that we will denote by the same symbol,

$$J : (T^*Q)/\mathbb{S}^1 \to \mathbb{R}.$$

Since $\mathbb{S}^1$ acts freely on Q the quotient space, $B = Q/\mathbb{S}^1$ inherits the structure of a smooth manifold for which the quotient projection

$$\pi : Q \to Q/\mathbb{S}^1 = B$$

is a submersion. This submersion gives Q the structure of a principal circle bundle over the base space B. We might call the base B "shape space" in honor of the planar three-body problem.

Here is a simple structure formula for the circle quotient of phase space:

$$(T^*Q)/\mathbb{S}^1 \cong T^*B \times \mathbb{R}. \tag{B.3}$$

Under this diffeomorphism the momentum map J becomes the projection onto the $\mathbb{R}$ factor. It follows from Equation (B.2) that the angular momentum J is a "Casimir": It Poisson commutes with all functions on $(T^*Q)/\mathbb{S}^1$. The Poisson leaves of our Poisson structure are the level sets of J, so are diffeomorphic to T^*B. However, as we now describe, these leaves are typically *not* symplectically equivalent to T^*B.

The diffeomorphism (B.3) depends on the choice of a connection for our circle bundle $Q \to B$. We recall a bit of the theory of connections on circle bundles now. Set

$$V = span \frac{\partial}{\partial\theta} = ker(d\pi) \subset TQ.$$

Call V the vertical distribution and vectors in V "vertical." Then V is a trivial line subbundle of TQ. A connection is, by definition, an $\mathbb{S}^1$-invariant complement H, called the "horizontal" to our vertical distribution V. Thus

$$TQ = H \oplus V. \tag{B.4}$$

(Apologies for the double use of the symbol H: for horizontal and for Hamiltionian.) So, for each $q \in Q$ we have a hyperplane $H_q \subset T_qQ$. The dual of the horizontal-vertical splitting is

$$T^*Q = V^\perp \oplus H^\perp, \tag{B.5}$$

where we write $\perp$ for annihilators, so that $V^\perp$ is the vector subbundle consisting of those covectors p that annihilate the vertical vectors $v \in V$. The dual splitting is also $\mathbb{S}^1$-invariant so descends to the quotient. The structure formula (B.3) now follows directly from Lemma B.5.

Lemma B.5 *(A)* $H^\perp$ *is a one-dimensional trivial line. (B)* $V^\perp = J^{-1}(0)$. *(C)* $V^\perp/\mathbb{S}^1 = T^*B$ *canonically.*

Proof of (A) That $H^\perp$ is one-dimensional and trivial as a line bundle follows immediately from the fact that H has codimension one and its complement V is trivial. $V \cong H^\perp$ is trivial since $\frac{\partial}{\partial\theta}$ provides a global section of V.

We will need some terminology to continue with the proof. A unique global section $A: Q \to H^\perp \subset T^*Q$ of the trivial line bundle of part (A) can be chosen by imposing the normalization

$$A\left(\frac{\partial}{\partial\theta}\right) = 1. \tag{B.6}$$

Definition B.6 The one-form A defined in Equation (B.6) is called the "connection one-form" for our choice of horizontal for the circle bundle $Q \to B$.

Then A determines H by $H = ker(A)$. And H plus the normalization condition determines A. Note that the connection one-form A is $\mathbb{S}^1$-invariant because H is $\mathbb{S}^1$-invariant. Translating by multiples of A induces the above splitting, Equation (B.5),

$$(q, p) \mapsto (p - J(q, p)A(q), J(q, p)A(q)) \in V^\perp \oplus H^\perp. \tag{B.7}$$

Proof of part (B) of Lemma B.5 Now $V^\perp = J^{-1}(0) \subset T^*Q$ since $J(q, p) = p(\frac{\partial}{\partial\theta}q)$ and $\frac{\partial}{\partial\theta}$ spans V. Observe that $J(q, A(q)) = 1$ by the normalization of the connection and that $J(q, p)$ is linear in the fiber p. It follows that $J(q, J(q, p)A(q)) = J(q, p)$ and that $(q, p - J(q, p)A(q)) \in J^{-1}(0)$.

It can be helpful to replace the vector bundle summand $H^\perp$ of the isomorphism (B.7) by the trivial line bundle $Q \times \mathbb{R}$ over Q. Then the isomorphism becomes the vector bundle map $T^*Q \cong V^\perp \times \mathbb{R} = J^{-1}(0) \times \mathbb{R}$, which sends (q, p) to $(\beta(q, p), J(q, p))$ where $\beta(q, p) = (q, p - J(q, p)A(q))$ and where $J(q, p) \in \mathbb{R}$. This isomorphism is an $\mathbb{S}^1$-equivariant bundle isomorphism where $\mathbb{S}^1$ acts trivially on the $\mathbb{R}$-factor.

Proof of part (C) of Lemma B.5 We will give two proofs of part (C). The first proof requires recalling a bit of background regarding circle bundles.

Circle bundles admit local trivializations. These are coverings of the base B by open sets U along with local diffeomorphisms $U \times \mathbb{S}^1 \to \pi^{-1}(U)$ that make the circle bundle over U look like the trivial bundle $U \times \mathbb{S}^1$. We can take the U's to be coordinate neighborhoods. Thus, if $x^i, i = 1, \ldots, n-1$ are coordinates on B defined within a locally trivializing neighborhood and if θ is the standard circle variable, we get induced coordinates x^i, θ on Q such that in these coordinates $\pi(x^i, \theta) = (x^1, \ldots, x^{n-1})$ and such that the infinitesimal generator of the circle action is the coordinate vector field $\frac{\partial}{\partial\theta}$. The coordinates x^i induce the usual momentum coordinates p_i on T^*B and the x^i, θ induce coordinates p_i, p_θ on T^*Q. In terms of these coordinates on T^*Q we have that $J(\theta, x^i, p_\theta, p_i) = p_\theta$, the momentum dual to rotation.

Coordinate proof of part (C) Locally, in the above coordinates, we have $J^{-1}(0) = \{p_\theta = 0\}$ and so x^i, θ, p_i coordinatize $J^{-1}(0)$. Forming the quotient by $\mathbb{S}^1$ gets rid of θ so that the quotient map $J^{-1}(0)/\mathbb{S}^1$ is, in these coordinates, $(x^i, \theta, p_i, 0) \mapsto (x^i, p_i)$, yielding canonical coordinates on T^*B. Some thought about how coordinates and local trivializations fit together shows this computation yields a global map and finishes the proof.

Intrinsic proof of part (C) Take a covector $\alpha \in T_b^* B$. Then $\pi^* \alpha \in T^* Q$ is a covector that annihilates V; thus $\pi^* \alpha \in J^{-1}(0)$. There is an ambiguity as to where $\pi^* \alpha$ lives. To make it a covector attached to a particular $q \in \pi^{-1}(b)$ we set $p = (\pi^* \alpha)_q = d\pi_q^* \alpha$. Now p annihilates $V_q = ker(d\pi_q)$ so that $J(q, p) = 0$. The ambiguity in where p is attached washes out upon forming the quotient, since the $\mathbb{S}^1$ orbit of any point is the whole π-fiber. It follows that the map $(b, \alpha) \mapsto (q, \pi^* \alpha_q)(mod\mathbb{S}^1)$ is a well-defined map $T^* B \to J^{-1}(0)/\mathbb{S}^1$. The coordinates can be used to verify it is smooth. To go the other way, take any $\beta_q \in J^{-1}(0) \cap T_q^* Q$. We define a covector $\alpha = [\beta_q] \in T_b^* B$ as follows. Choose $w \in T_b B$ and any $W_q \in T_q Q$ such that $d\pi_q W_q = w$. Set $\alpha(w) = \beta_q(W_q)$. Because β_q annihilates V_q this resulting number is independent of the choice of W_q projecting onto w, for if $d\pi_q W_q = d\pi_q \tilde{W}_q$ then $W_q - \tilde{W}_q \in V_q$. QED

We now identify the Poisson structure under our identification (B.3) of the reduced space $(T^* Q)/\mathbb{S}^1$. This is done in terms of the curvature of the connection. The curvature is a two-form F on the base B defined by

$$dA = \pi^* F.$$

In terms of our locally trivializing coordinates x^i, θ, we have that

$$A = d\theta + \Sigma A_i(x) dx^i$$

and

$$F = d(\Sigma A_i(x) dx^i),$$

where $\Sigma A_i(x) dx^i$ is a one-form on the trivializing neighborhood of B.

Theorem B.7 *Under the connection-induced diffeomorphism* (B.3), *the induced Poisson structure on* $T^* B \times \mathbb{R}$ *is characterized by*

$$\{x^i, x^j\} = 0, \{x^i, p_j\} = \delta_j^i, \{x^i, J\} = \{p_i, J\} = 0,$$

and

$$\{p_i, p_j\} = J F_{ij}$$

when the x^i, p_i *are a standard set of canonical coordinates on* $T^* B$ *induced by choosing coordinates* x^i *on* B, *and* $F_{ij} = F\left(\frac{\partial}{\partial x^i}, \frac{\partial}{\partial x^j}\right)$ *are the components of the curvature* F *of the connection with respect to the* x^i.

Proof Write the isomorphism (B.7) in the coordinates $x^i, \theta, p_i, p_\theta$ on $T^* Q$ described in the coordinate proof of Lemma B.5, part (C), and in terms of the coordinates $x^i, \theta, p^i, p_\theta$ that they induce on $T^* Q/\mathbb{S}^1$. The coordinates x^i on B also induce coordinates on $T^* Q$, which, for the purposes of the proof, we write as $x^i = X^i, P_i$. In terms of these coordinates our isomorphism is given by

$$X^i = x^i, \tag{B.8}$$

$$P_i = p_i - p_\theta A_i(x), \tag{B.9}$$

$$J = p_\theta. \tag{B.10}$$

To finish the proof, we compute Poisson brackets of these new functions. That $\{X_i, X_j\} = 0$ and $\{X_i, P_j\} = \delta^i_j$ is almost immediate. The only tricky bracket to compute is $\{P_i, P_j\}$. We compute $\{P_i, P_j\} = \{p_i - p_\theta A_i, p_j - p_\theta A_j\} = -p_\theta \{A_i, p_j\} - p_\theta \{p_i, A_j\}$. Since $\{f, p_j\} = \frac{\partial f}{\partial x^j}$ for any function f of the x's alone, this last expression equals $p_\theta \left(\frac{\partial A_j}{\partial x^i} - \frac{\partial A_i}{\partial x^j} \right)$. But $F = dA = \Sigma \left(\frac{\partial A_j}{\partial x^i} - \frac{\partial A_j}{\partial x^i} \right) dx^i \wedge dx^j$, which leads to $\{P_i, P_j\} = J F_{ij}$. Since X^i, P_i are the coordinates refered to as x^i, p_i in the statement of the theorem, this completes the proof. QED

Remark B.8   In case $Q = B \times \mathbb{S}^1$ is the trivial $\mathbb{S}^1$ bundle we can take the connection to be the flat connection corresponding to $H_{(b,\theta)} = T_b B \times 0 \subset T_{b,\theta} Q$. This connection has zero curvature, F_{ij}, and so in this case $\{p_i, p_j\} = 0$.

Remark B.9   The symplectic leaves for the reduced Poisson structure $T^*B \times \mathbb{R}$ are the level sets $J = const.$; that is, the leaves are the hypersurfaces $T^*B \times \{J_0\}$ with J_0 constant. Each leaf is diffeomorphic to T^*B but might not be symplectomorphic to T^*B with its canonical two-form ω_B. Indeed, the leaf symplectic form is $\omega_B - J_0 pr^* F$, where F is the curvature two-form on B and $pr: T^*B \rightarrow B$ is the cotangent projection. In the coordinates defined by Equations (B.8)–(B.10), this curvature-corrected symplectic form is $\Sigma dX^i \wedge dP_i - J_0 \Sigma F_{ij} dX^i \wedge dX^j$ and the cotangent projection is $pr(X^i, P_i) = X^i$.

B.5 Application to Natural Mechanical Systems

A natural mechanical system (Q, ds^2_Q, V) with $\mathbb{S}^1$ symmetry has a connection with horizontal defined by

$$H = V^\perp, \tag{B.11}$$

where $\perp$ here is with respect to the $\mathbb{S}^1$-invariant metric ds^2_Q. This connection induces splittings $TQ = V \oplus H$ and $T^*Q = V^\perp \oplus H^\perp$ (see Equations (B.4, B.5), which are mapped to each other by the Legendre transformation (the lowering index metric isomorphism) $TQ \rightarrow T^*Q$ so our two uses of $\perp$, one metric, the other by duality, should not lead to any confusion.

Definition B.10   The connection defined in Equation (B.11) is called the "natural mechanical connection."

Lemma B.11   *The connection one-form for the natural mechanical connection is given by*

$$A(q)(v) = \frac{1}{I(q)} J(q, v),$$

where $I(q) = \langle \frac{\partial}{\partial \theta}(q), \frac{\partial}{\partial \theta}(q) \rangle_q$ and where $J: TQ \rightarrow \mathbb{R}$ is the angular momentum J pulled back by the Legendre transformation

$$J(q, v) = \left\langle v, \frac{\partial}{\partial \theta} \Big|_q \right\rangle. \tag{B.12}$$

Proof Since $H = V^\perp$ and V is spanned by $\frac{\partial}{\partial\theta}$, it is clear that $ker\, J_q = H$. It follows that A_q and J_q are multiples of each other: $A_q = f(q)J_q$ for some non-vanishing scalar function f. The multiplier f is determined by the normalization $A\left(\frac{\partial}{\partial\theta}\right) = 1$. We have $J\left(q, \frac{\partial}{\partial\theta}\right) = I(q)$, which yields $f = \frac{1}{I}$. QED

In the case of the planar N-body problem, $J = \Sigma m_a q_a \wedge v_a$ is the usual angular momentum and $I(q) = \|q\|^2$ is the usual moment of inertia.

Metric on the Base

The restriction of $d\pi_q$ to H_q is a linear isomorphism $H_q \to T_b B$, where $b = \pi(q)$. The horizontal space H_q inherits an inner product from ds_Q^2 by restriction. Declare this linear isomorphism to be an isometry so as to define an inner product on $T_b B$. Since the circle acts by isometries, this inner product on the base tangent space is independent of the choice of $q \in \pi^{-1}(b)$. In this way, we arrive at a Riemannian metric ds_B^2 on the base space B which, in deference to the N-body problem, we refer to as the "shape metric."

In terms of the horizontal-vertical split we can now write

$$TQ = \pi^* TB \oplus \mathbb{R}, \tag{B.13}$$

with corresponding metric split as

$$ds_Q^2 = \pi^* ds_B^2 \oplus I d\theta^2. \tag{B.14}$$

A word or two is in order regarding $\pi^* TB$. This is the vector bundle over Q that assigns to $q \in Q$ the vector space $T_{\pi(q)} B$. Then $H \cong \pi^* TB$ with isomorphism $d\pi_q$ restricted to H_q. The inverse of this vector bundle isomorphism is given fiberwise as

$$h_q : T_b B \to H_q$$

and is called the horizontal lift operator.

The dual of our metric splitting (Equations (B.13) and (B.14)) yields the $\mathbb{S}^1$-invariant vector bundle splitting

$$T^* Q \cong \pi^* T^* B \oplus \mathbb{R},$$

with corresponding kinetic energy splitting as

$$K = K_B^* \oplus \frac{1}{2}\frac{J^2}{I}.$$

In local coordinates we have

$$K_B^*(x, p) = \frac{1}{2}g^{ij}(x)p_i p_j,$$

where g^{ij} is the inverse matrix to $ds_B^2 = \Sigma g_{ij} dx^i dx^j$. The total Hamiltonian $H = K + V$ has the shape

$$H = K_B + \frac{1}{2}\frac{J^2}{I} + V := K_B^* + V_{eff}(x),$$

where we have introduced the "effective potential"

$$V_{eff}(x) = V_{eff,J}(x) := \frac{1}{2}\frac{J^2}{I(x)} + V(x).$$

We will use the Poisson bracket formalism as summarized by Theorem B.7 to compute the equations of motion; that is, the circle-reduced Newton's equations. In order to do this, compute $\dot{f} = \{f, H\}$ as f varies over $f = x^i, p_i, J$. Now $\{x^i, V_{eff,J}(x)\} = 0$ since $\{x^i, f(x)\} = 0$ for any function of the x^j's alone. So

$$\dot{x}^i = \{x^i, K_B\} = g^{ij}(x)p_j.$$

We move on to the momentum evolution equations,

$$\dot{p}_i = \{p_i, H\} = \{p_i, K_B\} + \{p_i, V_{eff}(x)\},$$
$$\dot{J} = 0.$$

Now, $\{p_i, V_{eff}(x)\} = -\frac{\partial V_{eff}}{\partial x^i}$ while two terms arise out of $\{p_i, K_B\}$, one from the fact that $\{p_i, x^j\} = -\delta_i^j$ and the other from $\{p_i, p_j\} = JF_{ij}$. We compute that $\{p_i, K_B\} = -(\frac{\partial}{\partial x^i}\frac{1}{2}g^{km})p_k p_m + g^{km}p_k JF_{im}$. From the $\dot{x}^i$ equation, this last term equals $JF_{ik}\dot{x}^k$. Thus,

$$\dot{p}_i = -\frac{\partial}{\partial x^i}\left(\frac{1}{2}g^{km}(x)\right)p_k p_m + JF_{ik}\dot{x}^k - \frac{\partial V_{eff}}{\partial x^i}.$$

In order to frame these evolution equations in coordinate-free language, set the last two terms in the equation for $\dot{p}_i$ to zero. Observe that what remains is the geodesic equations for the shape metric. The two terms we just set to zero represent the deviation from geodesy, and so can be interpreted as forces. These forces are the coordinate expression of the one-form $JF(\dot{b}, \cdot) - dV(b)$ where $b \in B$ has coordinates x^i and its derivative $\dot{b}$ has coordinates $\dot{x}^i$. Raising indices using the metric and recalling that the contravariant version of the geodesic equation is $\nabla_{\dot{b}}\dot{b} = 0$, we see that the full system of equations is equivalent to

$$\nabla_{\dot{b}}\dot{b} = -\nabla V_{eff,J}(b) - JF(\dot{b}, \cdot)^\#, \qquad (B.15)$$

where J is a constant and where #: $T^*Q \to TQ$ is the metric-induced index-raising map, that is, the inverse of the Legendre transformation. These are the circle-reduced Newton's equations. They are a one-parameter family of ODEs parameterized by the value of J.

Reduction of the Planar N-Body Problem

The planar N-body problem forms a natural mechanical system with circle symmetry. We form the quotient by translations (and, conceptually speaking, by boosts) when we fix the center of mass to be the origin. After deleting total collision we arrive at the configuration space $Q = \mathbb{E}_0 \setminus \{0\} = \mathbb{C}^{N-1} \setminus \{0\}$, on which the circle $\mathbb{S}^1$ of planar rotations acts freely.

We described the metric quotient $Q/\mathbb{S}^1$ in Equation (0.63). We recall the salient features. Metrically speaking, $Q/\mathbb{S}^1 = Cone(\mathbb{CP}^{N-2}) \setminus \{0\}$, where 0 is the cone point and corresponds to total collision. Spherical coordinates provide a diffeomorphism $Q \cong \mathbb{S}^{2N-3} \times \mathbb{R}^+$ where the $\mathbb{R}^+$ factor is invariant under rotations and coordinated by $\|q\| = \sqrt{I(q)}$. Under these identifications the quotient projection $\pi: Q \to Q/\mathbb{S}^1$ becomes the map $\mathbb{S}^{2N-3} \times \mathbb{R}^+ \to \mathbb{CP}^{N-2} \times \mathbb{R}^+$, which is the Hopf fibration $\pi_{Hopf}: \mathbb{S}^{2N-3} \to \mathbb{CP}^{N-2}$ on the first factor and the identity on the size factor $\mathbb{R}^+$.

The natural mechanical connection for $Q \to Q/\mathbb{S}^1$ has connection one-form

$$A(Z) = \frac{1}{\Sigma_i |Z_i|^2} Im(\Sigma \bar{Z}_i dZ_i). \tag{B.16}$$

See Lemma B.11. Here, the Z_i are normalized Jacobi vectors yielding the isomorphism $\mathbb{E}_0 \cong \mathbb{C}^{N-1}$, the expression $Im(\Sigma \bar{Z}_i dZ_i)$ is the angular momentum (viewed as a one-form), and $I = \Sigma |Z_i|^2$ is the moment of inertia written in Jacobi vectors. This is the standard connection for the Hopf fibration, trivially extended by the $\mathbb{R}^+$ action. The curvature form F of this connection is the Fubini–Study Kahler form on $\mathbb{CP}^{N-2}$, trivially extended by the $\mathbb{R}^+$ factor. This two-form is invariant under the group of unitary transformations $U(N-1)$ acting on $\mathbb{CP}^{N-2}$ and, up to scale, is the unique two-form invariant under this group. The scale factor for F is fixed by the condition that $\int_{\mathbb{CP}^1} F = 2\pi$, where $\mathbb{CP}^1 \subset \mathbb{CP}^{N-2}$ is any standard complex projective line, for example, the line defined by setting all but the first of the $Z_i = 0$.

Planar Three-Body Reduction

We combine the perspectives of the earlier parts of this section to obtain the promised reduced three-body equations given in Equation (0.71). This derivation is simply a matter of translating Equation (0.71) over into the w-coordinates we use on shape space.

Recall the notation and set-up from Section 0.4.2. (Alternatively consult Appendix C or my article [150].) For the planar three-body problem we have $N = 2$ (in the earlier part of this section) so that $\mathbb{CP}^{N-2} = \mathbb{CP}^1 = \mathbb{S}^2$. This two-sphere is the shape sphere, identified as the unit sphere in shape space $\mathbb{R}^3$. We use the global coordinates $w = (w_1, w_2, w_3)$ on shape space.

The only tricky term to translate from Equation (0.71) is the final curvature term. The curvature F of the natural mechanical connection, the Fubini–Study form made scale-invariant, is the two-form

$$F = \frac{1}{2} \frac{w_1 dw_2 \wedge dw_3 + w_3 dw_1 \wedge dw_2 + w_2 dw_3 \wedge dw_1}{\|w\|^3}, \tag{B.17}$$

which is one-half the usual solid-angle form on $\mathbb{R}^3 \setminus \{0\}$. The one-form appearing in the reduced equations (B.15) is given by

$$F(w)(\dot{w}, \cdot) = \frac{1}{2|w|^3}(w_1 \dot{w}_2 dw_3 + w_3 \dot{w}_1 dw_2 + w_2 \dot{w}_3 dw_1). \tag{B.18}$$

We are to raise the indices of this form relative to the shape metric $ds_B^2 = \frac{\|dw\|^2}{2\|w\|}$ in order to convert this one-form to a vector field. Raising can be done by first raising indices relative to the Euclidean metric and then muliplying the result by the conformal factor $2\|w\|$. Euclidean index raising yields the vector field $-\frac{1}{2|w|^3} w \times \dot{w}$, where $w \times \dot{w}$ is the usual cross-product between w and $\dot{w}$. Multiplying by the conformal factor then yields $F(\dot{w}, \cdot)^\# = \frac{1}{|w|^2} w \times \dot{w}$ so that the curvature term is $-\frac{J}{|w|^2} w \times \dot{w}$.

The remaining terms of Equation (B.15) are easily translated to w-coordinates. Add them to the curvature term to obtain Equation (0.71).

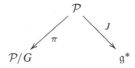

Figure B.1 The diagram defining reduction.

Remark B.12   The curvature two-form (B.17) is that of a magnetic monopole with source the origin. To see this, translate the two-form into a vector field using the standard duality between two-forms and vector fields on $\mathbb{R}^3$ induced by the volume form $dw_1 \wedge dw_2 \wedge dw_3$, and we arrive at the magnetic field B of a Dirac monopole of strength $1/2$ at the origin: $B = \frac{1}{2} \frac{w}{|w|^3}$.

B.6 Symplectic Reduction

The momentum map (Equation (A.7)) allows us to construct a refined version of Poisson reduction known as symplectic reduction. Consider the diagram shown in Figure B.1.

Fix a $\mu \in \mathfrak{g}^*$. Form the momentum level set $J^{-1}(\mu) \subset \mathcal{P}$ and project it to the Poisson quotient to form $\pi(J^{-1}(\mu))$.

Definition B.13   The symplectic reduced space at the value $\mu \in \mathfrak{g}^*$ is $\pi(J^{-1}(\mu))$.

This definition is equivalent to the original definition of Meyer [122] and Marsden and Weinstein [118].

Definition B.14   The symplectic reduced space is $J^{-1}(\mu)/G_\mu$, where $G_\mu = \{g \in G : Ad^*_g \mu = \mu\}$ is the isotropy subgroup of the frozen μ.

To see that these two definitions are equivalent, imagine the general situation of a Lie group G acting on a manifold Z with quotient map $\pi : Z \to Z/G$. Take a subset $X \subset Z$. Note that $\pi(X) = \pi(GX) = GX/G$. Form the subgroup $G_X = \{g \in G : gX = X\}$, where the $gX = X$ means equality as sets. I would like to say that $X/G_X = GX/G$. See Figure B.2. This equality is true in the case that $X = F^{-1}(\mu)$, where $F : Z \to \mathbb{V}$, $\mu \in \mathbb{V}$, and F is a G-equivariant map into a G-space $\mathbb{V}$. The relevant facts are that in this case $G_X = G_\mu := \{g \in G : g \cdot \mu = \mu\}$ and that if $p \in X$ and $g \in G$ then we have the equivalence $p' = gp \in X \iff g \in G_\mu$.

Remark B.15   If G is Abelian, then the co-adjoint action is trivial so $G_\mu = G$ regardless of μ. It follows that the reduced spaces all have the form $J^{-1}(\mu)/G$.

The reduced space can fail to be a manifold for several reasons. The subvariety $X = J^{-1}(\mu)$ may not be a manifold. The quotient of a manifold by a Lie group action need not be a manifold. We can guarantee that the reduced space is a manifold in a specific case.

Theorem B.16   *If μ is a regular value for J and if G_μ acts properly and freely on $J^{-1}(\mu)$, then the reduced space $J^{-1}(\mu)/G_\mu$ is a symplectic manifold. If it is connected, then it forms a symplectic leaf within $\mathcal{P}/G$.*

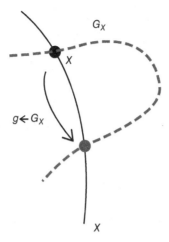

Figure B.2 Why $J^{-1}(\mu)/G_\mu = \mathcal{P}/G$. Take $X = J^{-1}(\mu)$ as indicated by the solid curve. Its G-orbit GX is all of $\mathcal{P}$, the whole background in the figure. The G-orbit, Gx (dotted curve), of a point $x \in X$ re-intersects X at the points $G_X x$, where $G_X \subset G$ is the stabilizer of the set X. The quotient space GX/X is thus equal to X/G_X.

When the G action is not free both the Poisson and symplectic reduced spaces may be singular algebraic varieties. All kinds of interesting complications come into play. See [47] and [109]. This situation arises in the N-body problem in d dimensions when $d \geq 3$ for any $N \geq 3$.

B.7 Linear Symplectic Reduction

A symplectic vector space $\mathbb{V}$ is a real vector space endowed with a constant symplectic form ω_0. If $\mathbb{L} \subset \mathbb{V}$ is a linear subspace, then by its reduction we mean the quotient space $\mathbb{L}/ker(\omega_\mathbb{L})$, where $\omega_\mathbb{L}$ is the restriction of ω to $\mathbb{L}$ and its kernel consists of all $v \in \mathbb{L}$ such that $\omega(v,w) = 0$ for all $w \in \mathbb{L}$. Equivalently, $ker(\omega_\mathbb{L}) = \mathbb{L} \cap \mathbb{L}^\perp$, where by $\mathbb{L}^\perp \subset \mathbb{V}$ we mean the ω-perpendicular to $\mathbb{L}$; that is, the set of all $v \in \mathbb{V}$ such that $\omega(v,w) = 0$ for all $w \in \mathbb{L}$. ($\mathbb{L}^\perp$ is a subspace of $\mathbb{V}$ whose dimension equals the codimension of $\mathbb{L}$, but, unlike with inner product perpendiculars, this symplectic perpendicular need not form a complementary subspace. Indeed $\mathbb{L} = \mathbb{L}^\perp$ can happen.) The form ω induces a linear symplectic form on the quotient in the standard way.

B.8 Local Slices and Nonlinear Symplectic Reduction

Moving on to the nonlinear world, if $(\mathcal{P}, \omega)$ is a symplectic manifold and $S \subset \mathcal{P}$ a submanifold, and if we write $j: S \to P$ for the inclusion then $\omega_S = j^*\omega$ is a closed two-form on S whose rank may vary. If this rank is constant, then the kernel of ω_S defines an involutive distribution on S, so we can form the local leaf space $S/ker\omega_S$. This local quotient space is the local symplectic reduction of S.

Words are in order regarding the meaning of the "local quotient" of a manifold M by an involutive distribution D. The Frobenius normal theorem asserts the following normal form for D: Every point of M is contained in a coordinate neighborhood $U \cong \mathbb{R}^k \times \mathbb{R}^\ell$, where k is the rank of D and under which the distribution D corresponds to the $\mathbb{R}^k$ directions. Thus, there are coordinates $u^a, v_i, a = 1, \ldots, k, i = 1, \ldots, \ell$ with respect to which D is defined by $dv_i = 0, i = 1, \ldots, \ell$. The plaques (local leaves) are the integral manifolds $v_i = const_i$ for $D|_U$, which is to say the image of any one of the k-planes $\mathbb{R}^k \times pt$ under the Frobenius normal forms a diffeomorphism. Whenever a leaf of D intersects U it does so in a union of these plaques. The plaques themselves are the connected components of the sets obtained by intersecting a leaf with U. In particular, the quotient $U/D|_U$ consists of the space of plaques in U and is diffeomorphic to $\mathbb{R}^\ell$ as coordinatized by the v_i. Any one of these spaces is what we mean by a local quotient of M by D.

As a primary example of this reduction process, suppose that H is a smooth function on the phase space $\mathcal{P}$ and that $c \in \mathbb{R}$ is a regular value for H. Then the set $S := \{H = c\} \subset \mathcal{P}$ is a smooth embedded hypersurface and the kernel of ω_S is spanned by X_H. A local symplectic reduction of $\{H = c\}$ is a space whose points are local integral curves for X_H that lie within $\{H = c\}$. To form the local quotient, invoke the symplectic version of the straightening lemma,[1] which yields canonical coordinates $q^1, p_1, q^2, p_2, \ldots, q^n, p_n$ such that $H = p_1$ and those under which the local integral curves of X_H are the q^1-curves. It follows that $q^2, p_2, \ldots, q^n, p_n$ coordinatize the local symplectic quotient.

More viscerally, choose any disc $D \subset S = \{H = c\}$ transverse to X_H. This disc must be symplectic: The restriction of ω to the disc is symplectic, since $ker(\omega_S)$ is spanned by X_H. Use the flow of X_H to form a tubular neighborhood of D diffeomorphic to $D \times (-\epsilon, \epsilon)$ with the $(-\epsilon, \epsilon)$ direction corresponding to the X_H direction. This process realizes the local symplectic quotient $S/ker(\omega_S)$ as D. In Section 2.3.3 we presented this same construction in order to realize the Poincaré map at constant energy as a map of a symplectic disc to itself.

B.9   Symplectic Reduction, Redux

We now incorporate the above discussion of kernels of the restrictions of symplectic forms into our understanding of symplectic reduced spaces. We gave the definition of the symplectic reduced space above in Definition B.14. Write $i_\mu \colon J^{-1}(\mu) \to \mathcal{P}$ for the inclusion and $\pi_\mu \colon J^{-1}(\mu) \to J^{-1}(\mu)/G_\mu$ for the projection. Then the reduced symplectic form ω_μ on the quotient $J^{-1}(\mu)/G_\mu$, which is to say on the symplectic reduced space, is determined by

$$i_\mu^* \omega = \pi_\mu^* \omega, \tag{B.19}$$

where ω is the symplectic form on the ambient phase space $\mathcal{P}$. This relation assumes that $J^{-1}(\mu)$ is an embedded submanifold and that the quotient is a manifold with π_μ a submersion.

---

[1] See https://en.wikipedia.org/wiki/Straightening_theorem_for_vector_fields and [2, p.67].

There is a close relation between μ being a regular value for J and the G-action being free in a neighborhood of $J^{-1}(\mu)$. A group action of a Lie group G on a manifold $\mathcal{P}$ is called *locally free* at $z \in \mathcal{P}$ if the infinitesimal generator map $\mathfrak{g} \to T_x\mathcal{P}$ is injective.

Exercise B.17 Show that μ is a regular value of the momentum map J if and only if the G-action on $\mathcal{P}$ is locally free at points $z \in J^{-1}(\mu)$.
Show that "free" implies "locally free."

In particular, if G acts freely on $\mathcal{P}$ then each value $\mu \in \mathfrak{g}^*$ is a regular value for J and hence each $J^{-1}(\mu)$ is a smooth embedded submanifold of $\mathcal{P}$.

Exercise B.18 Show that the leaves of the kernel of the two-form $i_\mu^*\omega$ on $J^{-1}(\mu)$ consist of the connected components of the orbits of the stabilizer G_μ.

Now suppose that G_μ acts on $J^{-1}(\mu)$ in such a way that the quotient space $J^{-1}(\mu)/G_\mu$ is a manifold with the quotient projection a smooth submersion. It follows from Exercise B.18 and the considerations of Section B.8 that the coordinate charts of the quotient realize the local symplectic quotients and, in particular, are symplectic manifolds. The global two-form ω_μ on the quotient defined by Equation (B.19) is the form that realizes this local quotient. If G is compact and acts freely, then the same is true of G_μ and automatically this hypothesis holds. The symplectic reduced space $J^{-1}(\mu)/G_\mu$ is indeed a smooth symplectic manifold with symplectic form ω_μ.

Appendix C   The Three-Body Problem and the Shape Sphere

In this appendix we derive facts and formulae regarding the w-coordinates on the shape space for the planar three-body problem. These facts are used in Section 0.4.2 in the introductory chapter. We also give a seperate derivation of the reduced three-body equation (0.73) valid for the case when the angular momentum is zero. It is an edited version of my article [150], with a section on Heron's formula added and referrals to the text made in some places.

C.1 Complex Variables and Mass Metric

Consider the planar three-body problem with plane of motion P. Choosing xy axes for P allows us to identify the plane with the complex number line $\mathbb{C}$ by sending a point $(x, y) \in P$ to the complex number $q = x + iy \in \mathbb{C}$. *The advantage of complex notation is that rotation corresponds to multiplication by a complex number of unit modulus.* In other words, rotation is given by

$$q \mapsto uq, u = \exp(i\theta),$$

where u is a unit complex number so that θ is real. The number θ is the radian measure of the amount of rotation. The set u of all unit complex numbers forms the circle group, denoted S^1.

We are now in the realm of Euclidean plane geometry. The locations $q_i \in \mathbb{C}$ of the three masses form the vertices of a Euclidean triangle.

Definition C.1 (Configuration space; located triangles)   A planar triangle is represented by a vector

$$\mathbf{q} = (q_1, q_2, q_3) \in \mathbb{C}^3,$$

whose components are the locations of the three vertices in the plane. We call $\mathbf{q}$ a configuration or *located triangle* and $\mathbb{C}^3$ the space of *located triangles*, or the *configuration space* for the three-body problem.

Definition C.2 (Mass metric) The mass metric on configuration space is the Hermitian inner product

$$\langle \mathbf{v}, \mathbf{w} \rangle = m_1 \bar{v}_1 w_1 + m_2 \bar{v}_2 w_2 + m_3 \bar{v}_3 w_3. \tag{C.1}$$

Using the mass metric, we have that

$$K(\dot{\mathbf{q}}) = \frac{1}{2} \langle \mathbf{q}, \mathbf{q} \rangle := \frac{1}{2} \Sigma m_i |\dot{q}_i|^2 \tag{C.2}$$

is the usual kinetic energy of a motion. Here, $\mathbf{q} = (\dot{q}_1, \dot{q}_2, \dot{q}_3) \in \mathbb{C}^3$ is the vector representing the velocities of the three masses. We also consider the gravitational potential energy

$$V(\mathbf{q}) = -\left\{ \frac{m_1 m_2}{r_{12}} + \frac{m_2 m_3}{r_{23}} + \frac{m_1 m_3}{r_{13}} \right\}. \tag{C.3}$$

Then

$$H(\mathbf{q}, \mathbf{q}) = K(\mathbf{q}) + V(\mathbf{q}) \tag{C.4}$$

is the energy of a motion $\mathbf{q}(t)$. In addition to the energy H we will make use of several other important functions. The *moment of inertia*,

$$I(\mathbf{q}) = \langle \mathbf{q}, \mathbf{q} \rangle = \Sigma m_i |q_i|^2, \tag{C.5}$$

measures the overall *size* of a located triangle. The *angular momentum*,

$$J = Im(\langle \mathbf{q}, \mathbf{q} \rangle) = m_1 q_1 \wedge \dot{q}_1 + m_2 q_2 \wedge \dot{q}_2 + m_3 q_3 \wedge \dot{q}_3, \tag{C.6}$$

measures the instantaneous spin of a triangle. The linear momentum,

$$P = \langle \mathbf{q}, \mathbf{1} \rangle = \Sigma m_i \dot{q}_i \in \mathbb{C}, \tag{C.7}$$

measures the instantaneous rate of translation of the entire three-body system. The *center of mass* of the located triangle $\mathbf{q}$ is

$$q_{cm} = \langle \mathbf{q}, \mathbf{1} \rangle / \langle \mathbf{1}, \mathbf{1} \rangle = (m_1 q_1 + m_2 q_2 + m_3 q_3)/(m_1 + m_2 + m_3) \in \mathbb{C}. \tag{C.8}$$

In the formula for angular momentum (Equation (C.6)) we used the notation

$$(x + iy) \wedge (u + iv) = det \begin{pmatrix} x & y \\ u & v \end{pmatrix} = xv - yu, \tag{C.9}$$

which is also equal to $Im(\bar{z}w)$ for $z = x + iy, w = u + iv \in \mathbb{C}$. This wedge operation $z, w \mapsto z \wedge w$ is the planar version of the cross product. If $\times$ denotes the usual cross product of vectors in $\mathbb{R}^3$, then $(x, y, 0) \times (u, v, 0) = (0, 0, z \wedge w)$, so that J is the 3rd component of the usual angular momentum of physics. In the formulae for linear momentum (Equation (C.7)) and center of mass (Equation (C.8)) we used the constant vector $\mathbf{1} = (1, 1, 1) \in \mathbb{C}^3$, which generates translations.

Definition C.3 (Phase space) The phase space of the planar three-body problem is $\mathbb{C}^3 \times \mathbb{C}^3$. Its points are written $(\mathbf{q}, \dot{\mathbf{q}})$. The first vector $\mathbf{q} \in \mathbb{C}^3$ represents the located triangle, that is to say, the positions of its three vertices. The second vector $\dot{\mathbf{q}}$ represents the velocities of these three vertices.

Here, H and J are functions on phase space, and I and q_{cm} are also functions on phase space, but functions that are independent of $\dot{\mathbf{q}}$ so can be thought of as functions on configuration space alone. The linear momentum is another function on phase space, but now one that is independent of position $\mathbf{q}$.

Definition C.4 A function $f : \mathbb{C}^3 \times \mathbb{C}^3 \to \mathbb{R}$ or $F : \mathbb{C}^3 \times \mathbb{C}^3 \to \mathbb{C}$ is *conserved* if its value is constant along any solution $\mathbf{q}(t)$ to the system of ODEs (0.4). (Different solutions typically yield different values for this constant.)

Proposition C.5 (Conservation laws) *The energy H, the angular momentum J, and the linear momentum P are* **conserved**.

The mass metric formalism yields a simple proof of this proposition. A complex vector space such as $\mathbb{C}^3$ becomes a real vector space when we only allow scalar multiplication by real scalars. The real part $\langle \cdot, \cdot \rangle_{\mathbb{R}}$ of the Hermitian mass inner product $\langle \cdot, \cdot \rangle$ of Equation (C.1) is a real inner product on $\mathbb{C}^3$. A real inner product induces a gradient operator, which sends smooth real-valued functions $W : \mathbb{C}^3 \to \mathbb{R}$ to smooth real vector fields $\nabla W : \mathbb{C}^3 \to \mathbb{C}^3$ according to the rule

$$\frac{d}{d\epsilon} W(\mathbf{q} + \epsilon \mathbf{h})|_{\epsilon=0} = \langle \nabla W(\mathbf{q}), \mathbf{h} \rangle_{\mathbb{R}}. \tag{C.10}$$

In terms of real linear orthogonal (not neccessarily orthonormal!) coordinates ξ^j, $j = 1, \ldots, 6$ for $\mathbb{C}^3$, the gradient ∇W is a variation of the usual coordinate formula from vector calculus. Namely $(\nabla W)_j = \frac{1}{c_j} \frac{\partial W}{\partial \xi^j}$ where $c_j = \langle E_j, E_j \rangle$. Here, the linear coordinates ξ^j are related to an orthogonal basis E_j for $\mathbb{C}^3$ as per usual: $\mathbf{q} = \Sigma_{j=1}^6 \xi^j E_j$. We will take the ξ^j to come in pairs (x_j, y_j) as per $q_j = x_j + i y_j$, so that the c_j are then equal to the m_j in pairs and we get the components of our gradient:

$$(\nabla V)_j = \frac{1}{m_j} \left(\frac{\partial V}{\partial x_j}, \frac{\partial V}{\partial y_j} \right) = \frac{1}{m_j} \left(\frac{\partial V}{\partial x_j} + i \frac{\partial V}{\partial y_j} \right).$$

Exercise C.6

1. Show that Newton's equations (0.4) can be rewritten as

$$\ddot{\mathbf{q}} = -\nabla V(\mathbf{q}), \quad \text{where } \nabla = \text{gradient for the mass metric}. \tag{C.11}$$

2. Use part 2 to prove Proposition C.5.
3. Show that if $P = 0$ and $q_{cm}(0) = 0$ then $q_{cm}(t) = 0$ for all time t.
4. Show that the moment of inertia $I(t) = I(\mathbf{q}(t))$ evolves along a solution to Equations (0.4) according to the Lagrange–Jacobi equation

$$\ddot{I} = 4H - 2V(\mathbf{q}).$$

C.2 Shape Space: Main Theorem

We seek reduced equations for the planar three-body problem: ODEs encoding the problem as dynamics on the space of congruence classes of triangles. Elementary geometry asserts that this space is three-dimensional, with the triangle's side lengths serving as coordinates. So we expect ODEs in these lengths. However, the collinear triangles form the boundary of the space of congruence classes of triangles and as a result any ODE in side lengths alone will exhibit a singularity along the space of collinear triangles. But Newton's equations (0.4) exhibit no such problem. The Earth, Moon, and Sun line up without creating singularities in the dynamics. In order to achieve what we seek, we must strengthen the notion of congruence by excluding reflections as being allowed congruences.

Definition C.7 The group G of rigid motions of the plane is the group of orientation-preserving isometries of the plane.

Every element of G is a composition of a rotation and a translation.

Definition C.8 Two planar triangles (possibly degenerate) are *oriented congruent* if there is a rigid motion taking one triangle to the other.

Definition C.9 (Shape space) Shape space is the space of oriented congruence classes of planar triangles endowed with the quotient metric.

Replacing the equivalence relation of congruence by oriented congruence removes the boundary from the resulting space of equivalence classes (see Theorem C.10).

Some words are in order regarding the meaning of "quotient metric" in the definition of shape space. The mass metric gives our space $\mathbb{C}^3$ of located triangles a norm $\| \cdot \|$ under which the distance between located triangles $\mathbf{q}, \mathbf{q}'$ is $\|\mathbf{q} - \mathbf{q}'\|$. Here, G acts on $\mathbb{C}^3$ by isometries relative to this distance. Denote the result of applying $g \in G$ to $\mathbf{q} \in \mathbb{C}^3$ by $g\mathbf{q}$. We define the shape space metric d by

$$d([\mathbf{q}], [\mathbf{q}']) = \inf_{g_1, g_2 \in G} \|g_1 \mathbf{q} - g_2 \mathbf{q}'\|. \tag{C.12}$$

Here $[\mathbf{q}]$ and $[\mathbf{q}']$ denote the *shapes*, or oriented congruence classes, of the located triangles $\mathbf{q}$ and $\mathbf{q}' \in \mathbb{C}^3$.

Theorem C.10 *(See Figure C.1.) The shape space $Shape(2, 3)$ for the planar three-body problem is homeomorphic to $\mathbb{R}^3$. The quotient map*

$$\pi \colon \mathbb{C}^3 \to \mathbb{R}^3, \qquad \pi(q) = w = (w_1, w_2, w_3),$$

which assigns to each located triangle q its shape w is the composition $\pi = \pi^{rot} \circ \pi_{tr}$ of the complex linear projection $\pi_{tr} \colon \mathbb{C}^3 \to \mathbb{C}^2$ described by Equation (C.15) and the real quadratic homogeneous map $\pi^{rot} \colon \mathbb{C}^2 \to \mathbb{R}^3$ described by Equation (C.22). This quotient map enjoys the following properties:

(A) $\pi(q) = \pi(q') \iff q, q'$ are oriented congruent.
(B) π is onto.
(C) π projects the triple collision locus onto the origin.
(D) Up to a mass-dependent constant, w_3 is the signed area of the triangle with shape w, so that $w_3 = 0$ defines the locus of collinear triangles.
(E) Let $\sigma \colon \mathbb{R}^3 \to \mathbb{R}^3$ be the reflection across the collinear plane; that is, $\sigma(w_1, w_2, w_3) = (w_1, w_2, -w_3)$. Then the two triangles $q_1, q_2 \in \mathbb{C}^3$ are congruent if and only if either $\pi(q) = \pi(q')$ or $\pi(q) = \sigma(\pi(q'))$.
(F) $w_1^2 + w_2^2 + w_3^2 = (\frac{1}{2}I)^2$, where $I = \langle q, q \rangle$ (see Equation (C.5)).
(G) $R = \sqrt{I}$ is the shape space distance to triple collision.

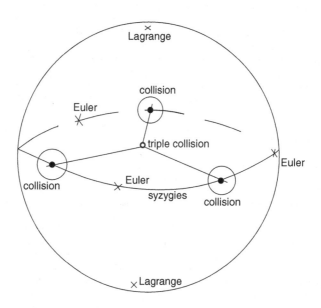

Figure C.1 The shape sphere, centered on triple collision. From [150].

Remark C.11   Parts (D) and (E) of Theorem C.10 say that the space of congruence classes of triangles can be identified with the closed half space $w_3 \geq 0$ of $\mathbb{R}^3$ and that the plane $w_3 = 0$ of collinear triangles forms its boundary, as claimed at the beginning of this section. The restriction of π^{rot} to the three-sphere $I = 1$ is the famous Hopf fibration $\mathbb{S}^3 \to \mathbb{S}^2$.

Iwai [86, 87, 88] was the first, as far as I am aware, to state an explicit version of this theorem in the context of the three-body problem. Earlier researchers certainly knew the theorem.

C.3 The Metric and the Shape Sphere

Although shape space is homeomorphic to $\mathbb{R}^3$ it is not isometric to $\mathbb{R}^3$: Shape space geometry is not Euclidean. However, the geometry does have spherical symmetry. Each sphere centered at triple collision is isometric to the standard sphere, up to a scale factor. We identify these spheres with the *shape sphere*.

Add scalings to the group G of rigid motions and we get the group of orientation-preserving similarities whose elements are compositions of rotations, translations, and scalings.

Definition C.12   Two planar triangles are *oriented similar* if there is an orientation-preserving similarity taking one to the other.

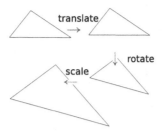

Figure C.2 Translating, rotating and scaling a triangle. From [150].

Definition C.13 The *shape sphere* is the resulting quotient space of the space of located triangles $\mathbb{C}^3 \setminus \mathbb{C}1$, after the triple collisions $\mathbb{C}1$ have been deleted.

In other words, the shape sphere is the space of oriented similarity classes of planar triangles where we do not allow all three vertices of the triangle to coincide. Now $\pi(\lambda \mathbf{q}) = \lambda^2 \pi(\mathbf{q})$ for λ real. It follows that the shape sphere can be realized as the space of rays through the origin in $\mathbb{R}^3$. This space of rays can in turn be identified with the unit sphere $\|\mathbf{w}\| = 1$ within shape space. Various special types of triangles, including the five families of solutions of Euler and Lagrange are encoded on this sphere as indicated in Figure C.1.

C.4 Forming the Quotient: Proving Theorem C.10

A vector in $\mathbb{C}^3$ represents a located triangle with its three components representing the triangle's three vertices. Translation of a triangle $\mathbf{q} = (q_1, q_2, q_3) \in \mathbb{C}^3$ by $c \in \mathbb{C}$ sends $\mathbf{q}$ to the located triangle $\mathbf{q} + c\mathbf{1}$, where $\mathbf{1} = (1, 1, 1)$. Rotation by θ radians about the plane's origin sends $\mathbf{q}$ to $e^{i\theta}\mathbf{q} = (e^{i\theta}q_1, e^{i\theta}q_2, e^{i\theta}q_3)$. Scaling the plane by a positive factor ρ corresponds to multiplication by the real number ρ and so sends the triangle $\mathbf{q}$ to $\rho\mathbf{q} = (\rho q_1, \rho q_2, \rho q_3)$. See Figure C.2.

Shape space is the quotient of $\mathbb{C}^3$ by the action of the group G generated by translation and rotation. We form this quotient in two steps: translation, then rotation.

C.5 Dividing by Translations

We divide by translations by using the isomorphism

$$\mathbb{C}^3/\mathbb{C}1 \cong \mathbb{C}1^\perp,$$

which is a special case of

$$\mathbb{E}/S \cong S^\perp,$$

and is valid for any finite-dimensional complex vector space $\mathbb{E}$ with a Hermitian inner product and any complex linear subspace $S \subset \mathbb{E}$. This isomorphism is a metric isomorphism. Here, $\mathbb{E}/S$ inherits a Hermitian inner product whose distance is given by Equation (C.12) with the group G replaced by S acting on $\mathbb{E}$ by translation, and with

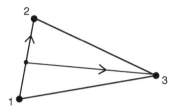

Figure C.3 Jacobi vectors (usually called Jacobi coordinates). From [150].

the elements q_i in that formula being elements of $\mathbb{E}$. In the isomorphism, the metric we use on $S^\perp$ is the restriction of the metric from $\mathbb{E}$.

In our situation S is the span of $\mathbf{1}$. Define

$$\mathbb{C}_0^2 := \mathbf{1}^\perp = \{\mathbf{q} : m_1 q_1 + m_2 q_2 + m_3 q_3 = 0\}$$

to be the set of planar three-body configurations whose center of mass is at the origin. This two-dimensional complex space represents the quotient space of $\mathbb{C}^3$ by translations.

C.6 Jacobi Coordinates: Diagonalizing the Mass Metric

We will need coordinates that diagonalize the restriction of the mass metric to $\mathbb{C}_0^2$. The quadratic form associated to the mass metric is the moment of inertia $I = \langle \mathbf{q}, \mathbf{q} \rangle$ of Equation (C.5). Thus we look for coordinates Z_1, Z_2 on $\mathbb{C}_0^2$ such that

$$I = |Z_1|^2 + |Z_2|^2 \quad \text{whenever } \mathbf{q} \perp \mathbf{1}. \tag{C.13}$$

These coordinates are traditionally attributed to Jacobi, despite having been found earlier, and explained more clearly by Lagrange [99, p. 292].

Exercise C.14 Show that $\mathbf{1} = (1,1,1), E_1 = \left(\frac{1}{m_1}, -\frac{1}{m_2}, 0\right)$, and $E_2 = \left(\frac{-1}{m_1+m_2}, \frac{-1}{m_1+m_2}, \frac{1}{m_3}\right)$ form an orthogonal (not orthonormal) basis relative to the mass metric on $\mathbb{C}^3$.

The corresponding coordinates $\langle \mathbf{q}, \mathbf{1} \rangle, \langle \mathbf{q}, E_1 \rangle, \langle \mathbf{q}, E_2 \rangle$ are orthogonal coordinates for $\mathbb{C}^3$.

Definition C.15 The coordinates $\langle q, E_1 \rangle = q_1 - q_2 := Q_{12}$ and $\langle q, E_2 \rangle = q_3 - \frac{m_1 q_1 + m_2 q_2}{m_1 + m_2}$ are *Jacobi coordinates* for $\mathbb{C}_0^2 := \{q \in \mathbb{C}^3 : q_{cm} = 0\}$.

Jacobi coordinates are indicated in Figure C.3.

Normalizing the Jacobi coordinates yields our desired unitary diagonalizing coordinates $Z_i = \langle \mathbf{q}, e_i \rangle, i = 1, 2$, for $\mathbb{C}_0^2$ where $e_i = E_i / \|E_i\|$. We compute

$$Z_1 = \mu_1(q_1 - q_2), \qquad Z_2 = \mu_2 \left(q_3 - \frac{m_1 q_1 + m_2 q_2}{m_1 + m_2}\right), \tag{C.14}$$

with $\frac{1}{\mu_1^2} = \|E_1\|^2 = \frac{1}{m_1} + \frac{1}{m_2}$ and $\frac{1}{\mu_2^2} = \|E_2\|^2 = \frac{1}{m_3} + \frac{1}{m_1+m_2}$. These normalized Jacobi coordinates define the complex linear projection

$$\pi_{tr} : \mathbb{C}^3 \to \mathbb{C}^2, \qquad \pi_{tr}(q_1, q_2, q_3) = (Z_1, Z_2), \tag{C.15}$$

which realizes the metric quotient of $\mathbb{C}^3$ by translations.

C.7 Dividing by Rotations

It remains to divide $\mathbb{C}_0^2$ by the action of the rotation group S^1. A rotation by θ radians acts on the triangle's vertices q_j by $q_j \mapsto e^{i\theta}q_j$, and so it acts on the normalized Jacobi coordinates by $(Z_1, Z_2) \mapsto (e^{i\theta}Z_1, e^{i\theta}Z_2)$. We want to understand the resulting equivalence classes under rotation.

The functions $Z_i \bar{Z}_j, i, j = 1, 2$, are invariant under rotation. Put these functions into a two by two Hermitian matrix,

$$\Phi(Z_1, Z_2) = \begin{pmatrix} |Z_1|^2 & Z_1 \bar{Z}_2 \\ \bar{Z}_1 Z_2 & |Z_2|^2 \end{pmatrix} = A, \tag{C.16}$$

or

$$\Phi(\mathbf{Z}) = \mathbf{Z}^T \bar{\mathbf{Z}}, \tag{C.17}$$

where $\mathbf{Z} = (Z_1, Z_2)$,

$$\mathbf{Z}^T = \begin{pmatrix} Z_1 \\ Z_2 \end{pmatrix} \text{ and } \bar{\mathbf{Z}} = (\bar{Z}_1, \bar{Z}_2).$$

From the factorization (C.17) we see that $\Phi(\mathbf{Z}) \mathbf{Z}^T = (|Z_1|^2 + |Z_2|^2) \mathbf{Z}^T$, while $\Phi(\mathbf{Z}) \mathbf{W}^T = 0$ for $\mathbf{W} \perp \mathbf{Z}$. Thus $\Phi(Z_1, Z_2)$ is the matrix of orthogonal projection onto the complex line spanned by $\mathbf{Z}$ multiplied by $\|\mathbf{Z}\|^2$. Now, two nonzero vectors $\mathbf{Z}, \mathbf{U}$ are related by rotation if and only if they span the same complex line and their lengths are equal. It follows that the image of Φ represents the quotient space $\mathbb{C}^2/S^1$, and Φ can be considered as the quotient map. We have just seen that the image of Φ consists of the Hermitian matrices of rank 1 whose nonzero eigenvalue is positive (corresponding to $\|\mathbf{Z}\|^2$), together with the zero matrix (corresponding to $\mathbf{Z} = 0$). In terms of the determinant and trace, these conditions on A are $\det(A) = 0$ and $tr(A) \geq 0$. Coordinatize Hermitian matrices as

$$A = \begin{pmatrix} w_4 + w_1 & w_2 + iw_3 \\ w_2 - iw_3 & w_4 - w_1 \end{pmatrix} \text{ for } w_j \text{ real}, \tag{C.18}$$

so that $\det(A) = w_4^2 - w_1^2 - w_2^2 - w_3^2$ and $tr(A) = 2w_4$. The discussion we have just had proves the following proposition.

Proposition C.16 *The image of the map Φ is the cone of two-by-two Hermitian matrices A (Equation (C.18)) satisfying*

$$w_4^2 - w_1^2 - w_2^2 - w_3^2 = 0 \tag{C.19}$$

and

$$w_4 \geq 0. \tag{C.20}$$

This cone realizes the quotient $\mathbb{C}^2/S^1$ with Φ implementing the quotient map $\mathbb{C}^2 \to \mathbb{C}^2/S^1$.

Now map the real four-dimensional space of Hermitian matrices onto $\mathbb{R}^3$ by projecting out the trace w_4:

$$(w_1, w_2, w_3, w_4) \mapsto pr(w_1, w_2, w_3, w_4) = (w_1, w_2, w_3).$$

The restriction of this linear projection to the cone given by Equations (C.19) and (C.20) is a homeomorphism onto $\mathbb{R}^3$. Indeed, solve the cone equations for w_4 to find $w_4 = +\sqrt{w_1^2 + w_2^2 + w_3^2}$, and hence the inverse of the restricted projection is

$$(w_1, w_2, w_3) \mapsto \left(w_1, w_2, w_3, \sqrt{w_1^2 + w_2^2 + w_3^2} \right).$$

We have proved the following proposition.

Proposition C.17 *The map*

$$\pi^{rot} = pr \circ \Phi \colon \mathbb{C}^2 \to \mathbb{R}^3, \tag{C.21}$$

given by

$$\pi^{rot}(Z_1, Z_2) = \left(\frac{1}{2}(|Z_1|^2 - |Z_2|^2), Re(Z_1 \bar{Z}_2), Im(Z_1 \bar{Z}_2) \right) = (w_1, w_2, w_3), \tag{C.22}$$

realizes $\mathbb{R}^3$ as the quotient space of $\mathbb{C}^2$ by the rotation group S^1.

Remark C.18 The restriction of the map (C.21) to the sphere $w_4 = 1$ is the famous *Hopf map* from the three-sphere to the two-sphere.

Proof of Theorem C.10 We form $\pi = \pi^{rot} \circ \pi_{tr}$ by composing the linear projection π_{tr} of Equation (C.15) with the map π^{rot} shown in Equation (C.21). The first map realizes the quotient by translations and the second realizes the quotient by rotations, so together they realize the full quotient by the group of rigid motions. This establishes parts (A) and (B) of the theorem. Part (C), that the only triangles sent to $0 \in \mathbb{R}^3$ are the triple collision triangles $\mathbf{q} = (q, q, q)$, follows directly from the formulae for π^{rot} and π_{tr}. Indeed, the only point of $\mathbb{C}^2$ mapped to $0 \in \mathbb{R}^3$ by π^{rot} is the origin 0, and the only points of $\mathbb{C}^3$ mapped to $0 \in \mathbb{C}^2$ by π_{tr} are the triple collision triangles.

(D) says that w_3 is a constant times the oriented area of the triangle. We have $w_3 = -Z_1 \wedge Z_2$. The wedge of Equation (C.9) represents the oriented area of the parallelogram whose edges are $z = x + iy$ and $w = u + iv$. Thus the oriented area of our triangle is $\frac{1}{2}(Q_{21}) \wedge (Q_{31})$, where we write $Q_{ij} = q_i - q_j$ for the edge connecting vertex j to vertex i. We have $Z_1 = \mu_1 Q_{12}$ and $Z_2 = \mu_2(p_1 Q_{31} + p_2 Q_{32})$ where $p_1 = m_1/(m_1 + m_2)$ and $p_2 = m_2/(m_1 + m_2)$ so that $p_1 + p_2 = 1$. Use $Q_{12} + Q_{23} + Q_{31} = 0$ and $Q_{ij} = -Q_{ji}$ to compute that $Z_2 = \mu_2(Q_{31} - p_2 Q_{12})$. The wedge operator satisfies $Q \wedge Q = 0$ for any Q. It follows that $w_3 = -\mu_1 \mu_2 \frac{1}{2} Q_{12} \wedge Q_{31} = +\mu_1 \mu_2 \frac{1}{2} Q_{21} \wedge Q_{31}$, as desired. (D) follows immediately.

To establish part (E) regarding the operation of reflection on triangles, observe that we can reflect triangle **q** by changing all vertices q_i to $\bar{q}_i$, which in turn changes (Z_1, Z_2) to its conjugate vector $(\bar{Z}_1, \bar{Z}_2)$. This conjugation operation leaves w_1 and w_2 unchanged and changes w_3 to $-w_3$; the oriented area flips sign.

Part (F) is a computation. Observe from Equations (C.16, C.18) that $w_4 = \frac{1}{2}I$ and recall the cone condition given by Equation (C.19): $w_4^2 = w_1^2 + w_2^2 + w_3^2$. For part (G), see Section C.12, Equation (C.32). QED

C.8 Heron's Formula and the Light Cone

Heron's formula expresses the area of a triangle in terms of its sidelengths. We show here that Heron's formula defines the light cone (Equation (C.19)).

What is the intrinsic meaning of the ambient Lorentzian vector space $\mathbb{R}^4$ containing our light cone? The coordinates w_i that we've been using for the ambient Lorentzian space are S^1-invariant polynomials on $\mathbb{V}_{tr}$. Now a single coordinate w on a real vector space $\mathbb{E}$ is simply a linear function $w\colon \mathbb{E} \to \mathbb{R}$, which is to say $w \in \mathbb{E}^*$. A full set of coordinates, such as our w_1, w_2, w_3, w_4 is then a basis for the dual vector space $\mathbb{E}^*$. So the *dual space* to our ambient vector space is the vector space $\mathcal{P}$ of all real quadratic S^1-invariant polynomials on $\mathbb{V}$. Using the theorem $\mathbb{E}^{**} = \mathbb{E}$ about the double dual, we see that our ambient vector space is intrinsically identified with $\mathcal{P}^*$. Now each $v \in \mathbb{V}_{tr}$ defines an element of $\mathcal{P}^*$ by evaluation:

$$\delta_v(Q) = Q(v) \in \mathbb{R}, Q \in \mathcal{P}, v \in \mathbb{V}_{tr}.$$

This assignment is the intrinsic version of the Hopf map,

$$\Phi(v) = \delta_v.$$

Exercise C.19 Verify that the Hopf map is the coordinate expression of the map $v \mapsto \delta_v$ from $\mathbb{V}_{tr}$ to $\mathcal{P}^*$ when we use coordinates (Z_1, Z_2) on $\mathbb{V}_{tr}$ and w_1, w_2, w_3, w_4 on $\mathcal{P}$.

There are other bases for $\mathcal{P}$ besides the w_i. The squared side lengths and the signed area

$$s_1 = r_{23}^2, s_2 = r_{13}^2, s_3 = r_{12}^2, \text{ and } \Delta$$

are quadratic polynomials on $\mathbb{C}^3$ invariant under translation and rotation, and hence define rotation invariant quadratic polynomials on $\mathbb{V}_{tr}$. They are linearly independent, and hence yield another basis for $\mathcal{P}$. We call $\{s_1, s_2, s_3, \Delta\}$ the *ancient basis*. Heron's formula reads $p(p - r_{12})(p - r_{23})(p - r_{31}) = \Delta^2$ where $p = \frac{1}{2}(r_{12} + r_{23} + r_{31})$ is half the triangle's perimeter. A bit of algebra shows that Heron's formula is equivalent to

$$16\Delta^2 = 2s_1 s_2 + 2s_3 s_1 + 2s_2 s_3 - \left(s_1^2 + s_2^2 + s_3^2\right).$$

This is the equation of the light cone, Equation (C.19), in the ancient basis.

C.9 Mechanics via Lagrangians

One of our goals is to write out reduced equations that encode Newton's equations (0.4) as ODEs on shape space. The strategy for achieving this goal is to push the least action principle for the three-body problem down from the space $\mathbb{C}^3$ of located triangles to our shape space $\mathbb{R}^3$. We begin by stating the least action principle.

A classical mechanical system can be encoded by its *Lagrangian L* (see Appendix A),

$$L = K - V, \tag{C.23}$$

the difference of its kinetic (K) and potential (V) energies. (The energy is the sum $K + V$.) Integrating the Lagrangian over a path c in the configuration space Q of the mechanical system defines that path's *action*:

$$A[c] = \int_c L dt = \int_a^b L(c(t), \dot{c}(t)) dt.$$

In this last expression, the time interval $[a, b]$ parameterizes so that $c \colon [a, b] \to Q$. The *principle of least action* asserts that the curve c satisfies Newton's equations if and only if c minimizes A among all paths $\gamma \colon [a, b] \to Q$ for which $c(a) = \gamma(a)$ and $c(b) = \gamma(b)$.

The principle is not a theorem! It is a guiding principle. To turn the principle to a theorem for the three-body problem requires careful wording and more hypotheses.

Theorem C.20 *If a curve* $c \colon [0, T] \to \mathbb{C}^3$ *is collision-free on the open interval* $(0, T)$ *and minimizes the action among all curves sharing its endpoints* $c(0)$ *and* $c(T)$, *then that curve solves Newton's equations on* $(0, T)$. *Conversely, if* c *satisfies Newton's equations, then* c *is a critical point (but not necessarily a minimizer) of the action functional restricted to all curves sharing its endpoints.*

One of the beauties of the action principle is that it is coordinate-independent. If a path minimizes the action, then it does not matter what coordinate system we use to express that path: The path still minimizes the action and so satisfies the Euler–Lagrange equations in that coordinate system. We refer to Appendix A on geometric mechanics for more on the principle of least action and the Euler–Lagrange equations.

C.10 Reducing the Least Action Principle

The curves competing in the least action principle are subject to boundary conditions. In the principle as we stated it, the curves connect two fixed points. Replace the fixed points by fixed oriented congruence classes to get new boundary conditions. If we remember that an oriented congruence class is represented by a point of shape space, we arrive at an action principle for shape space.

Remark C.21 (Shape space action principle)   Fix two shapes $\mathbf{w}_0, \mathbf{w}_1$ in the shape space $\mathbb{R}^3$. Suppose that the curve $\mathbf{q}(t) \in \mathbb{C}^3$ for $0 \le t \le T$ minimizes the standard action (C.23) among all curves in the space $\mathbb{C}^3$ of located triangles that join the corresponding oriented congruence classes $\Sigma_0 = \pi^{-1}(\mathbf{w}_0), \Sigma_1 = \pi^{-1}(\mathbf{w}_1) \subset \mathbb{C}^3$ in time T. Then we will say that its projected curve $\mathbf{w}(t) = \pi(\mathbf{q}(t)) \in \mathbb{R}^3$ minimizes the *shape space action* among all curves connecting the endpoints $\mathbf{w}_0, \mathbf{w}_1$ in time T.

Consider an analogous change of boundary conditions for the simplest action functional in the plane, the length functional. Instead of minimizing the length of curves among all curves connecting two fixed points in the plane, replace the two points by two concentric circles Σ_0 and Σ_1. We know that the minimizer will be a radial line segment, perpendicular to both Σ_0 and Σ_1 at its endpoints. More generally, for a Lagrangian on $\mathbb{R}^n$ of the general form given by Equation (C.23) , if we replace the fixed endpoint minimization problem with the problem of minimizing the action among all curves connecting two given *subspaces* $\Sigma_0, \Sigma_1 \subset \mathbb{R}^n$, then we induce a derivative condition at the endpoints. Namely, extremal curves, in addition to satisfying the Euler–Lagrange equations, must hit their targets orthogonally: $\dot{c}(0) \perp \Sigma_0$ at $c(0)$ and $\dot{c}(T) \perp \Sigma_1$ at $c(T)$. We call this added condition *first variation orthogonality*.

Returning to our situation, $\Sigma_0 = \pi^{-1}(\mathbf{w}_0), \Sigma_1 = \pi^{-1}(\mathbf{w}_1) \subset \mathbb{C}^3 = \mathbb{R}^6$. We will interpret first variation orthogonality in mechanical terms, so Σ_0 is formed by applying variable rigid motions $g \in G$ to any single point $\mathbf{q}_0 \in \Sigma_0$. Let $g = g(t)$ be any smooth path in G and form the corresponding path $g(t)\mathbf{q}_0$ in Σ_0. Differentiating this path, and then alternately taking $g(t)$ to be a curve of translations or a curve of rotations, we see that the tangent space $T_{\mathbf{q}_0}\Sigma_0$ to Σ_0 at $\mathbf{q}_0$ is spanned by two subspaces, $\{\dot{c}1, \dot{c} \in \mathbb{C}\}$ for translations, and $\{i\dot{\theta}\mathbf{q}_0 : \dot{\theta} \in \mathbb{R}\}$ for rotations:

$$
\begin{aligned}
T_{\mathbf{q}_0}\Sigma_0 &= \text{infinitesimal rigid motions} \\
&= \text{(translational)} + \text{(rotational)} \\
&= span_{\mathbb{C}}\mathbf{1} + span_{\mathbb{R}}(i\mathbf{q_0}).
\end{aligned}
\tag{C.24}
$$

The first variation orthogonality condition is thus the condition that our extremal $\mathbf{q}(t)$ be orthogonal to both the translation and rotational spaces: $\langle \mathbf{1}, \mathbf{q}(0) \rangle = 0$ and $\langle i\dot{\theta}\mathbf{q}(0), \mathbf{q}(0) \rangle_{\mathbb{R}} = 0$. But as we saw in Equations (C.6, C.7), these orthogonality conditions are equivalent to the assertions that the linear and angular momentum are zero at $\mathbf{q}_0$. (The inner product used for orthogonality is the real part of the Hermitian mass inner product and $Im(\langle \mathbf{q_0}, \mathbf{q} \rangle) = Re(\langle i\mathbf{q_0}, \mathbf{q} \rangle)$.) We summarize this in Lemma C.22.

Lemma C.22 *The curve $q(t)$ in $\mathbb{C}^3$ is orthogonal to the oriented congruence class through $q_0 = q(0)$ if and only if its linear and angular momentum are zero at $t = 0$. (See Equations (C.6) and (C.7).)*

Now, if the curve $\mathbf{q}(t)$ of Lemma C.22 is an extremal for our shape space action principle, then it must satisfy the Euler–Lagrange equations, which are Newton's equations. Since linear and angular momentum are conserved for solutions to Newton's equations, we have that the linear and angular momentum are identically zero all along the curve. Equivalently, if an extremal curve is orthogonal to the group orbit Σ_0 through one of its points $\mathbf{q}(0)$, then it is orthogonal to the group orbit Σ_t through every one of its points $\mathbf{q}(t)$. We have established Proposition C.23.

Proposition C.23 *The extremals for the shape space action principle are precisely those solutions to Newton's equations whose linear and angular momentum are zero.*

Proposition C.23 suggests a strategy for finding a Lagrangian L_{shape} on shape space whose action minimization is equivalent to the shape space action principle. Break up kinetic energy into

$$K = \text{translational part} + \text{rotational part} + \text{shape part.} \qquad \text{(C.25)}$$

We have just agreed that the translation and rotational part of the kinetic energy must be zero along our shape extremals, corresponding to the fact that they are orthogonal to G-orbits. Let us denote the last term, the shape term of the kinetic energy, as K_{shape}. Thus,

$$L_{shape} = K_{shape} - V \qquad \text{(C.26)}$$

is the shape Lagrangian. It remains to express K_{shape} in terms of shape coordinates w_i and their time derivatives $\dot{w}_i$ and V in terms of the w_i.

C.11 Shape Kinetic Energy

The decomposition (C.25) applied to velocities yields the *Saari decomposition*:

$$\mathbf{q} = (\text{translational part} + \text{rotational part}) + \text{shape part}$$
$$= T_q(Gq) \qquad\qquad \oplus (T_q(Gq))^\perp$$
$$= \text{vertical} \qquad\qquad \oplus \text{horizontal.} \qquad \text{(C.27)}$$

(See Section 0.3.7, regarding this decomposition. Also see Saari [183] and Chenciner [29, p. 331].) In the differential geometry of bundles, such a splitting of tangent vectors is known as a *vertical–horizontal* splitting. The group directions $T_q(Gq)$ form the *vertical space*. The orthogonal complement $T_q(Gq)^\perp$ to the vertical space forms the *horizontal space*. This vertical–horizontal decomposition, which depends on the base point $\mathbf{q}$ at which velocities are attached, is orthogonal and leads to Proposition C.24.

Proposition C.24 *Suppose that the center of mass of our located triangle is zero. Then the Saari decomposition of kinetic energy, Equation (C.25), is*

$$K = \frac{1}{2}\frac{\|\boldsymbol{P}\|^2}{M} + \frac{1}{2}\frac{J^2}{I} + \frac{1}{2}\frac{\|\dot{w}\|^2}{I},$$

where $\dot{w} = \frac{d}{dt}\pi(q(t))$, $P = P(\boldsymbol{q})$, *and* $J = J(\boldsymbol{q},\boldsymbol{q})$ *are the linear and angular momenta (Equations (C.6) and (C.7)). In particular,*

$$K_{shape} = \frac{1}{2}\frac{\|\dot{w}\|^2}{I} \qquad \text{where } I = 2\|w\|. \qquad \text{(C.28)}$$

Remark C.25 Apologies to readers of the original American Mathematical Monthly version [150] for a typographic error. There I had $I = 2\sqrt{\|\mathbf{w}\|}$ instead of the correct $I = 2\|\mathbf{w}\|$.

Proof of Proposition C.24 A real basis for the two-dimensional translational part of the motion consists of 1 and $i1$. A real basis for the one-dimensional rotational part is $i\mathbf{q}$. The rotational part is orthogonal to the translational part since the center of mass is

$\frac{1}{M}\langle\mathbf{q},\mathbf{1}\rangle$ and has been set equal to zero. Hence, $\mathbf{1}, i\mathbf{1}$, and $i\mathbf{q}$ is an orthogonal basis for the vertical part, $T_q(Gq)$. Normalize to get the real orthonormal basis

$$e_1, e_2, e_3 = 1/\sqrt{M}, i\mathbf{1}/\sqrt{M}, i\mathbf{q}/\sqrt{I}, \qquad \text{where } M = \langle\mathbf{1},\mathbf{1}\rangle = m_1 + m_2 + m_3,$$

for the vertical space. Let $\mathbf{q} \in \mathbb{C}^3$ be an arbitrary vector based at the located triangle $\mathbf{q} \in \mathbb{C}^3$. Expand this vector out as an orthogonal direct sum to get a quantitative form of the Saari decomposition equations (C.27),

$$\mathbf{q} = \langle\mathbf{q}, e_1\rangle_{\mathbb{R}} e_1 + \langle\mathbf{q}, e_2\rangle_{\mathbb{R}} e_2 + \langle\mathbf{q}, e_3\rangle_{\mathbb{R}} e_3 + (\text{shape}).$$

The first three terms form the vertical part of $\mathbf{q}$ in Equation (C.27) while the final (shape) part is, by definition, orthogonal to the first three terms and forms the horizontal part. Squaring lengths and using the orthonormality of e_1, e_2, e_3, we find that

$$\langle\mathbf{q},\mathbf{q}\rangle = |P|^2/M + J^2/I + \text{shape}^2.$$

It remains to show that $|\text{shape}|^2 = \frac{\|\dot{w}\|^2}{I}$. In other words, we need to show that

$$\|\dot{w}\|^2 = \|\mathbf{q}\|^2\|\mathbf{q}\|^2 \text{ if } P(\mathbf{q}) = 0, J(\mathbf{q},\mathbf{q}) = 0, \text{ and } \dot{w} = D\pi_{\mathbf{q}}(\mathbf{q}). \tag{C.29}$$

To this end, write out the map π^{rot} in real coordinates, using $Z_j = x_j + iy_j$, $\mathbf{Z} = (Z_1, Z_2) = (x_1, y_1, x_2, y_2)$. We have $\pi^{rot}(x_1, y_1, x_2, y_2) = \left(\frac{1}{2}\left(x_1^2 + y_1^2 - x_2^2 - y_2^2\right),\right.$ $\left. x_1 x_2 + y_1 y_2, x_2 y_1 - x_1 y_2\right)$. Compute the Jacobian,

$$D\pi^{rot}_{\mathbf{Z}} = \begin{pmatrix} x_1 & y_1 & -x_2 & -y_2 \\ x_2 & y_2 & x_1 & y_1 \\ -y_2 & x_2 & y_1 & -x_1 \end{pmatrix}.$$

Set

$$L = D\pi^{rot}_{\mathbf{Z}}, D\pi_{\mathbf{q}} = L \circ \pi^{tr}.$$

In the last equality, π denotes the quotient map of Theorem C.10 and we used the fact that $\pi = \pi^{rot} \circ \pi_{tr}$ with π_{tr} linear. The three rows of L are orthogonal and have length $\|\mathbf{Z}\|^2 = x_1^2 + y_1^2 + x_2^2 + y_2^2$. Now, $\mathbf{q} \in \mathbb{C}_0^2$ so that $\|\mathbf{Z}\|^2 = \|\mathbf{q}\|^2$ and π_{tr} is a linear isometry of $\mathbb{C}_0^2$ to $\mathbb{C}^2$. It follows that

$$LL^T = \|\mathbf{q}\|^2 Id = D\pi_{\mathbf{q}} D\pi_{\mathbf{q}}^T.$$

The kernel of $D\pi = D\pi_{\mathbf{q}}$ is the vertical space, the span of e_1, e_2, e_3, since π is invariant under rotations and translations. Thus the image of $D\pi^T$ is the horizontal space, which is the orthogonal complement to e_1, e_2, e_3 and is the space called "(shape)" above. Consequently, any vector $\mathbf{q}$ of the form required in Equation (C.29) can be written $\mathbf{q} = D\pi^T \mathbf{v}$ for some $\mathbf{v} \in \mathbb{R}^3$. Thus,

$$\begin{aligned}
\|\mathbf{q}\|^2 &= \langle D\pi^T\mathbf{v}, D\pi^T\mathbf{v}\rangle \\
&= \langle\mathbf{v}, D\pi D\pi^T\mathbf{v}\rangle \\
&= \langle\mathbf{v}, \|\mathbf{q}\|^2\mathbf{v}\rangle \\
&= \|\mathbf{q}\|^2\|\mathbf{v}\|^2. \tag{C.30}
\end{aligned}$$

Moreover, $\dot{w} = D\pi(\mathbf{q})$ so that $\dot{w} = D\pi D\pi^T(\mathbf{v}) = \|\mathbf{q}\|^2\mathbf{v}$. We get that $\|\dot{w}\|^2 = \|\mathbf{q}\|^4\|\mathbf{v}\|^2 = \|\mathbf{q}\|^2\|\mathbf{q}\|^2$, which is the required Equation (C.29).

C.12 Shape Space Metric

Definition C.26 (Shape space metric)   The shape space metric is twice the shape space kinetic energy K_{shape}, when viewed as a Riemannian metric on shape space.

We saw in Proposition C.24 that the shape space metric is given by

$$ds^2_{shape} = \frac{dw_1^2 + dw_2^2 + dw_3^2}{2\sqrt{w_1^2 + w_2^2 + w_3^2}}. \tag{C.31}$$

Define the *length* ℓ of a path c in shape space to be $\ell(c) = \int_c ds_{shape} := \int_a^b \sqrt{2K_{shape}}dt$. Define the distance between two points of shape space to be the infimum of the lengths of all paths joining the two points. In other words, the shape space length is the action relative to the Lagrangian $\sqrt{2K_{shape}}$, and the shape space distance between two points is realized by a length-minimizing curve joining them. We call such a length-minimizing curve a *geodesic*.

Recall the Cauchy–Schwarz inequality,

$$\int f(t)g(t)dt \leq \sqrt{\int f(t)^2 dt}\sqrt{\int g(t)^2 dt},$$

with equality if and only if $f(t) = cg(t)$ (a.e.), with c a constant. Applied to $f = 1, g(t) = \sqrt{2K_{shape}(t)}$, we get $\ell(\gamma) \leq \sqrt{b-a}\sqrt{2\int K_{shape}dt}$ with equality if and only if K_{shape} is constant along the curve γ. Reparameterizing a curve γ does not change its length, and we can always parameterize γ so that its square speed $2K_{shape}$ is constant. It follows from Cauchy–Schwarz that the curves that minimize $\int_\gamma K_{shape}dt$ are precisely the constant speed geodesics. The shape space action principle holds just as well for K in place of $K - V$, and the reduced Lagrangian for K is K_{shape}. The geodesics for K are straight lines in $\mathbb{C}^3$. Putting together these observations, we have proved the assertions of the first two sentences Theorem C.27.

Theorem C.27   *The distance function defined by the shape space metric agrees with the shape space distance of Equation (C.12). Its geodesics are the projections by $\pi\colon \mathbb{C}^3 \to \mathbb{R}^3$ of horizontal lines in $\mathbb{C}^3$. Each plane $\Pi\colon Aw_1 + Bw_2 + Cw_3 = 0$ through the origin is totally geodesic: A geodesic that starts on Π initially tangent to Π, lies completely in the plane Π. The restriction of the Riemannian metric to such a plane Π, when expressed in standard Euclidean polar coordinates (r, θ) on that plane, has the form*

$$ds^2_{shape}|_\Pi = dr^2 + \frac{1}{4}r^2 d\theta^2.$$

In order to finish the proof of this theorem, let $\ell(t) = \mathbf{q} + t\mathbf{v}$ be a horizontal line passing through the point $\mathbf{q} \in \mathbb{C}^2_0 \subset \mathbb{C}^3, \mathbf{q} \neq 0$, with horizontal tangent vector $\mathbf{v}$. There are two possibilities for $\mathbf{v}$: Either $\mathbf{v}$ is a multiple of $\mathbf{q}$ or $\mathbf{v}$ is linearly independent from $\mathbf{q}$. In the first case, we may assume that $\mathbf{v} = \mathbf{q}$ is the radial vector. Then ℓ is a radial line and $\pi(\ell)$ is the ray connecting the triple collision point 0 to $\mathbf{w} = \pi(\mathbf{q})$ (traversed twice). The distance from 0 to $\mathbf{w}$ along this ray is the radial variable

$$r = dist(0, \mathbf{w}) = \|q\| = \sqrt{I} = \sqrt{2\|\mathbf{w}\|}. \tag{C.32}$$

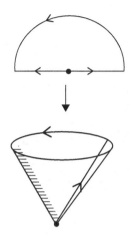

Figure C.4 Folding a half sheet of paper makes the desired cone. From [150].

In the second case, $\mathbf{q}$ and $\mathbf{v}$ span a real horizontal two-plane P in $\mathbb{C}^3$ that passes through 0 and contains the line ℓ. One computes that the projection $\Pi := \pi(P) \subset \mathbb{R}^3$ is a plane (relative to the coordinates w_i) passing through 0. However, the projection $\pi(\ell)$ is not a line (relative to the linear coordinates w_i)!

We can understand the geodesic $\pi(\ell)$ in shape space by understanding the restriction $ds^2_{shape}|_\Pi$ of the shape space metric to the plane $\pi(P)$. Here is what we know so far about this metric. Radial lines are geodesics. The distance along a radial line from the triple collision point 0 to $\mathbf{w} \in \pi(P)$ is r as given by Equation (C.32). To dilate the metric by a factor $t > 0$, we multiply $\mathbf{w} \in \mathbb{R}^3$ by t^2, since $\mathbf{q} \mapsto t\mathbf{q}$ corresponds to $\mathbf{w} \mapsto t^2\mathbf{w}$ under π. Finally, the metric on Π is rotationally symmetric, since the expression (C.31) is rotationally invariant. From all of this information, we deduce that the restricted metric has the form

$$ds^2_{\pi(P)} = dr^2 + c^2 r^2 d\theta^2, \tag{C.33}$$

where (r, θ) are polar coordinates on the plane and c is a constant. It remains to show that $c = 1/2$. With this in mind, consider the circle $r = 1$ in the plane Π. Its circumference computed from Equation (C.33) is $2\pi c$. But we can also compute its length by working up on $P \subset \mathbb{C}^3$. Take $\mathbf{q}$ and $\mathbf{v}$ to both be unit length and orthogonal, so that they form an orthonormal basis for P. Then the corresponding horizontal circle on P is $\cos(s)\mathbf{q} + \sin(s)\mathbf{v}$, for $0 \leq s \leq 2\pi$. But $\pi(\mathbf{q}) = \pi(-\mathbf{q})$, since $-\mathbf{q} = e^{i\pi}\mathbf{q}$, so that the π-projection of this circle closes up once we have gone half way around, from $s = 0$ to $s = \pi$. Thus the projected circle on $\pi(P) = \Pi$ has the length of half a unit circle, or π. Comparing lengths, we see that $c = 1/2$.

Any metric of form (C.33) is that of a cone. We can form our $c = 1/2$ cone by taking a sheet of paper and marking the midpoint of one edge to be the cone point (see Figure C.4). Fold that edge up so the two halves touch each other and we have a paper model of the required cone. Note that the circle of radius r about the cone point has circumference πr, as required.

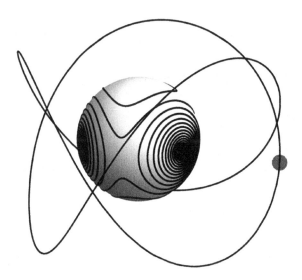

Figure C.5 The sphere here represents the shape sphere with contour level sets for the normalized potential drawn. Surrounding it is the shape curve of the figure eight orbit as it looks in shape space. From [150].

C.13 Potential on Shape Space

We need a formula for the three triangle side lengths r_{ij} in terms of the shape coordinates w_i's in order to express the potential (Equation (C.3)) as a function on shape space. Let $\mathbf{b}_{ij}$ be the point on the shape sphere $\{\|\mathbf{w}\| = 1\}$ that represents the ij collision: $r_{ij} = 0$. The desired equation is

$$r_{ij}^2 = \frac{m_i + m_j}{m_i m_j}(\|\mathbf{w}\| - \mathbf{w} \cdot \mathbf{b}_{ij}). \tag{C.34}$$

An important geometric fact underlies this computation. We let $d_{ij}(\mathbf{w}) = d(\mathbf{w}, \mathbb{R}^+\mathbf{b}_{ij})$ denote the shape space distance (Equation (C.12)) from the shape space point $\mathbf{w}$ to the ij binary collision ray $\mathbb{R}^+\mathbf{b}_{ij}$. Then

$$d_{ij} = \sqrt{\mu_{ij}} r_{ij}, \tag{C.35}$$

where $\mu_{ij} = m_i m_j/(m_i + m_j)$. For proofs of this formula or Equation (C.34) see [33, 148].

The potential is homogeneous of degree -1 so that

$$V(r, \theta, \varphi) = \frac{1}{r}\tilde{V}(\theta, \varphi),$$

where $\tilde{V}$ is the restriction of V to the unit sphere $r = 1$, and φ, θ are standard spherical coordinates: $\mathbf{w} = \frac{r^2}{2}(\sin(\varphi)\cos(\theta), \sin(\varphi)\sin(\theta), \cos(\varphi))$. A contour plot of V is indicated in Figure C.5.

C.14 Reduced Equations of Motion

Having written the shape space kinetic energy (Equation (C.28)) and potential energy (Equations (C.3) and (C.34)) in terms of shape variables w_i, we have the shape space Lagrangian

$$L_{shape} = \frac{1}{2} \frac{\dot{w}_1^2 + \dot{w}_2^2 + \dot{w}_3^2}{2\sqrt{w_1^2 + w_2^2 + w_3^2}} + \frac{c_{12}}{r_{12}} + \frac{c_{23}}{r_{23}} + \frac{c_{13}}{r_{13}}, \qquad (C.36)$$

with constants $c_{ij} = (m_i m_j)^{3/2}/\sqrt{m_i + m_j}$. Its Euler–Lagrange equations (see Equation (A.2))

$$\frac{d}{dt}\left(\frac{\partial L_{shape}}{\partial \dot{w}_i}\right) = \frac{\partial L_{shape}}{\partial w_i}, \qquad \text{for } i = 1,2,3, \qquad (C.37)$$

are our desired reduced equations and are the ODEs for the zero angular momentum three-body equations as written in shape space.

C.15 Infinitely Many Syzygies

The shape space Lagrangian (Equation (C.36)), combined with the realization (Equation (C.35)) is that of a point mass moving in $\mathbb{R}^3$, endowed with metric (C.31), subject to the attractive force generated by the pull of the three binary collision rays. These rays lie in the collinear plane $w_3 = 0$. Consequently, the point is always attracted toward the collinear plane. This physical analogy suggests that the point must oscillate back and forth, crossing that plane infinitely often.

Theorem C.28 (See [148]) *If a solution with negative energy and zero angular momentum does not begin and end in triple collision, then it must cross the collinear plane $w_3 = 0$ infinitely often.*

Sketch of proof of Theorem C.28 The physical analogy just described led to us to discover a differential equation of the form $\frac{d}{dt}(f\frac{d}{dt}z) = -gz$ for a normalized height variable $z = w_3/\tilde{I}$, where $\tilde{I} = r_{12}^2 + r_{23}^2 + r_{31}^2$. Here, f is a positive function on shape space and g is a non-negative function of the w_i and $\dot{w}_i$, which is positive away from the Lagrange homothetic solution. The theorem follows from the differential equation by a Sturm–Liouville argument and the fact that near-triple collision behavior is governed by behavior near the Lagrange homothetic solution.

C.16 Finale

We end with another theorem whose conception and proof relies on the shape space reformulation of the three-body problem.

Theorem C.29 (See [33]) *There is a periodic solution to the equal mass zero angular momentum three-body problem in which all three masses chase each other around the same figure of eight shaped curve.*

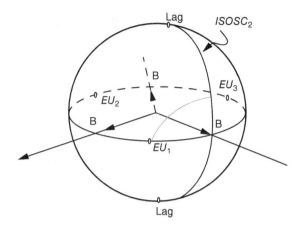

Figure C.6 The shape sphere. The three binary collision rays pierce the shape sphere at the three points (B) that lie on the collinear equator, which in turn also contains the three Euler points (EU). One isosceles great circle ($ISOSC_2$) is drawn. Like all isosceles circles, it joins the two Lagrange points (Lag), and passes through a binary collision point. The minimizer out of which the eight is built is drawn in a light gray and connects an Euler point to $ISOSC_2$. First published in Notices Amer. Math. Soc. 48 (May 2001), published by the American Mathematical Society. © 2001.

A version of the proof sketched below can be found in Chapter 3, Section 3.3.3. That version was built to allow more direct generalizations to choreographies with more bodies.

Sketch of proof of Theorem C.29   There are three types of isosceles triangles depending on which mass forms the vertex. Each type defines a *longitude* – a great circle on the shape sphere passing through the north and south poles. (See Figure C.6) Recall that these poles represent the equilateral triangles, or Lagrange points. Each longitude intersects the equator of collinear triangles at two antipodal points, one of which is a binary collision point. In the equal mass case, the other point of intersection is an Euler point, the binary collision points and Euler points are equally spaced along the equator, and the three isosceles great circles together with the equator divide the sphere up into 12 congruent spherical triangles with angles 90–90–60. One key to the proof is that this apparent 12-fold symmetry is indeeed a symmetry group of the differential equations.

   Viewed in shape space, instead of on the shape sphere, each isosceles great circle represents a plane of isosceles triangles, and each Euler point represents a ray of Euler collinear triangles. Label these planes $ISOSC_i$ for $i = 1, 2, 3$, and label these rays EU_i for $i = 1, 2, 3$. For example, $ISOSC_2$ is defined by the equation $r_{12} = r_{23}$. Consider the problem of minimizing the shape space action $\int_c L_{shape} dt$ among all paths $c : [0, T] \to \mathbb{R}^3$ connecting EU_1 to $ISOSC_2$. Suppose such a minimizer, call it γ_*, exists and is collision-free. First variation orthogonality implies that γ_* must hit both the Euler ray and the isosceles plane orthogonally. (See Figure C.6). The minimizer γ_* will form one-twelfth of the figure eight solution.

To build the rest of the eight from γ_*, we use equality of masses and the consequent 12-fold symmetry group. This order 12 group is generated by reflections about the isosceles planes and the equator, and each of its elements are symmetries of the kinetic energy, the potential energy, the shape space Lagrangian, and consequently of the reduced equations. Reflection R_E about the equator is induced by reflecting all three masses about any fixed line in the inertial plane and is a symmetry no matter what the masses are. The half-twist σ_{13} about the binary collision ray $r_{13} = 0$ is induced by the operation $(q_1, q_2, q_3) \mapsto (q_3, q_2, q_1)$ of interchanging masses 1 and 3 and is a symmetry provided $m_1 = m_3$. Reflection R_2 about $ISOSC_2$ is the composition of R_E with σ_{13}. Applying the symmetries in turn to our minimizer γ_*, we obtain 12 congruent curves that fit together continuously to "wrap" twice around the sphere. Due to first-variation orthogonality they also fit together smoothly! For example, γ_* and its reflection $R_2(\gamma_*)$ about $ISOSC_2$ share the same derivative since both are orthogonal to the isosceles plane. In other words, $R\gamma_*(2T - t)$ is a reduced solution to Newton's equations whose shape space point and shape space velocity agree with those of γ_* at $t = T$. By unique dependence of solutions to ODEs on their initial conditions, it follows that in order to continue the solution γ_* past $t = T$ through the isosceles plane, we simply concatenate $\gamma_*(t)$ with its reflected image $R\gamma_*(2T - t)$. Continuing in this manner with reflections or half-twists we see see that the concatenation of these 12 congruent arcs taken together forms one smooth periodic solution to the reduced Newton's equation. The horizontal lift to $\mathbb{C}^3$ of this periodic solution is automatically a solution to Newton's equations. With extra work we can show that this horizontal lift is itself periodic of period $12T$ (i.e. the solution is periodic, not just periodic modulo rigid motions) and is also a *choreography*.

Definition C.30   An N-body choreography of period T is a solution to N-body problem that has the particular form $\mathbf{q}(t) = (q_1(t), q_2(t), \dots, q_N(t))$, where $q_i(t) = s(t - iT/N)$ for $i = 2, 3, \dots, N$, for some fixed T periodic curve $s(t)$ in the plane (or space).

This curve is the figure eight. See Figure C.5 for the curve γ, the eight projected to shape space.

We have skipped the difficult part of the proof, which is showing that the minimum γ_* exists and is collision-free. For this we refer the reader to [33].

Appendix D   The Orthogonal Group and Its Lie Algebra

In this appendix we gather some facts regarding the orthogonal group and its Lie algebra, $so(d) = o(d) = \wedge^2 \mathbb{R}^d$, which are used in Chapter 0 and occasionally in other chapters. One set of facts described here and not easily found in other sources is our comparison of various norms on $so(d)$.

The orthogonal group $O(d)$ is the group of linear maps $g : \mathbb{R}^d \to \mathbb{R}^d$ that preserve the inner product $x, y \to x \cdot y$ on $\mathbb{R}^d$. If we write $gl(d)$ for the space of all linear maps on $\mathbb{R}^d$, that is, the vector space of real $d \times d$ matrices, then $O(d) \subset gl(d)$ can be defined by the matrix equation $gg^t = Id$. Equivalently, $O(d)$ consists of the $g \in gl(d)$ for which $gx \cdot gy = x \cdot y$ holds for all $x, y \in \mathbb{R}^d$. Fix x, y and differentiate this equation respect to g at $g = Id$ by setting $g = Id + t\xi + O(t^2)$, $\xi \in gl(d)$, to arrive at the condition $\xi x \cdot y + x \cdot \xi y = 0$, valid for all $x, y \in \mathbb{R}^d$. This condition asserts that $\xi = -\xi^T$ and identifies the skew-symmetric linear maps as the Lie algebra $o(d)$ of $O(d)$.

The rotation group $SO(d) \subset O(d)$ is the identity component of $O(d)$ and so its Lie algebra $so(d)$ equals that of $o(d)$. We can verify this fact by hand by noting that $SO(d) \subset gl(d)$ is defined by imposing the equation $det(g) = 1$ in addition to $gg^t = Id$. The derivative of the determinant with respect to the curve $g = I + t\xi + \cdots$ is $tr(\xi)$, and $\xi = -\xi^T \implies tr(\xi) = 0$, verifying that $O(d)$ and $SO(d)$ have the same Lie algebra.

The Lie bracket for $o(d)$ is the usual bracket of matrices $[\xi_1, \xi_2] = \xi_1 \xi_2 - \xi_2 \xi_1$. The matrix exponential $\omega \mapsto exp(\omega)$ is the Lie group exponential and maps $o(d)$ onto $SO(d)$. The matrix exponential of a line $t\xi, t \in \mathbb{R}$ through the origin in $o(d)$ yields a one-parameter subgroup $t \mapsto exp(t\xi)$ of rotations in the rotation group $SO(d)$.

Identify $\wedge^2 \mathbb{R}^d$ with $o(d)$ by sending the bivector $e \wedge f \in \wedge^2 \mathbb{R}^d$ to the linear map $v \mapsto \langle e, v \rangle f - \langle f, v \rangle e$, and then extending this map to all of $\wedge^2 \mathbb{R}^d$ by linearity. Here $e, f \in \mathbb{R}^d$. When $d = 2$ and e_1, e_2 is the standard orthonormal basis for $\mathbb{R}^2$ we find

$$e_1 \wedge e_2 = \begin{pmatrix} 0 & 1 \\ -1 & 0 \end{pmatrix},$$

which corresponds to multiplication by i upon identifying $\mathbb{R}^2$ with $\mathbb{C}$ in the standard way.

The case $SO(3)$. When $d = 3$ the linear map $e_1 \wedge e_2$ corresponds to the cross product by e_3, which leads to the isomorphism $o(3) \cong \mathbb{R}^3$ by which $\omega \in \mathbb{R}^3$ corresponds to the skew-symmetric operator of cross-product by ω. The one-parameter group $t \mapsto exp(t\omega)$ generated by $\omega \in \mathbb{R}^3$ is rotation about the axis ω with $exp(t\omega)$ being counterclockwise rotation by $t|\omega|$ radians.

The linear isomorphism $\Phi : \wedge^2 \mathbb{R}^d \cong o(d)$ just described is $SO(d)$-equivariant, meaning that $\Phi(g\omega) = g\Phi(\omega)g^t$ where $g \in SO(d)$ acts on bivectors by $e \wedge f \mapsto ge \wedge gf$. An orthonormal basis $e_a, a = 1, \ldots, d$ for $\mathbb{R}^d$, induces the basis $e_a \wedge e_b, a < b$ for $\wedge^2 \mathbb{R}^d$. *Declare this basis for $\wedge^2 \mathbb{R}^d$ to be orthonormal to define an inner product on $o(d)$.* One easily verifies that this inner product is independent of the choice of initial orthonormal basis for $\mathbb{R}^d$. *This is the inner product we use on $o(d)$.* One verifies without difficulty that

$$\langle \omega(q), v \rangle_{\mathbb{R}^d} = \langle \omega, q \wedge v \rangle_{so(d)}.$$

Exercise D.1   Up to scale, there is only one $O(d)$ invariant inner product on $o(d)$. Since $(\xi, \omega) = tr\xi^t \omega$ is another such inner product, we must have that $\langle \xi, \omega \rangle_{so(d)} = ctr(\xi^t \omega)$. Use the normal form theorem described next to compute that $c = \lfloor d/2 \rfloor$.

The normal form theorem for $o(d)$ asserts that for each $\omega \in o(d)$ there exists an orthonormal basis $e_1, f_1, e_2, f_2, \ldots, e_\ell, f_\ell$ for the range of ω, a subspace of $\mathbb{R}^d$, and nonzero constants $\omega_i \in \mathbb{R}$ such that

$$\omega = \Sigma \omega_i e_i \wedge f_i.$$

In particular, the range of ω is even-dimensional, and if d is odd, ω must have a kernel. Note that

$$\langle \omega, \omega \rangle = \Sigma \omega_i^2.$$

This normal form theorem is a version of the maximal torus theorem for $SO(d)$. The maximal torus theorem says that a compact connected Lie group admits a maximal torus and any two such tori are conjugate in the group. To get a maximal torus for $SO(d)$ decompose $\mathbb{R}^d$ orthogonally into $k = \lfloor d/2 \rfloor$ two-planes, $P_1, P_2, \ldots, P_k$, and an additional orthogonal line ℓ if d is odd. Thus $\mathbb{R}^d = P_1 \oplus P_2 \oplus \cdots \oplus P_k$ for $d = 2k$ even and $\mathbb{R}^d = P_1 \oplus P_2 \oplus \cdots \oplus P_k \oplus \ell$ for $d = 2k + 1$ odd. Choose an orthonormal basis e_i, f_i as above for P_i. Then $exp(\Sigma \omega_i e_i \wedge f_i)$ preserves this decomposition of $\mathbb{R}^d$ and acts on the plane P_i by rotating it by ω_i radians.

Operator Norm versus Inner Product

Recall that the operator norm of a linear operator $\xi : \mathbb{R}^d \to \mathbb{R}^d$ is given by $\|\xi\|_{op} = sup_{\{v \in \mathbb{R}^d, \|v\|=1\}} \|\xi v\|_{\mathbb{R}^d}$. If $\omega \in o(d)$ is given by the normal form we compute that $\langle \omega x, \omega x \rangle_{\mathbb{R}^d} = \Sigma \omega_i^2 (x_i^2 + y_i^2)$, where the coefficents x_i, y_i of x are defined by expanding out $x = \Sigma x_i e_i + y_i f_i + k$, and where k is in the kernel of ω which is perpendicular to the range of ω. Thus $\|x\|^2 = \Sigma(x_i^2 + y_i^2) + \|k\|^2$. It follows immediately that $\|\xi\|_{op} = max_i |\omega_i|$. Conversely, $\|\xi\|_{o(d)} = \sqrt{\Sigma \omega_i^2}$. Thus we find that

$$\|\xi\|_{op} \le \|\xi\|_{o(d)}.$$

When $d = 2, 3$, we have equality since there is only one ω_i: $\|\xi\|_{op} = \|\xi\|_{so(d)} = |\omega_1|$.

Appendix E Braids, Homotopy, and Homology

In this appendix we collect introductory material on the braid group used in Chapter 3. Good general references are [20] and [76].

Two closed curves in a space X are called *freely* homotopic if we can deform one into the other while staying within X. When defining the fundamental group $\pi_1(X)$ of X we must fix a base point $* \in X$ and then insist that all loops pass through $*$ at the base time $t = 0 \in \mathbb{S}^1$. An element of the fundamental group is a homotopy class of such based loops.

In forming free homotopy classes, neither the loops nor the homotopies need pass through the base point. The set of all free homotopy classes of loops $S^1 \to X$ is denoted $[S^1, X]$ and is not a group, but it can be canonically identified with the space of *conjugacy classes* of elements of the fundamental group $\pi_1(X)$ when X is path-connected. We construct this identification by taking a given free loop $c\colon S^1 \to X$ and adding a leg L to it that connects the base point $*$ to $c(0)$. Travel the leg up to $c(0)$, then travel the loop c, and finally return down the leg to get a based loop $\tilde{c} = L^{-1} * c * L$ with $\tilde{c}(0) = *$. Concatenation with the leg acts effectively as conjugation. See Figure E.1. In this way we get a composition of surjective maps

$$\pi_1(X) \to [S^1, X] \to H_1(X). \tag{E.1}$$

Here, $H_1(X)$ is the 1st homology group of X with integer coefficients. The composition of these two maps yields the map $\pi_1(X) \to H_1(X)$, which is the surjective group homomorphism called the "Hurewicz map" and which "Abelianizes" the fundamental group by sending every commutator $aba^{-1}b^{-1}$ to $0 \in H_1(X)$.

If $X = \mathbb{C}^N \backslash \Delta$ is the configuration space of the collision-free planar N-body problem then its fundamental group $\pi_1(X)$ is called the pure braid group on N strands, and will be denoted here as P_N. The sequence above (Equation (E.1)) becomes

$$P_N \to [S^1, X] \to \mathbb{Z}^{\binom{N}{2}}.$$

Here, P_N has $\binom{N}{2}$ standard generators, one for each pair ab of bodies, denoted A_{ab} by Birman [20], and here sometimes simply by (ab). The generator (ab) projects to $[\mathbb{S}^1, X]$ to a loop realized by an $a - b$ *tight binary*, meaning a loop in which masses a and b make one full counterclockwise revolution around each other while all the other masses stay still and very far away from the two moving masses. Equivalently, such a loop in

264

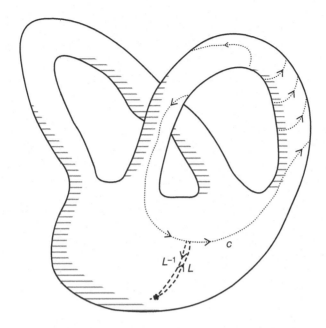

Figure E.1 A leg L joining the base point $*$ to a point on a loop implements the realization between conjugacy classes of elements in the fundamental group of a space and elements of the set of free homotopy classes of loops of that space.

$\mathbb{C}^n$ encircles the binary collision plane $\Delta_{ab} \subset \mathbb{C}^N$ once, staying far away from all the other binary collision planes that make up Δ. Mapped to homology, these generators form a standard integer basis for $H_1(\mathbb{C}^N \setminus \Delta) \cong \mathbb{Z}^{\binom{N}{2}}$ and the winding numbers of edges $q_a - q_b$ realize the coefficients of the homology class of a loop $q(t)$ relative to this basis.

When people speak of *the* braid group they usually do not mean P_N but another group known as the Artin braid group B_N, which is related to P_N via an exact sequence

$$1 \to P_N \to B_N \to S_N \to 1,$$

where S_N denotes the symmetric group on N letters, and S_N acts freely on $X = \mathbb{C}^N \setminus \Delta$ by permuting mass labels. The quotient space $Y = X/S_N$ is the configuration space of N indistinguishable but non-colliding particles and $B_N = \pi_1(Y)$. Generators of B_N can be obtained by taking the square roots σ_{ab} of the standard generators A_{ab}. The σ_{ab} map to the element $(ab) \in S_N$ which transposes a and b. It is extravagant to take all of the σ_{ab} (or A_{ab}) to generate, just as we do not need all of the (ab) to generate S_N. In the usual representation of the braid group, one takes the list of generators to be $\sigma_{i,i+1}, i = 1, \dots, N - 1$.

Rotate any collision-free configuration $* \in X$ one full revolution by using the action of the rotation group. The resulting loop in X commutes with all elements of the braid group.

Proposition E.1 *The center of Artin's braid group B_N lies in P_N and is a copy of the integers $\mathbb{Z}$ whose generator is obtained by rotating the base loop $*$ one full revolution using the action of the rotation group $\mathbb{S}^1$.*

See [20, p. 28, corollary 1.8.4] or [28, corollary 3.4] for proofs. Note that this center $\mathbb{Z}$ lies in P_N.

Definition E.2   The projective pure braid group, denoted $\mathbb{P}P_N$, is the quotient $P_N/\mathbb{Z}$ of the pure braid group P_N by its center $\mathbb{Z}$, this center being generated by applying a full rotation to the initial configuration of N points.

Take its quotient of X by the group $SE(2)$ of rigid motions to get collision-free shape space $Shape(2,N) \setminus \Delta := X/SE(2)$. (See Definition 0.31.) The fibration $SE(2) \to X \to Shape(2,N) \setminus \Delta$ is homotopic to the collision-free Hopf fibration $\mathbb{S}^1 \to \mathbb{S}^{2N-3} \setminus \Delta \to \mathbb{CP}^{N-2} \setminus \Delta$. (See the discussion in Section 0.4.) This *exact sequence* of spaces induces an exact sequence between their fundamental groups: $\mathbb{Z} \to P_N \to \mathbb{P}P_N := P_N/\mathbb{Z}$, where $\mathbb{Z} = \pi_1(S^1)$ embeds as the center of P_N by Proposition E.1. Deleting any one of the binary collision planes allows us to trivialize the Hopf fibration so that we have a diffeomorphism $\mathbb{S}^{2N-3} \setminus \Delta \cong \mathbb{S}^1 \times (\mathbb{CP}^{N-2} \setminus \Delta)$. It follows that

$$P_N = \mathbb{P}P_N \times \mathbb{Z}.$$

(See [28, corollary 3.4] for an alternative derivation of this fact.) It follows that for collision-free shape space Y, the exact sequence, Equation (E.1), becomes

$$\mathbb{P}P_N \times \mathbb{Z} \to [\mathbb{S}^1, Y] \to H_1(Y) = \mathbb{Z}^{\binom{N}{2}-1}, \qquad (E.2)$$

where we have used that $\mathbb{Z}^{\binom{N}{2}}/\mathbb{Z} = \mathbb{Z}^{\binom{N}{2}-1}$.

The Case of Three Bodies

The Lagrange solution, projected to the shape sphere, is trivial homotopically since it is a constant loop, As per Proposition E.1, this loop generates the center of the fundamental group of the collision-free configuration space. Now a loop in the shape sphere S^2 represents a *relative* periodic curve: a curve that closes up modulo a rotation. If that loop avoids the three binary collision points Δ, then the curve upstairs in configuration space is collision free and so represents a conjugacy class within $\mathbb{P}P_3$.

Projected to the shape sphere, the tight binary loop representing the generator A_{ij} is a small circle surrounding the binary collision point B_{ij} of the shape sphere. Denote by (ij) the corresponding element of the fundamental group of the collision-free shape sphere. The three (ij)'s generate this fundamental group and are subject to the single relation $(12)(23)(31) = Id$. (This exhibits $\pi_1(\mathbb{S}^2 \setminus \Delta)$ as the free group on two letters, namely, any two of the (ij)'s.) We can write an eclipse sequence in terms of generators. For example, $123123 = (12)(31)(23)$. To see where such an element maps in homology, let the generators commute: $(12)(31)(23) = (12)(23)(31)$, which is trivial, being the class of the identity. This computation shows that the figure eight's class, represented by 123123, is homologically trivial, a fact that can also be seen by drawing the loop traced out by the eight on the shape sphere.

Appendix F   The Jacobi–Maupertuis Metric

This appendix is only used in passing in the body of the text, mostly as notes and remarks in Chapter 3. It is used in an essential way in Appendices G and H.

We can reformulate most problems in classical mechanics as problems of finding geodesics. Almost. The geodesics are for a family of metrics known as the Jacobi–Maupertuis (JM) metric. We refer to the reformulation as the JM trick.

The JM trick works for any natural mechanical system. (See Section A.2.) Recall that such a system is specified by a Riemannian metric $\langle \cdot, \cdot \rangle$, which we also denote by ds_K^2, and a function V that we call the potential, both defined on a manifold Q that we call configuration space. Following our N-body convention we set

$$U = -V.$$

Newton's equations are

$$\nabla_{\dot{q}} \dot{q} = \nabla U(q),$$

where ∇ is the Levi-Civita connection of the Riemannian metric. We saw in Appendix A that Newton's equations are equivalent to Euler–Lagrange equations for the Lagrangian $L = K + U$ and to Hamilton's equations for the Hamiltonian $H = K - U$. Here K is the kinetic energy for the metric. In coordinates $K = \frac{1}{2}\Sigma g_{ij}(x)v^i v^j = \frac{1}{2}\Sigma g^{ij}(q)p_i p_j$, where $ds_K^2 = \Sigma g_{ij}(x)dx^i dx^j$, and where the v^i and p_i are the corresponding coordinate-induced velocity and momentum coordinates, which are fiber linear coordinates on TQ and on T^*Q. The momenta and velocities are related by the Legendre transformation $p_i = \Sigma g_{ij}(x)v^j$.

Energy is conserved along solutions to Newton's equations so that Hamilton's equations induce a flow on the constant energy hypersurface

$$\{(q, p) : H(q, p) = E\} = \Sigma_E \subset T^*Q.$$

Since $K \geq 0$ and $K = E + U$ on Σ_E, we have that the projection of Σ_E to the configuration space Q forms the open subset

$$\Omega_E := \{q : E + U(q) \geq 0\} \subset Q. \tag{F.1}$$

An energy E solution $q(t)$ to Newton's equations must lie in Ω_E.

Definition F.1   The region Ω_E (Equation (F.1)) is called the Hill region for energy E. The JM metric on this Hill region is the metric

$$ds_E^2 = \lambda_E \, ds_K^2, \text{ with conformal factor } \lambda_E(q) := 2(E + U(q)).$$

The JM metric is Riemannian on the interior of the Hill region $\{0 < \lambda_E < \infty\}$. The metric degenerates to zero on the Hill boundary $\partial\Omega_E : \{q : E + U(q) = 0\} \subset \Omega_E$ and becomes infinite at the collision locus $\{q : U(q) = +\infty\}$. The "interior of the Hill region" will refer to the Hill region minus the Hill boundary and the collision locus.

When an energy E solution $q(t)$ to Newton's equations hits the Hill boundary $\{E + U = 0\}$, its kinetic energy is zero, so its velocity is zero. The solution has "braked" – it has instantaneously stopped. We say the solution has a brake and refer to it as a brake orbit.

Proposition F.2   *Geodesics for the JM metric ds_E^2 at energy E that lie in the interior of the Hill region are reparameterizations of collision-free, brake-free energy E solutions to Newton's equations. The reparameterizaton is given by $ds = \lambda_E(q(t))dt$ where t is Newtonian time t and s is JM arclength.*

The proof follows from a simple but powerful lemma in symplectic geometry.

Lemma F.3   *Let (P, ω) be a symplectic manifold. Suppose that $H, F: P \to \mathbb{R}$ are two smooth functions on P that share a common level set,*

$$\{H = c_H\} = \{F = c_F\} = \Sigma,$$

where the constants c_H, c_F are regular values for the respective functions so that the shared level set is a smooth hypersurface in P. Then, on Σ, the solutions to Hamilton's equations for H and for F are reparameterizations of each other.

Proof of lemma   Recall that ω, being non-degenerate, defines a fiber linear isomorphism $T^*P \to TP$. The Hamiltonian vector field X_F of a function F on P is obtained by applying this isomorphism to the one-form dF, according to $\omega(X_F, \cdot) = dF$. It follows that if $dF(\zeta) = c \, dH(\zeta)$ at some point $\zeta \in P$ and for some constant c, then $X_F(\zeta) = c X_H(\zeta)$.

Since the constants c_F, c_H are regular values for F and H, the hypersurface Σ is a smooth hypersurface whose tangent space at $\zeta \in \Sigma$ is the kernel of $dF(\zeta)$, which equals the kernel of $dH(\zeta)$. It follows that $dF(\zeta) = \lambda \, dH(\zeta)$ for some nonzero constant λ. Letting ζ vary, we get a smooth non-vanishing function $\lambda: \Sigma \to \mathbb{R}$ such that $dF = \lambda \, dH$ holds along Σ. It follows that the identity $X_F = \lambda X_H$ holds along Σ, which proves the lemma, with the function λ supplying the reparameterization. QED

Proof of proposition F.2   Apply Lemma F.3 with F being the Hamiltonian on T^*Q whose geodesic flow is that of the JM metric and H the Hamiltonian for our natural mechanical system. Since F is defined by the inverse metric to ds_{JM}^2, we have that $F = \frac{1}{\lambda_E} K$ or $F = \frac{1}{2} \frac{1}{\lambda_E} \|p\|^2$, where $K = \frac{1}{2} \|p\|^2$ denotes the original kinetic energy on T^*Q. We now take constants $c_H = E$ for H and $c_F = 1/2$ corresponding to parameterizing JM-curves by arclength and we verify the hypothesis of the lemma:

$$F(q, p) = \frac{1}{2} \iff \|p\|^2 = \lambda_E(q)$$
$$\iff \|p\|^2 = 2(E - V(q))$$
$$\iff K(q, p) = E - V(q)$$
$$\iff K(q, p) + V(q) = E$$
$$\iff H(q, p) = E.$$

It remains to verify that the conformal factor λ_E defines the reparameterization. So let $(q(t), p(t)) = \zeta(t)$ be an energy E solution for H and $(q(s), p(s)) = \zeta(s)$ the same solution curve reparameterized by arclength according to $F = 1/2$. According to what we just proved, we have that $\frac{d}{dt}\zeta = f\frac{d}{ds}\zeta$ for some function f and the problem is to find f. From Hamilton's equations, we have $\frac{d}{dt}q = g^{-1}p$ where $g^{-1} = g^{-1}(q)$ is the inverse of the kinetic energy metric. (In index notation $\dot{q}^i = g^{ij}p_j$.) Conversely, from the geodesic equations, rewritten as Hamilton's equations for F, the Hamiltonian associated to the JM metric, we have that $\frac{d}{ds}q = \frac{1}{\lambda_E}g^{-1}p$. Comparing these two ODEs for q, we see that $\frac{d}{ds} = \frac{1}{\lambda_E}\frac{d}{dt}$ or $\lambda_E dt = ds$. QED

Proposition F.2 asserts that JM geodesics are *almost* in bijection with the solutions to Newton's equations having energy E. The bijection fails when we try to proceed past brake points or collision points occuring on brake orbits and collision orbits. For one thing, it is no longer even clear what we mean by being a "geodesic" when we pass through such a point. Giving meaning to the notion is not difficult by using the JM arclength to give the Hill region the structure of a metric "length-space" as outlined in Burago et al's [23]. Travel along the Hill boundary costs nothing, so we must collapse the Hill boundary to a single point if we are to continue to work in the context of a metric space. This done, the notion of "geodesic" now makes sense.

In [154] I show for standard Newtonian gravity in dimensions $d > 1$ that the completion of the interior of the Hill region with its JM metric coincides with the Hill region with its boundary collapsed to a point, endowed with this JM-length induced metric. I show that brake solutions fail to be geodesic past their Hill instant. This is no surprise. Brake solutions back up and retrace themselves after the Hill instant. A JM version of the Marchal lemma (Lemma 3.2.3) shows that it is impossible to extend solutions beyond collisions in such a way that they continue to be geodesics. So the brake and collision solutions of standard gravity cannot be continued beyond their interactions with the Hill boundary and the collision locus and remain geodesic.

We have arrived at a strange dichotomy between solutions to Newton's equations and metric geodesics. The solutions to Newton's equations extend with no problem through brake instants. They just reverse their path. This natural Newtonian extension fails to be a metric geodesic. Can we extend a brake solution beyond its Hill point in such a way that it remains geodesic? Certainly! Simply take *any other brake solution* for this energy E and leave the Hill point using this alternate brake solution. The result will be a metric geodesic passing through the Hill point. Up in configuration space, it hits the Hill boundary, travels along the Hill boundary for free building up no length, then leaves the Hill boundary along a different brake solution. (In the context of the Kepler problem these are the discontinuous solutions of Todhunter [208, 209]. See also [153].)

Open Problems

1. Consider a minimizing JM geodesic connecting a point on the collision locus Δ to a point on the Hill boundary. Is U strictly monotone decreasing along such a geodesic?

2. Background. (i) By the virial theorem (proven by averaging the Lagrange–Jacobi equation over an orbit) any bounded energy $-E$ solution, with $-E < 0$ must repeatedly intersect the locus $U = 2E$ "half-way" between the Hill boundary $U = E$ and the collision locus $U = \infty$. See [154] for details. (ii) In the case of the planar three-body problem, the only solutions defined for all time and for which $U = 2E$ are the relative equilibria of Euler and Lagrange. This is the $N = 3$ version of the Saari conjecture, which Moeckel has established [136].

(A) Prove that for the standard (i.e. $\alpha = 1$) equal mass planar three-body problem, there is an $\epsilon > 0$ with the following property. If a bounded energy $-E$ solution is defined for all time and lies in the region $\{2(E - \epsilon) < U < 2(E + \epsilon)\}$, then that solution is a central configuration solution. (Some small eccentricities are allowed in the Kepler orbits that yield the central solution, the eccentricities bounded by ϵ.)

(B) We can project the JM metric to shape space and work on the shape space Hill region, indicated in Figure 0.6 (Chapter 0), drawn there as an unbounded domain within the shape space $\mathbb{R}^3$ for the planar three-body problem. The corresponding shape space JM geodesics correspond, away from brake and collision, to solutions to the zero angular momentum fixed energy problem. Modify (A) for the zero angular momentum, negative energy, equal mass standard planar three-body problem as follows. There is an $\epsilon > 0$ with the following property. No bounded solution exists for this problem, which is defined for all time and lies in the region $\{2(E - \epsilon) < U < 2(E + \epsilon)\}$.

(C) Continue in the spirit of (B). If we increase ϵ, eventually there will be such solutions, for example, the figure eight solution. Show that the figure eight at this energy $-E$ realizes the infimum of the ϵ's such that such a periodic orbit lying in the region $\{2(E - \epsilon) < U < 2(E + \epsilon)\}$ exists for the problem.

Appendix G   Regularizing Binary Collisions

G.1 Levi-Civita Regularization

The Levi-Civita transformation is

$$z^2 = q; \qquad d\tau = \frac{dt}{|q|} \tag{G.1}$$

for $z \in \mathbb{C}$. The transformation converts the planar Kepler problem

$$\ddot{q} = -\mu \frac{q}{|q|^3} \tag{G.2}$$

with energy $E = \frac{1}{2}|\dot{q}|^2 - \frac{\mu}{|q|}$, to the isotropic planar harmonic oscillator

$$z'' = \frac{E}{2} z \tag{G.3}$$

with Hooke's constant $E/2$. Here the dots over q indicate derivatives with respect to the Kepler time t and the primes over z indicate derivatives with respect to the oscillator time τ, with τ and t being related by the second part of the Levi-Civita transformation (G.1). The main point of the transformation is that the Kepler problem has a singularity at the origin while the oscillator has no singularities. Levi-Civita's transformation "regularizes" the collision singularity.

We have used the word oscilator loosely. Equation (G.3) is that of an oscilator when $E < 0$. However, when $E > 0$ it is the equation of a particle under the influence of a repulsive "Hooke's law," while for $E = 0$ the equation is that of a free particle whose solutions are straight lines in the z-plane. It is a pleasant surprise of complex analysis that squaring takes lines in the plane to parabolas with focus at the origin, and ellipses centered at the origin to ellipses one of whose foci is at the origin. See Figure G.1. The transformation takes Kepler collision-ejection solutions, which reflect off the origin with infinite velocity, and unwraps them so they pass through the origin with finite velocity tracing out line segments (at negative E) centered at the origin. See Figure G.1 again.

Bohlin, nine years before Levi-Civita, discovered the same transformation (G.1) taking the harmonic oscillator to Kepler's problem. So Arnol'd calls this fact "Bohlin's theorem." See Arnol'd [12, p. 96, theorem 2]. The transformation had been discovered multiple times in the previous centuries. See [9] for a history. Perhaps the first to

271

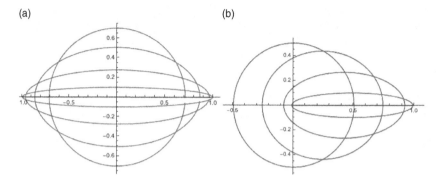

Figure G.1 The Levi-Civita transformation maps solutions to (a) the harmonic oscillator, to those of (b) the Kepler problem.

uncover Levi-Civita's transformation was Maclaurin [115], who published the result in 1742, about 260 years before Levi-Civita. See section 451 of [115] for the statement and section 875 for the proof. It is worth noting that Maclaurin found the full "Bohlin duality," valid for all power law potentials, and described in Section G.6.

G.2 Jacobi–Maupertuis Explains Levi-Civita

One can verify by direct computation that the Levi-Civita transformation converts the oscillator to the Kepler problem. This does not explain the transformation.

The JM (Jacobi–Maupertuis) metric reformulation of dynamics helps explain the transformation. Recall that the Kepler JM metric at fixed energy E is

$$ds_E^2 = 2 \left(E + \frac{\mu}{|q|} \right) |dq|^2.$$

Under $q = z^2$, we have $|q| = |z|^2$ while $|dq|^2 = 4|z|^2|dz|^2$. Since $(E + \frac{\mu}{|q|})|z|^2 = E|z|^2 + \mu$, we have, in the variable z,

$$ds_E^2 = 8(\mu + E|z|^2)|dz|^2,$$

which is the JM metric for a harmonic oscillator whose potential is $-4E|z|^2$. This computation shows that $z \mapsto q = z^2$ maps "oscillator geodesics" to "Kepler geodesics."

In order to verify the time reparameterization, recall that in Proposition F.2 we showed that in going from the Newtonian to the JM metric reformulation of dynamics, the JM arclength and Newtonian time are related by the conformal factor $\lambda_E :=$ $2(E - V(q))$. In our case, $\lambda = 2(E + \frac{\mu}{r})$ for Kepler → JM while $\lambda = 4(E|z|^2 + \mu)$ for going from the harmonic oscillator to its JM. Write t_K for Kepler time, t_{osc} for oscillator time, and s for JM arclength. Use $|z|^2 = r$. It follows that

$$\frac{dt_K}{dt_{osc}} = \frac{dt_K}{ds}\frac{ds}{dt_{osc}}$$

$$= 2\left(E + \frac{\mu}{r}\right)\frac{1}{4(Er + \mu)}$$

$$= \frac{1}{2}\frac{\frac{1}{r}(Er + \mu)}{(Er + \mu)}$$

$$= \frac{1}{2}\frac{1}{r}, \tag{G.4}$$

which, up to a constant, is the Levi-Civita time change of Equation (G.2).

G.3 On to the N-Body Problem

Heggie [77] describes an elegant method for regularizing binary collisions in the planar N-body problem. He proceeds, in essence, by applying a Levi-Civita transformation to each pair of relative position vectors $q_{ab} := q_a - q_b$. Rewrite the equations in terms of the $\binom{N}{2}$ variables q_{ab}. Introduce the complex variables Z_{ab} by $Z_{ab}^2 = q_{ab}$ and the new time τ by $dt = \Pi_{a<b}r_{ab}d\tau$. In these new variables the flow becomes analytic in a neighborhood of the isolated binary collisions.

Heggie actually wrote his paper for the spatial N-body problem, not the planar one. Instead of applying the Levi-Civita transformation to each $q_a - q_b$ he uses its three-dimensional variant known as the Kuustanheimo–Steifel transformation. In this transformation, the configuration part $z \mapsto q = z^2$ of the Levi-Civita transformation is replaced by the same Hopf map $\mathbb{C}^2 = \mathbb{R}^4 \to \mathbb{R}^3$ central to constructing shape space. See Equation (C.28). See the book by Stiefel and Scheifele [199] for details regarding the K–S transformation.

For the specific case of the three-body problem, see the earlier work of Lemaitre [106, 107], which both regularizes and reduces the planar three-body problem, leading to ODEs on a regularized shape space. Similar work in this vein can be found in Waldvogel [211], Moeckel and Montgomery [139], and references therein.

For large N-body simulations, regularizing binary collisions can be essential. Bodies come close enough to near collision to render huge forces requiring some kind of regularization in order to proceed with accurate numerical integration. Variants of the Kuustanheimo–Steiffel or Levi-Civita regularization are the methods of choice for numerical implementations. See Aarseth's book [1].

G.4 Failure to Regularize More Collisions

The simultaneous collision of three or more bodies cannot be regularized although they can be blown up. What's the difference between regularizing and blow-up? In regularization, solutions pass smoothly *through* the collision locus, while in blow-up the collision locus becomes an invariant submanifold and hence non-collision solutions cannot pass through collisions. See Easton [50] for a discussion of obstructions to regularization.

Rather strangely, simultaneous binary collisions among two pairs of bodies cannot be analytically regularized. The best we can do is get Holder 8/3 smoothness. See Duignan and Dullin [45] and Martinez and Simó [119]. Dullin and Duignan relate the "8" and "3" of 8/3 to eigenvalues of equilibria arising upon blowing up the simultaneous binary collision locus in McGehee style [45, section 1.4].

G.5 Other Regularizations

Moser [159] formulated an alternative type of regularization of the Kepler problem in any dimension d that is now called Moser regularization. See also [126]. Knauf describes yet another regularization in chapter 11.3.1 of [96], which we also recommend for an all-round discussion of regularizations.

All Kepler regularizations extend to perturbed Kepler problems

$$\ddot{q} = -\mu \frac{q}{|q|^3} + f(q,t),$$

with f bounded and Lipshitz in a neighborhood of $q = 0$. Near an isolated binary collision, the full N-body problem looks like a perturbed Kepler problem – the forces of the far-away bodies act like the perturbation f. Thus any of these other types of regularization can, in principle, be scaled up to work for the full N-body problem. As far as I know, no one has done this work.

G.6 Regularizing Other Power Laws and Bohlin's Theorem

Besides the Newtonian $1/r$ potential of Kepler, the power law potentials with pair potential $1/r^\alpha$ can be regularized provided that

$$\alpha = 2 - \frac{2}{n+1}, \quad n = 1, 2, 3, \dots . \tag{G.5}$$

The regularizing transformation is $q = z^{n+1}$ and maps solutions of the $1/r^\alpha$ central force problem to solutions to a central force problem with power law potential r^{2n}.

To see why and how this works, return to the JM trick used to explain the Levi-Civita transformation, but now applied to any central force potential $V = c/r^\alpha$, c a constant, in place of the Kepler potential. The JM metric is

$$ds^2 = 2(E + c/r^\alpha)|dq|^2.$$

Under the substitution $q = z^\beta$ we get $|dq| = \beta |z|^{\beta-1}|dz|$ and $r = |z|^\beta$. Now tune the exponent β so that the conformal factor $|z|^{2(\beta-1)}$ occuring in $|dq|^2 = \beta^2 |z|^{2(\beta-1)}|dz|^2 = |dq|^2$ cancels the $\frac{1}{r^\alpha}$ singularity. One computes that this tuning requires that

$$2 - \frac{2}{\beta} = \alpha$$

and converts the JM metric into $ds^2 = 2(\beta^2 E|z|^\gamma + k)|dz|^2$ where $\gamma = 2\beta - 2$ and k is a constant. This later metric is the JM metric for a power law potential $V(z) = C|z|^\gamma$ in the z-plane, where C is a constant. One computes that the exponents γ and α are related by

$$\left(1 - \frac{\alpha}{2}\right)\left(1 + \frac{\gamma}{2}\right) = 1.$$

We have established Maclaurin duality. See [9, 12, 71, 115]. This duality asserts that the central force problems $V(r) = 1/r^\alpha$ and $V(r) = r^\gamma$ are dual to each other in that the multivalued transformation $z \mapsto z^\beta = q$ maps solutions of the latter central force problem to those of the former, provided α, β, γ are related algebraically as above.

To understand the quantization condition (G.5) on the power law exponent α, note that a fractional power β yields a multivalued transformation $z \mapsto z^\beta$ of the complex plane. For this transformation to be single-valued β must be an integer. Write this integer as $n + 1$ and recover the quantization condition (G.5).

We could work out Heggie-type versions that regularize power law N-body problems with the exponents satisfying the quantization condition (G.5). So far no one has felt the need to do so. These same special exponents α described by Equation (G.5) play an important yet curious role in scattering. See [98], where Knauf and Krapf show that the associated scattering map yields a degree n map $\mathbb{S}^1 \to \mathbb{S}^1$. Also see the comments at the end of Appendix H.

Appendix H   One Degree of Freedom and Central Scattering

H.1 Radial Motion as a One-Degree-of-Freedom System

In this appendix we gather together some well-known facts regarding one-degree-of-freedom motions and apply them to scattering a particle off a central potential, which is essentially the same problem as the two-body scattering problem. Results from this appendix are used in Chapter 4, particularly the last sections of that chapter.

By a one-degree-of-freedom system, we mean the flow defined by a one-dimensional Newton equation $\ddot{x} = -V'(x)$ where $x \in \mathbb{R}$. The associated energy $E = \frac{1}{2}\dot{x}^2 + V(x)$ is constant along solutions. In the phase space whose coordinates are $(x, \dot{x})$, the solutions to this ODE trace out the contour level sets of energy.

One-degree-of-freedom systems arise within two-body problems where they have $x = r = r_{12}$ the distance between the two bodies. If we fix the angular momentum J and work in the center-of-mass frame then the two-body energy is

$$E = \frac{1}{2}\dot{r}^2 + \frac{1}{2}\frac{J^2}{r^2} - f(r), \tag{H.1}$$

where $f(r)$ is the negative of the two-body potential. The corresponding central force equation for the evolution of $q = q_1 - q_2$ is $\ddot{q} = f'(r)\frac{q}{r}$. See Section 0.2. The function

$$V_J(r) := \frac{1}{2}\frac{J^2}{r^2} - f(r) \tag{H.2}$$

is called the effective potential and plays the role of the one-degree-of-freedom potential $V(x)$. In particular, r evolves by $\ddot{r} = -V_J'(r)$.

We can qualitatively understand everything about a one-degree-of-freedom system by simply graphing the potential $V(x)$ versus x. We then fix any energy value E of interest, draw the horizontal line $V = E$, and use the graph to mark out the intervals of the x-line where $V(x) \le E$. We call these the Hill intervals associated to the energy E. Since $\dot{x}^2 \ge 0$ and E is constant, the solutions $x(t)$ having energy E are constrained to move within these Hill intervals. The endpoints of a Hill interval satisfy $V(x) = E$, which means that $\dot{x} = 0$ there. If we choose the time origin so that $t = 0$ when $\dot{x} = 0$ then one has $x(-t) = x(t)$ and $\dot{x}(-t) = -\dot{x}(t)$, and for this reason the Hill endpoints are also called *turning points*. If a particular Hill interval is compact and neither endpoint is a critical point for $V(x)$, then the solution $x(t)$ shuttles back and

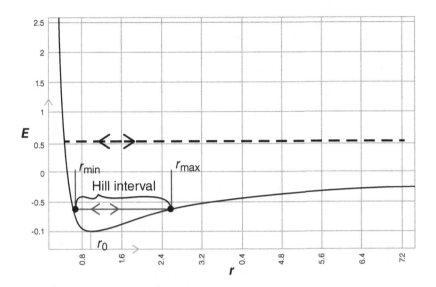

Figure H.1 The graph of the effective potential $V_J(r) = 1/r^2 - 2/r$ when $J^2 = 2$ for the Kepler problem. The horizontal lines represent fixed energy levels. The gray line is negative and r_0 is the circular orbit for this angular momentum. The dashed line represents a positive energy value with corresponding unbounded, or scattering orbit.

forth between these endpoints. The solution is periodic and its period is twice the time required to pass from one endpoint to the other. When the Hill interval is a half-line $[a, \infty)$ with $V(a) = E$ and $V'(a) < 0$, then the solution comes in from infinity, hits the turning point $x = a$, and turns around, retracing its steps to infinity.

Let us see how this one-degree-of-freedom analysis pans out for the Kepler problem. Then $V_J = \frac{1}{2}\frac{J^2}{r^2} - \frac{\mu}{r}$. The graph of V_J is shown in Figure H.1 with $J^2 = 2$ and $\mu = 2$. The effective potential has a global minimum $V_{J,min}$ when $J \neq 0$, this minimum occuring at $r = r_0$. If $E < V_{J,min}$, then the Hill interval is empty. When $E = V_{min}$ the Hill interval is the single point r_0, which is a stable equilibrium for the one-dimensional dynamics and represents the circular Kepler orbit at that energy. For $V_{J,min} < E < 0$, the Hill interval is a bounded interval containing the circular radius r_0 and the turning points are the perihelion and aphelion radii r_{min} and r_{max} of the corresponding Kepler ellipse. When $E \geq 0$ the Hill interval is a half-line, representing parabolic or hyperbolic motion.

Exercise H.1 Show that the radial motions for power law potentials $f(r) = 1/r^\alpha$ having $0 < \alpha < 2$ and $J \neq 0$ are qualitatively the same as those of Kepler. That is, when $J \neq 0$ the negative energy solutions $r(t)$ are oscillatory, with r shuttling back and forth between pericenter r_{min} and apocenter r_{max}, while when $E \geq 0$ the solutions *scatter*: They come in from infinity, hit pericenter (closest point of approach), and return to infinity.

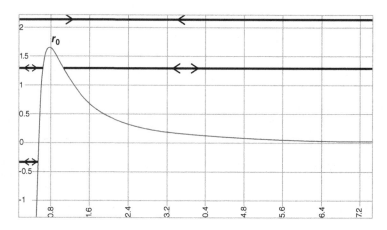

Figure H.2 The graph of the effective potential $V_J(r) = 2/r^2 - .6/r^4$ arising in the $\alpha = 4$ strong central force problem when $\mu = .6$ and $J^2 = 2$. The horizontal lines represent various energy levels with corresponding solutions indicated. The top line goes from infinity to collision, or the reverse. The intermediate energy level splits into two Hill intervals, one ending in collision and bounded, the other coming in from infinity and returning to infinity. The bottom energy level is bounded and ends in collision.

For the strong force case of $\alpha > 2$ the situation is, in a way, the reverse of the previous $0 < \alpha < 2$ case. In Figure H.2 we graph the effective potential for such a case, namely $\alpha = 4$, so that $f(r) = r^{-4}$, $\mu = .6$, and $J^2 = 2$. When $\alpha > 2$ and when $J \neq 0$, V_J now has a unique global maximum $V_{J,max} = V(r_0)$ instead of a unique global minimum, and it is unbounded below instead of above. Only for energy $E = V_{J,max}$ are the solutions bounded and collision free. Here the Hill interval is the single point r_0 which is an unstable fixed point for the one-dimensional dynamics, representing a circular orbit. If $E > V_{J,max}$, all solutions come in from infinity and hit total collision $r = 0$, or, are the time reversal of such an orbit. One can verify that they spiral infinitely often on the way to collision. If $E < 0$, then the Hill interval is of the form $(0, a]$ and all solutions end at total collision and reach a finite size corresponding to $a > 0$. For $0 < E < V_{J,max}$ there are two Hill intervals, one ending in collision at $r = 0$ and the other of scattering type, with a finite nonzero perihelion.

The case $\alpha = 2$ is special among the power law potentials. Here the two terms of V_J balance. Indeed, by choosing $J^2 = \mu$ they cancel and we get a family of circular periodic orbits all having zero energy.

H.2 Scattering

Here we substantiate the 3rd claim of Section 4.3 regarding two-body scattering for power law potentials by using some one-degree-of-freedom tricks and the Jacobi–Maupertuis (JM) metrics for zero energy.

Proposition H.2 *Consider positive energy dynamics for a power law potential with exponent* α *satisfying* $0 < \alpha < 1$. *Then the image of a beam under the scattering map for the corresponding central force law* $\ddot{q} = -\frac{q}{r^{\alpha+2}}$ *in the plane is a circular arc of arclength* $2\pi\alpha/(2-\alpha)$ *centered on the "undeflected" ray, which travels in the beam direction. If* $1 < \alpha < 2$, *the scattered image of the beam covers the entire circle and then some: Some directions are hit more than once. As* $\alpha \to 2$ *the circle is covered more and more times, so that in the limit* $\alpha \to 2$ *each direction is covered infinitely many times.*

The angular momentum for a central force problem, after normalizing units, satisfies $J = r^2\dot{\theta}$, so that

$$\frac{d\theta}{dt} = \frac{J}{r^2}.$$

Solve the energy equation (H.1) for the radial velocity to get

$$\frac{dr}{dt} = \pm\sqrt{2(E - V_J(r))},$$

with the $+$ sign valid for the branch of $r(t)$ receding from the origin and the $-$ sign for the branch coming in towards the origin. Dividing the equation for dr/dt by that for $d\theta/dt$ we get an equation for $dr/d\theta$ valid along one branch or the another. Integrating this equation for $dr/d\theta$, and paying attention to the signs, we get, in the unbounded case,

$$\Delta\theta(E, J) = 2\int_{r_{min}}^{\infty} \frac{J/r^2}{\sqrt{2(E - V_J(r))}}dr \tag{H.3}$$

for the total change in angle $\Delta\theta = \Delta\theta(E, J)$ suffered in traveling along the trajectory labeled by E and J from $t = -\infty$ to $t = +\infty$. In the integral, r_{min} is the pericenter or turning point, that is, the unique value of r for which $E - V_J(r) = 0$. Note that $\Delta\theta$ is the angle between the two asymptotic positions, $\lim_{t\to-\infty}(q(t)/r(t))$ and $\lim_{t\to+\infty}(q(t)/r(t))$. In particular, $\Delta\theta = \pi$ corresponds to straight line motion. In the Kepler case, this angle $\Delta\theta$ is the (signed) angle between the asymptotes of Keplerian hyperbola, with both asymptotes oriented to point outwards.

Recall Rutherford scattering (Section 4.1) and the meaning of the impact parameter b used to describe it. If we turn the force off, angular momentum is still conserved, and from this we see by a quick sketch, or by an inspection of the relation $J = q \wedge \dot{q}$ as q tends to infinity along a line, that

$$J = b.$$

Henceforth we use b in place of J. Choose the x-axis along the beam direction, with rays pointed toward the origin, coming in from infinity, heading in the direction $(-1, 0)$ of the negative x-axis. Reflection about the x-axis maps solutions to solutions, preserving the beam direction and hence maps the entire beam's worth of trajectories to itself, acting on the impact parameter by $b \to -b$. Reflectional symmetry implies that $\Delta\theta(E, -b) = -\Delta\theta(E, b)$. As $b \to \infty$ the corresponding trajectories tend to straight lines, undeflected by the central force providing $f(r) \to 0$ sufficiently fast as $r \to \infty$. (The power law potentials with $\alpha > 0$ all have sufficiently fast decay.) Since $\Delta\theta = \pi$ for straight lines, this yields $\lim_{b\to\pm\infty} \Delta\theta(E, b) = \pi$. Reflectional symmetry implies that the arc A swept out by the scattering map is reflectionally symmetric, so of the form

$(\pi - \Delta\varphi, \pi + \Delta\varphi)$. The midpoint $\theta = \pi$ of the arc A corresponds to undeflected lines arising when $b = \pm\infty$. The total change of angle of an undeflected line not through the origin is π. Our goal is to compute the half-width $\Delta\varphi = \Delta\varphi(\alpha)$ of the arc A. We will do this by relating this half-width to $\lim_{b\to 0^+} \Delta\theta(E, b)$.

The ray $b = 0$ corresponds to the solution with angular momentum zero and is the unique trajectory in the beam that ends in collision when $f(r) = r^{-\alpha}$ with $0 < \alpha < -2$. This collision trajectory associated to $b = 0$ splits the beam into two connected half-beams, one having $b > 0$ and the other having $b < 0$. Reflection symmetry shows that if the scattered image of one half beam is an arc $(\pi, \pi + \Delta\varphi)$, then the other half beam has scattered image equal to the arc $(\pi - \Delta\varphi, \pi)$.

If, as with power laws, the potential $f(r)$ is strictly monotonic in r, then the scattering angle $\Delta\theta(E, b)$ is a strictly monotonic function of b in either half-beam, a fact that can be verified by inspecting the integral expression (H.3). It follows that the half-width of the scattered arc is characterized by

$$\Delta\varphi(E) = \lim_{b\to 0^\pm} |\Delta\theta(E, b) - \pi|. \tag{H.4}$$

We will see momentarily that for power law potentials this half-width is independent of the energy E as long as $E > 0$.

Knauf and Krapf [98] compute

$$\lim_{b\to 0^+} \Delta\theta(E, b) = \frac{2\pi}{2 - \alpha}, \qquad f(r) = r^{-\alpha}, \tag{H.5}$$

by taking the limit of the integral expression (H.3) for $\Delta\theta(E, b)$, with $U = f = r^{-\alpha}$ to obtain V_J. (See their equation (4.6).) From Equation (H.4), we have that $\Delta\varphi = \lim_{b\to 0^+} \Delta\theta(E, b) - \pi$, so it follows from this last limit and a tiny bit of algebra that

$$\Delta\varphi(\alpha) = \pi\frac{\alpha}{2 - \alpha}. \tag{H.6}$$

Doubling this last angle, we obtain the overall width $2\Delta\varphi = 2\pi\frac{\alpha}{2-\alpha}$ of the scattered arc, as claimed.

$\diamond$

In the next few pages we will compute the half-width $\Delta\varphi(\alpha)$ by a means different from that of Knauf and Krapf. We will use conical geometry and the fact that the zero-energy JM metric is conical for power-law potentials. An understanding of geodesics on cones leads directly to the same answer given in Equation (H.5) for the limit of $\Delta\theta$, and hence for $\Delta\varphi$. Figure H.3 summarizes the situation described in the last few paragraphs for the particular case of $\alpha = 1/2$ for which $\Delta\varphi(1/2) = \pi/3$. The computation is broken into three steps.

● **Step one: Scaling to reduce to $E = 0$**

In the first step of our computation we use scaling to trade $b \to 0$ with $E \to 0$. Recall the scaling symmetry $q(t) \mapsto \lambda q(\lambda^{-\beta}t)$ where $\beta = 1 + \alpha/2$. (See Section 0.1.8.) Scaling a solution does not change its limiting direction but changes its energy and angular momentum scale according to $(E, J) = (\lambda^{-\alpha}E, \lambda^{1-\alpha/2}J)$. It follows that

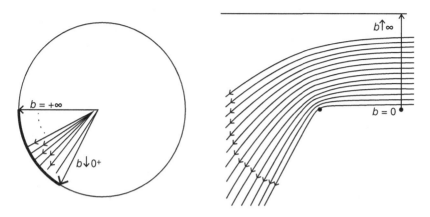

Figure H.3 The right-hand panel shows a half beam being scattered by an attractive central force power law $f(r) = r^{-1/2}$. The left-hand panel shows the arc of the circle resulting from the scattering map applied to this half beam, its image being half of the scattered image of the full beam.

$$\Delta\theta(\lambda^{-\alpha}E, \lambda^c b) = \Delta\theta(E, b), \qquad c = 1 - \frac{\alpha}{2}.$$

It follows immediately from this scaling law that the width of the scattered arc A does not depend on energy as long as the energy is positive. Scaling symmetry yields that $\Delta\theta(1, \lambda^c) = \Delta\theta(\lambda^\alpha, 1)$, or, more generally, $\Delta\theta(E_*, \lambda^c b_*) = \Delta\theta(\lambda^\alpha E_*, b_*)$ for any positive constants E_*, b_*. Now let $\lambda \to 0$ and use that $\alpha, c > 0$ to conclude that

$$\lim_{E \to 0^+} \Delta\theta(E, b_*) = \lim_{b \to 0} \Delta\theta(E_*, b)$$

for any $E_*, b_* > 0$. This common limit is the desired angle of Equation (H.5).

Now in the scaling limit $E \to 0$ our entire beam (now with spatial scale λ growing according to the relation $\lambda^{-\alpha} = E \to 0$) limits to a beam of trajectories for the zero energy problem. At zero energy all solutions except the collision solution $b = 0$ suffer the same overall deflection and this angle is the desired limiting angle $\lim_{b \to 0} \Delta\theta(E, b)$ we want to compute.

- **Step two: Reduction to conical geometry**

In this step we use conical geometry to compute $\lim_{b \to 0} \Delta\theta(E, b)$.

Lemma H.3 *For a power law central potential $-r^\alpha$ with $0 < \alpha < 2$ the Jacobi–Maupertuis metric at zero energy is isometric to the cone metric*

$$d\rho^2 + c^2\rho^2 d\theta^2, \quad with \quad c = 1 - \frac{\alpha}{2}. \tag{H.7}$$

Figure H.4 shows the metric of a cone over a circle of radius c. (See the discussion of the second paragraph following Equation (0.66) for a definition and description of metric cones.)

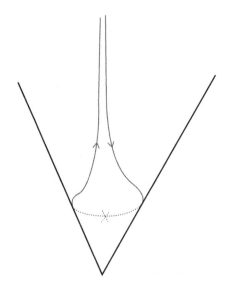

Figure H.4 The Kepler cone: The JM metric for the Kepler problem is depicted, with one of its geodesics.

Proof In standard r, θ polar coordinates, the JM metric is

$$ds_{JM}^2 = r^{-\alpha}(dr^2 + r^2 d\theta^2) = r^{-\alpha}dr^2 + r^{2-\alpha}d\theta^2.$$

We bring this metric into the form $d\rho^2 + f(\rho)^2 d\theta^2$ by making a change of variables $r \to \rho$ obtained by imposing $r^{-\alpha}dr^2 = d\rho^2$, which is to say $d\rho = r^{-\alpha/2}dr$. Integrating, we find that

$$\rho = \frac{1}{c}r^c,$$

with c as in the lemma. Then $\rho^2 = \frac{1}{c^2}r^{2c} = \frac{1}{c^2}r^{2-\alpha}$ or $r^{2-\alpha} = c^2\rho^2 = f(\rho)^2$. Finally, to verify that this is the metric over a circle of radius c, set $\rho = 1$ and compute the length of the resulting circle. QED

• **Flattening the cone. Glue and Paste**

The change of variables $\psi = c\theta$ puts the cone metric (Equation (H.7)) into the form

$$ds_{JM}^2 = d\rho^2 + \rho^2 d\psi^2$$

of a flat Euclidean metric on a plane endowed with polar coordinates (ρ, ψ). These new variables are related to the old (r, θ) variables by

$$\frac{1}{c}r^c = \rho, \tag{H.8}$$

$$c\theta = \psi. \tag{H.9}$$

Remark H.4 The change of variables is, up to the multiplicative constant c, the inverse of the Maclaurin transformation described in Section G.6. Write $q = re^{i\theta}$ and $w = \rho e^{i\psi}$, then, ignoring the multiplicative constant $1/c$ in front of w, we have that $w = q^c$ or $q = w^{1/c}$. The variable w is the Bohlin variable.

We can build the cone ((Equation (H.7)) by a standard cut and paste construction. Cut out the sector $2\pi(1 - c) \le \psi \le 2\pi$ from the standard (Bohlin!) Euclidean plane endowed with polar coordinates (ρ, ψ). The remaining sector is $0 \le \psi \le 2\pi c$ and is bound by the two rays $\psi = 0$ and $\psi = 2\pi c$. Glue these two bounding rays together by identifying points on them with the same value of ρ. We get a metric cone in which the locus $\rho = 1$ is isometric to a circle of radius $2\pi c$ and so it is our cone. See Appendix C, Figure C.4.

• **Step three: analysis of geodesics on the cone**

The group of rotations and dilations acts on our cone. Rotations act by $(\rho, \theta) \mapsto (\rho, \theta + \theta_0)$ and are isometries. Dilations act by $(\rho, \theta) \mapsto (\lambda \rho, \theta)$, $\lambda > 0$, and scale distances by λ. The collision geodesics are the curves $\theta = const$. They all hit the cone point $\rho = 0$, and are equivalent to each other under rotation. We also use *ray, radial geodesic,* or *generator of the cone* as synonyms for collision geodesic.

We claim that all non-collision geodesics are the same up to rotation and scaling. Indeed, take any such geodesic and consider the point $p_* = (\rho_*, \theta_*)$ of its closest approach to the cone point. Necessarily, the geodesic is orthogonal to the ray $\theta = \theta_+$. By scaling we can force $\rho_* = 1$. By rotation we can force $\theta_* = \pi/2$. In this way, we have taken any non-collision geodesic to the unique geodesic orthogonal to the ray $\theta = \pi/2$ at the given point $p_* = (1, \pi/2)$.

What does a non-collision geodesic look like? How does its shape change as we vary c? We answer these questions by describing the set of rays our geodesic intersects. Normalize the measure of the set of all rays to be that of the standard circle, 2π, corresponding to the total variation of θ from 0 to 2π. Now consider the angular measure of the set of all rays our geodesic hits, counting a ray twice if our geodesic hits that ray twice, thrice if the geodesic hits the ray three times, and so on. When $c = 1$, the conical metric is the standard Euclidean metric whose geodesics are the standard straight lines. A straight line not passing through the origin hits exactly half the set of all rays, hitting them once. Its angular measure is π.

Lemma H.5 *The angular measure of a non-collision geodesic on the cone* (H.7) *is* π/c. *This angular measure coincides with* $\Delta\theta$.

Proof Represent the geodesic in the w-plane which is Euclidean and where the polar coordinates are ρ, ψ. Being a straight line there, its angular measure relative to these coordinates is $\Delta\psi = \pi$. But $\theta = \psi/c$, so $\Delta\theta = \pi/c$. To see that this angular measure coincides with $\Delta\theta$ simply realize that rays in the conical metric coincide with rays through the origin in the standard configuration space $\mathbb{R}^2$. QED

Since $c = 1 - \alpha/2$ we have that $\pi/c = 2\pi/(2 - \alpha)$, which is the scattering angle of Equation (H.5). With this lemma, we have completed our alternative derivation of this scattering angle.

H.3 Commentary

Let us consider how the geodesics change as we decrease c from its Euclidean value $c = 1$. The function π/c is monotonically, increasing so the angular measure is monotonically increasing: We hit more and more rays as c decreases. At the value of $c = 1/2$, corresponding to Kepler ($\alpha = 1$), we have $\Delta\theta = 2\pi$ corresponding to the fact that the non-collision geodesics hit every ray exactly once, except for one ray, namely the ray the geodesic is asympotic to in positive and negative time. As we decrease c below $1/2$, our non-collision geodesics begin to intersect some of the rays more than once. For example, if $c = 1/4$ then any non-collision geodesic winds exactly twice around the cone, returning to infinity asymptotically along the same ray that it came in on.

As we continue to decrease c, we pass through $\Delta\theta = 4\pi$, then 6π, and so on, with $2n\pi$ corresponding to the degree n-scattering maps. The integer multiples of π correspond to the regularizable exponents and, thus, the magic exponent values for α given in Equation (G.5) in Section G.6. As $\alpha \to 2$ we find $c \to 0$ with geodesics winding more and more times around the cone point, the limit corresponding to infinitely many windings.

Remark H.6   Notes on strong forces: $\alpha \geq 2$

(A) The $\alpha = 2$ metric is cylindrical.

   Setting $c = 0$ in the metric (Equation (H.7)) when $\alpha = 2$ is not the correct metric for $\alpha = 2$. The correct way to understand this case is to return to the initial JM metric for $\alpha = 2$. Then $ds_{JM}^2 = \frac{1}{r^2}(dr^2 + r^2 d\theta^2) = (\frac{dr}{r})^2 + d\theta^2$. Set $\rho = \log(r)$ so that $d\rho = dr/r$. We have put the JM metric into the form of the cylindrical metric $d\rho^2 + d\theta^2$.

(B) Collision and infinity switch places when $\alpha > 2$.

   If $\alpha > 2$ the metric is still conical but collision and infinity switch places. The collision point $r = 0$ of the usual configuration space is at infinite JM distance from any other point of configuration space and corresponds to $\rho = \infty$. These other points are a finite distance away from $r = \infty$ so that $r = \infty$ corresponds to $\rho = 0$.

   The role-switching of the cone point with infinity when we go from $\alpha < 2$ to $\alpha > 2$ arises from the integral defining ρ in terms of r. When $\alpha < 2$ the integral converges for $r \to 0$ and diverges for $r \to \infty$, while for $\alpha \geq 2$ the integral diverges as $r \to 0$ and converges for $r \to \infty$. This divergence as $r \to 0$ for $\alpha > 2$ is the same divergence that allowed us to exclude collision paths (see Lemma 3.13) from consideration when applying the direct method of the calculus of variations applied to strong force potentials.

References

[1] Aarseth, S. (2003) **Gravitational N-Body Simulations**, Cambridge Monographs on Mathematical Physics, Cambridge University Press.

[2] Abraham, R. and Marsden, J. E. (1978) **Foundations of Mechanics**, 2nd ed., Benjamin/Cummings.

[3] Albouy, A. (2002) *Lectures on the two-body problem*. In **Classical and Celestial Mechanics. The Recife Lectures**, H. Cabral and F. Diacu (eds.), Princeton University Press, pp. 63–116.

[4] Albouy, A. (1996) *The symmetric central configurations of four equal masses.* Contemp. Math. **198**, 131–136.

[5] Albouy, A. and Dullin R. (2020) *Relative equilibria of the 3-body problem in $\mathbb{R}^4$*, Journal of Geometric Mechanics **12**, 323–341. www.aimsciences.org/article/doi/10.3934/jgm.2020012.

[6] Albouy, A., Cabral H., and Santos A. (2012) *Some problems on the classical N-body problem.* Celest. Mech. Dyn. Astron. **113**, 369–375.

[7] Albouy, A. and Kaloshin V. (2012) *Finiteness of central configurations of five bodies in the plane.* Ann. Math. **176**, 535–588.

[8] Albouy, A. and Chenciner, A. (1998) *Le probléme des N corps et les distances mutuelles.* Invent. Math. **131**, 151–184.

[9] Albouy, A. and Zhao, L. (2022) *Darboux Inversions of the Kepler Problem.* Reg. Chaotic Dyn. **27**, no. 3, 253–280.

[10] Arnol'd, V. I. (1983) *Singularities of systems of rays.* Uspehy Math. Nauk **38**, no. 2, 77–147.

[11] Arnol'd, V. I. (1989) **Mathematical Methods in Classical Mechanics**, translated by A. Weinstein and K. Vogtman, 2nd ed., Springer-Verlag.

[12] Arnol'd, V.I. (1990), **Huyghens & Barrow, Newton & Hooke**, Birkhauser Verlag.

[13] Aubin, T. (1982) **Nonlinear Analysis on Manifolds, Monge-Ampére Equations**, Grundlehren der math. **252**, Springer-Verlag.

[14] Bahri, A. and Rabinowitz, P. (1991) *Periodic solutions of hamiltonian systems of 3-body type.* Annales de l'I. H. P. Analyse non lineaire **8**, no. 6, 561–649.

[15] Benettin, G., Fassò, F., and Guzzo, M. (1998) *Nekhoroshev-stability of L4 and L5 in the spatial restricted three-body problem.* Reg. Chaotic Dyn. **3**, no. 3, 56–72.

[16] Bernshtein, D.N. (1975) *The number of roots of a system of equations*. Fun. Anal. Appl. **9**, 183–185.

[17] Bierstone, E. (1980) **The Structure of Orbit Spaces and the Singularities of Equivariant Mappings**, Instituto de Matemática Pura e Aplicada.

[18] Binney, J. and Tremaine, S. (1987) **Galactic Dynamics**, Princeton University Press.

[19] Birkhoff, G. D. (1927) **Dynamical Systems**, American Mathematical Society.

[20] Birman, J. (1974) **Braids, Links, and Mapping Class Groups**, Princeton University Press.

[21] Borisov, A., Mamaev, I., and Kilin, A. (2004) *Two-body problem on a sphere. Reduction, stochasticity, periodic orbits*. Reg. Chaotic Dyn. **9**, no. 3, 265–279.

[22] Broucke R. (1975) *On relative periodic solutions of the planar general three-body problem*. Celest. Mech. **12**, 439–462.

[23] Burago, D., Burago, Y., and Ivanov, S. (2001) **A Course in Metric Geometry**, Graduate Studies in Mathematics **33**, American Mathematical Society.

[24] Chandrasekhar, S. (1995) **Newton's Principia: For the Common Reader**, Oxford University Press.

[25] Chazy, J. (1922) *Sur l'allure du mouvement dans le problème des trois corps quand le temps croît indéfiniment*. Ann. Sci. École Norm. Sup. **39**, 29–130.

[26] Chen, K-C. (2003) *Variational methods on periodic and quasi-periodic solutions for the N-body problem*. Ergod. Th. Dyn. Sys. **23**, 1691–1715.

[27] Chen, K-C., Ouyang, T., and Xia, Z. (2012) *Action-minimizing periodic and quasi-periodic solutions in the N-body problem*. Math. Res. Lett. **19**, no. 02, 483–497.

[28] Chen, Y. Q., Glover, H. H., and Jensen, C. A. (2008) *The center of some braid groups and the Farrell cohomology of certain pure mapping class groups*. Algebraic and Geometric Topology **7**, 1987–2006.

[29] Chenciner, A. (1997) *À l'infini en temps fini*. In **Seminaire Bourbaki, 1996/97**, 323–353.

[30] Chenciner, A. (2002) *Action minimizing solutions of the Newtonian N-body problem: From homology to symmetry*. International Congress of Mathematicians.

[31] Chenciner, A. (2011) *The Lagrange reduction of the N-body problem, a survey*. arXiv:1111.1334.

[32] Chenciner, A. and Jiménez-Pérez, H. (2013) *Angular momentum and Horn's problem*. Moscow Math. Jour. **13**, 612–630.

[33] Chenciner, A. and Montgomery, R. (2000) *A remarkable periodic solution of the three body problem in the case of equal masses*. Ann. Math. **152**, 881–901.

[34] Chenciner A. and Venturelli, A. (2000) *Minima de l'intégrale d'action du problème de 4 corps de masses égales dans* **R**^{3}: *orbites hip-hop*. Cel. Mech. Dyn. Ast. **77**, 139–152.

[35] Chenciner, A., Gerver, J., Montgomery, R., and Simó, C. (2002) *Simple choreographies of N bodies: A preliminary study*. In **Geometry, Mechanics, and Dynamics**, P. Newton, P. Holmes, and A. Weinstein (eds.), Springer-Verlag.

[36] Chierchia, L. and Pinzari, G., (2011) *The planetary N-body problem: Symplectic foliation, reductions and invariant tori*. Invent. Math. **186**, no. 1, 1–77.

[37] Chirikov, B.V. and Shepelyansky, D. (2008) *Chirikov Standard Map*. Scholarpedia 3, no. 3, 3550. www.scholarpedia.org/article/Chirikov_standard_map.

[38] Clarke, A., Féjoz, J., and Guardia, M. (2022) *Why are inner planets not inclined?* arXiv:2210.11311.

[39] Deprit, A. and Deprit-Bartholome, A. (1967) *Stability of the triangular Lagrangian points*. Astron. J. **72**, 173–179.

[40] Derezínski, J. and Gérard, C. (1997) **Scattering Theory of Classical and Quantum N? Particle Systems**, Texts and Monographs in Physics, Springer.

[41] Devaney, R. (1980) *Triple collision in the planar isosceles three-body problem*. Invent. Math. **60**, 249–267.

[42] F. Diacu (2013) *The curved N-body problem: Risks and rewards*. Math. Intelligencer **35**, no. 3, 24–33.

[43] Diacu, F., Perez-Chavela, and E., Santoprete, M. (2012) *The n-body problem in spaces of constant curvature*. J. Nonlinear Sci. **22**, no. 2, 247-266. arXiv:0807 .1747v6

[44] Dirac, P. A. M. (1950) *Generalized Hamiltonian dynamics*. Canad. J. Math. **2**, 129–148.

[45] Duignan, N. and Dullin, H. (2021) *On the $C^{8/3}$-regularisation of simultaneous binary collisions in the planar four-body problem*. Nonlinearity **34**, no. 7, 4944–4982.

[46] Duignan, N., Moeckel, R., Montgomery, R., and Yu, G. (2020) *Chazy-type asymptotics and hyperbolic scattering for the n-body problem*. Archive Rat. Mech. Anal. **238**, no. 1, 255–297.

[47] Duistermaat, J. J. (2004) *Symplectic Geometry*, Course Notes, Summer Course, Utrecht University.

[48] Dullin, H. and Scheurle, J. (2020) *Symmetry reduction of the 3-body problem in $\mathbb{R}^4$*. J. Geom. Mech. **12**, no. 3, 377–394.

[49] Dumas, H. S. (2014) **The KAM Story: A Friendly Introduction to the Content, History, and Significance of Classical Kolmogorov-Arnold-Moser Theory**, World Scientific.

[50] Easton, R. (1971) *Regularization of vector fields by surgery*. J. Diff. Eq. **10**, no. 1, 92–99.

[51] Easton, R. (1971) *Some topology of the 3-body problem*. J. Diff. Eq. **10**, 371–377

[52] Einstein, A. (1945) **The Meaning of Relativity**, 2nd ed., Princeton University Press.

[53] Elbialy, M. S. (1990) *Collision singularities in celestial mechanics*. SIAM J. Math. Anal. **21**, 1563–1593.

[54] Euler, L. (1767) *De motu rectilineo trium corporum se mutuo attrahentium*. Novi Commentarii academiae scientiarum Petropolitanae **11**, 144–151.

[55] Fasso, F. Guzzo, M., and Benettin, G. (1998) *Nekhoroshev-stability of elliptic equilibria of Hamiltonian systems*. Comm. Math. Phys. **197**, 347–360.

[56] Fathi, A. (2014) *Weak KAM theory: The connection between Aubry–Mather theory and viscosity solutions of the Hamilton–Jacobi equation*. Proc. International Congress of Mathematicians **III**, 597–621.

[57] Féjoz, J. (2002) *Quasi periodic solutions in the planar three-body problem*. J. Diff. Eq. **183**, no. 2, 303–341.

[58] Féjoz, J. (2004) *Démonstration du "théorème d'Arnol'd" sur la stabilité du système planétaire (d'après Herman)*. Erg. Th. Dyn. Sys. **24**, no. 5, 1521–1582.

[59] Féjoz, J., Knauf, A., and Montgomery, R. (2017) *Lagrangian relations and linear point billiards*. Nonlinearity **30**, no. 4, 1326.

[60] Féjoz, J, Knauf, A. and Montgomery, R. (2021) *Classical N-body scattering with long-range potentials*. Nonlinearity **34**, no. 11, 8017–8054.

[61] Ferrario D. and Terracini S. (2004) *On the existence of collisionless equivariant minimizers for the classical n-body problem*. Invent. Math. **155**, no. 2, 305–362.

[62] Feynman, R. P. (1942) *The principle of least action in quantum mechanics*. PhD thesis, Princeton University. Published as **Feynman's Thesis: A New Approach to Quantum Theory**, L. M. Brown (ed.), World Scientific (2005).

[63] Feynman, R. and Wheeler, J. (1949) *Classical electrodynamics in terms of direct interparticle action*. Rev. Mod. Phys. **21**, no. 3, 425–433.

[64] Feynman, R., Leighton, R.B., and Sands, M. (1963) **The Feynman Lectures on Physics**, Addison-Wesley.

[65] Fusco, G., Gronchi, G. F., and Negrini P. (2011) *Platonic polyhedra, topological constraints and periodic solutions of the classical N-body problem*. Invent. Math. **185**, 283–332.

[66] Gelfand, I. and Fomin S. (2000) **Calculus of Variations**, Dover.

[67] Gerver, J., Huang, G., and Xue, J. (2022) *A new mechanism for noncollision singularities*. arXiv:2202.08534.

[68] Glimm, J. (1964) *Formal stability of Hamiltonian systems*. Comm. Pure Appl. Math. **XVI**, 509–526.

[69] Gordon, W. (1975) *Conservative dynamical systems involving strong forces*. Trans. AMS **204**, 113–135.

[70] Gordon, W. (1977) *A minimizing property of Keplerian orbits*. Am. J. Math. **99**, no. 5, 961–971.

[71] Grandati, Y., Bérard, A., and Mohrbach, H. (2008) *Bohlin–Arnold–Vassiliev's duality and conserved quantities*. https://arxiv.org/pdf/0803.2610.pdf.

[72] Guardia, M., Martín P., Paradela, J., Seare T. M., (2022) *Hyperbolic dynamics and oscillatory motions in the 3 body problem*. arXiv:2207.14351.

[73] Hale, J. (1969) **Ordinary Differential Equations**, Wiley-Interscience.

[74] Hampton, M. (2019) *Planar N-body central configurations with a homogeneous potential*. Celest. Mech. Dyn. Astr. **131**, 20. https://doi.org/10.1007/s10569-019-9898-0.

[75] Hampton, M. and Moeckel R. (2006) *Finiteness of relative equilibria of the four-body problem*. Invent. Math. **163**, 289–312.

[76] Hansen, V. L. (1989) **Braids and Coverings**. London Mathematical Society Student Texts **18**, Cambridge University Press.

[77] Heggie, D. (1974) *A global regularisation of the gravitational N-body problem*. Celest. Mech. **10**, 217–241.

[78] Heggie, D. (2000) *A new outcome of binary-binary scattering*. Mon. Not. R. Astron. Soc. **318**, 61–63.

[79] Heggie, D. and Hut, P. (2003) **The Gravitational Million-Body Problem**, Cambridge University Press.

[80] Hénon M (1976) *A family of periodic solutions of the planar three-body problem, and their stability*. Celest. Mech. **13**, 267–285.

[81] Herman, M. (1998) *Some open problems in dynamical systems.* Proc. Int. Congress of Math. **II**, 797–808.

[82] Hsiang, Wu-Yi (1994) *Geometric study of the three-body problem I*, report -620, Center for Pure and Applied Math, University of California at Berkeley.

[83] Hsiang, Wu-Yi (1997) *Kinematic geometry of mass-triangles and reduction of Schrödinger's equation of three-body systems to partial differential equations solely defined on triangular parameters.* Proc. Nat. Acad. Sci. U.S.A. **94**, no. 17, 8936–8938.

[84] Hsiang, Wu-Yi and Straume, E. (1995) *Kinematic geometry of triangles with given mass distribution*, report PAM-636, Center for Pure and Applied Math, University of California at Berkeley.

[85] Hut, P. and Bahcall, J. N. (1983) *Binary-single star scattering. I. Numerical experiments for equal masses*, Astrophys. J. **268**, 319–341.

[86] Iwai, T. (1987) *A gauge theory for the quantum planar three-body problem.* J. Math. Phys. **26**, 964–974.

[87] Iwai, T. (1987) *A geometric setting for classical molecular dynamics.* Ann. Inst. Henri Poincaire Phys. Th. **47**, no. 2, 199–219.

[88] Iwai, T. (1987) *A geometric setting for internal motions of the quantum three-body system.* J. Math. Phys. **28**, 1315–1326.

[89] Kadowaki, K. (2018) *A Note on Saari's treatment of rotation curve analysis.* Astroph. J. **869**, 160.

[90] Kapela, T. and Simó, C. (2007) *Computer assisted proofs for nonsymmetric planar choreographies and for stability of the Eight.* Nonlinearity **20**, 1241–1255.

[91] Katok, A. and Hasselblatt, B. (1995) **Introduction to the Modern Theory of Dynamical Systems**, Cambridge University Press.

[92] Kelley, A. (1967) *The stable, center-stable, center, center-unstable and unstable manifold.* J. Diff. Eq. **3**, no. 4, 546–570.

[93] Kendall, D. G. (1984) *Shape manifolds, procrustean metrics, and complex projective space.* Bull. London Math. Soc. **16**, 81–121.

[94] Khesin, B. A. and Tabachnikov, S. L. (eds.) (2014) **Arnold: Swimming Against the Tide**, American Mathematical Society.

[95] Klein, M. and Knauf, A. (1992) **Classical Planar Scattering by Coulombic Potentials**, Lecture Notes in Physics 13, Springer-Verlag.

[96] Knauf, A. (2012) **Mathematical Physics: Classical Mechanics**, Springer.

[97] Knauf, A. (2018) *Asymptotic velocity for four celestial bodies.* Phil. Trans. R. Soc. A. **376**, 20170426.

[98] Knauf, A. and Krapf, M. (2008) *The non-trapping degree of scattering.* Nonlinearity **21**, 2023–2041.

[99] Lagrange, J. (1772) *Essai sur le problème des trois corps.* In **Prix de l'Académie Royale des Sciences de Paris, IX**, pp. 229–331.

[100] Lagrange, J. (1788) **Mécanique Analytique**, 1st ed., Veuve Desaint.

[101] Landau, I. and Lifshitz, E. (1976) **Mechanics**, Pergamon Press.

[102] Laskar, J. (1996) *Large scale chaos and marginal stability in the solar system.* Celest. Mech. Dyn. Astron. **64**, 115–162.

[103] Laskar, J. (2013) *Is the solar system stable?* Prog. Math. Phys. **66**, 239–270.

[104] Laskar, J. (2014) *Michel Hénon and the Stability of the Solar System.* arXiv:1411.4930.

[105] Laskar, J. (2024) Stability of the solar system. Scholarpedia 19, no. 4, 5216. www.scholarpedia.org/article/Stability_of_the_solar_system#Laplace-Lagrange_stability_of_the_Solar_System.

[106] Lemaitre, G. (1952) *Coordonnées symétriques dans le problème des trois corps.* Bull. Cl. Sc. Acad. Belg. **38**, no. 5, 582–592, 1218–1234.

[107] Lemaitre, G. (1955) *Regularization of the three body problem.* Vistas Astron. **1**, 207–215.

[108] Levi-Civita, T. (1920) *Sur la régularisation du problème des trois corps.* Acta Math. **42**, no. 1, 99–144.

[109] Lerman, E., Montgomery, R., and Sjamaar, R. (1993) *Examples of Singular Reduction.* In **Symplectic Geometry**, D. Salamon (ed.), Cambridge University Press, pp. 127–155.

[110] Li, X. and Liao, S. (2018) *Collisionless periodic orbits in the free-fall three-body problem.* arXiv:1805.07980.

[111] Li, X. and Liao, S. (2021) *Movies of the collisionless periodic orbits in the free-fall three-body problem in real space or on shape sphere.* https://numericaltank.sjtu.edu.cn/free-fall-3b/free-fall-3b-movies.htm.

[112] Lyapunov, M. (1947) **Probléme Général de la Stabilité du Mouvement**, Princeton University Press.

[113] Maderna, E. and Venturelli, A. (2020) *Viscosity solutions and hyperbolic motions: A new PDE method for the N-body problem.* Ann. Math. **192**, 499–550.

[114] Mañe, R (1990) **Global Variational Methods in Conservative Dynamics**, Colóquio Brasileiro de Mathematica, IMPA.

[115] Maclaurin, C. (1742), **Treatise of Fluxions: In Two Books**, Ruddimans.

[116] Marchal, C. (1990) **The Three-Body Problem**, Elsevier.

[117] Marchal C. (2002) *How the method of minimization of action avoids singularities.* Celest. Mech. Dyn. Astron. **83**, 325–353.

[118] Marsden, J. E. and Weinstein, A. (1974) *Reduction of symplectic manifolds with symmetry.* Rep. Math. Phys. **5**, 121–130.

[119] Martinez, R. and Simó, C (1999) *Simultaneous binary collisions in the planar four-body problem.* Nonlinearity **12**, no. 4, 903–930.

[120] McGehee, R. (1973) *A stable manifold theorem for degenerate fixed points with applications to celestial mechanics.* J. Diff. Eq. **14**, 70–88.

[121] McGehee, R. (1974) *Triple collision in the collinear three-body problem.* Invent. Math. **27**, 191–227.

[122] Meyer, K. (1973) *Symmetries and integrals in mathematics.* In **Dynamical Systems**, M Peixoto (ed.), Academic Press, pp. 259–272.

[123] Meyer, K. (1999) **Periodic Solutions of the N-body Problem**, Lecture Notes in Math., Springer.

[124] Meyer, K. and Hall, G. (1991) **Introduction to Hamiltonian Dynamical Systems and the N-Body Problem**, Applied Math. Sciences, series 90, 1st ed., Springer-Verlag.

[125] Meyer, K. and Schmidt, D. (1986) *The stability of the Lagrange triangular point and a theorem of Arnol'd.* J. Diff. Eq. **62**, no. 2, 222–236.

[126] Milnor, J. (1973) **Morse Theory**, Annals of Math. Studies, Princeton University Press.

[127] Minton, G. (n.d.), *choreo.2.3.js.* http://gminton.org/#choreo

[128] Moczurad, M. and Zgliczyński, P. (2019) *Central configurations in planar n-body problem for n = 5,6,7 with equal masses.* Celest. Mech. Dyn. Astron. **131**, 46–74.

[129] Moeckel, R. (1981) *Orbits of the three-body problem which pass infinitely close to triple collision.* Amer. J. Math. **103**, no. 6, 1323–1341.

[130] Moeckel, R. (1983) *Orbits near triple collision in the three-body problem.* Indiana U. Math. J. **32**, 221–239.

[131] Moeckel, R. (1984) *Heteroclinic phenomena in the isosceles three-body problem.* SIAM J. Math. Anal. **15**, 857–876.

[132] Moeckel, R. (1988) *Some qualitative features of the three-body problem.* Contemp. Math. **81**, 1–21.

[133] Moeckel, R. (1989) *Chaotic dynamics near triple collision.* Arch. Rat. Mech. **107**, no. 1, 37–69.

[134] Moeckel, R. (2005) *A variational proof of the existence of transit orbits in the restricted three-body problem.* Dyn. Sys. **20**, 45–58.

[135] Moeckel, R. (2007) *Symbolic dynamics in the planar three-body problem.* Reg. Chaotic Dyn. **12**, no. 5, 449–475.

[136] Moeckel, R. (2008) *A proof of Saari's conjecture for the three-body problem in* $\mathbb{R}^d$. Dis. Cont. Dynam. Sys. **1**, no. 4, 631–646.

[137] Moeckel, R. (2014) *Central configurations.* Scholarpedia 9(4), 10667. www .scholarpedia.org/article/Central_configurations.

[138] Moeckel, R. (2014) *Lectures on central configurations.* www-users.cse.umn .edu/~rmoeckel/notes/CentralConfigurations.pdf.

[139] Moeckel, R. and Montgomery, R. (2013) *Symmetric regularization, reduction, and blow-up of the planar three-body problem.* Pac. J. Math. **262**, no. 1, 129–189.

[140] Moeckel, R. and Montgomery, R. (2015) *Realizing all reduced syzygy sequences in the planar threebody problem.* Nonlinearity **28**, 1919–1935.

[141] Moeckel, R. and Montgomery, R. (2023) *No infinite spin for the planar N-body problem.* arXiv:2302.00177

[142] Montaldi, J. (n.d.) *n-body choreographies.* https://personalpages.manchester.ac .uk/staff/j.montaldi/Choreographies/

[143] Montaldi, J. and Steckles, K. (2013) *Classification of symmetry groups for planar n-body choreographies.* Forum of Mathematics, Sigma **1**, e5 doi:10.1017/fms.2013.5.

[144] Montgomery, R. (1996) *The geometric phase of the three-body problem.* Nonlinearity **9**, no. 5, 1341–1360.

[145] Montgomery, R. (1998) *The N-body problem, the braid group, and action-minimizing periodic orbit.* Nonlinearity **11**, no. 2, 363–376.

[146] Montgomery, R. (2000) *Action spectrum and collisions in the planar three-body problem.* Contemp. Math. **282**, 173–184.

[147] Montgomery, R. (2002) **A Tour Of Subriemannian Geometries, their Geodesics, and Applications**, Mathematical Surveys and Monographs **91**, American Mathematical Society.

[148] Montgomery, R. (2002) *Infinitely many syzygies.* Arch. Rat. Mech. Anal. **164**, no. 4, 311–340.

[149] Montgomery, R. (2007) *The zero angular momentum three-body problem: All but one solution has syzygies.* Erg. Th. Dyn. Sys. **27**, no. 6, 1933–1946.

[150] Montgomery, R. (2015) *The three-body problem and the shape sphere.* Amer. Math. Monthly **122**, no. 4, 299–321.

[151] Montgomery, R. (2018) Blow-Up, Homotopy and Existence for Periodic Solutions of the Planar Three-Body Problem. In **Geometrical Themes Inspired by the N-body Problem**, L. Hernández-Lamoneda, H. Herrera, and R. Herrera (eds.), Springer, pp. 49–89.

[152] Montgomery, R. (2019) *The three body problem.* Scientific American **321**, no. 2, 67–73.

[153] Montgomery, R. (2020) *Minimizers for the Kepler problem.* Qual. Theory Dyn. Sys. **19**, no. 31. https://doi.org/10.1007/s12346-020-00363-8

[154] Montgomery, R. (2023) *Brake orbits fill the N-body Hill region.* Reg. Chaotic Dyn. **28**, 374–394.

[155] Montgomery, R. (2023) *Dropping bodies.* Math. Intell. **45**, 168–174.

[156] Montgomery, R. (2023) *Lyapunov Instability.* Unpublished notes, available at https://peopleweb.prd.web.aws.ucsc.edu/~rmont/papers/unpublished/Lyapunov_Instability.pdf.

[157] Moore, C. (1993) *Braids in classical gravity.* Phys. Rev. Lett. **70**, 3675–3679.

[158] Moore, C. and Nauenberg, M. (2005) *New periodic orbits for the N-body problem.* J. Comput. Nonlinear Dynam. **1**, no. 4, 307–311.

[159] Moser, J. (1973) **Stable and Random Motion**, Princeton University Press.

[160] Moulton, F. R. (1910) *The straight line solutions of the problem of n bodies.* Ann. Math. **12**, 1–17.

[161] Musso, M. (2022) *Bubbling blow-up in critical elliptic and parabolic problems.* Notices Amer. Math. Soc. **69**, no. 10, 1700–1706.

[162] Nauenberg M. (2001) *Periodic orbits for three particles with finite angular momentum.* Phys. Lett. **292**, 93–99.

[163] Newton, I. (1667) **Philosophiæ Naturalis Principia Mathematica**, translated by Andrew Motte. Available at http://en.wikisource.org/wiki/The_Mathematical_Principles_of_Natural_Philosophy_%281846%29.

[164] Newton, I. (1667) **Philosophiae Naturalis Principia Mathematica**. www.thelatinlibrary.com/newton.scholium.html

[165] Niederman, L. (1998) *Nonlinear stability around an elliptic equilibrium point in a Hamiltonian system.* Nonlinearity **11**, 1465–1479.

[166] Ouyang, T. and Xie, Z. (2017) *A continuum of periodic solutions to the four-body problem with various choices of masses.* J. Diff. Eq. **264**, no. 7, 4425–4455.

[167] Palais, R. (1993) *The principle of symmetric criticality.* Comm. Math. Phys. **69**, 19–30.

[168] Palmore J. (1973) *Classifying relative equilibria, I.* Bull. Amer. Math. Soc. **79**, 904–908.

[169] Poincaré, H. (1892) **Les Methodes Nouvelles de la Mécanique Céleste**, vol. 1, ch. 3, Gauthier-Villars et fils. See also **New Methods of Celestial Mechanics**, translated by D. Goroff. Introduction, p. I (1993).

[170] Poincaré, H. (1896) *Sur les solutions périodiques et le principe de moindre action*, C.R.A.S. t. **123**, 915–918, in Oeuvres, tome VII.

[171] Pollard, H. (1966) **Celestial Mechanics**, Prentice Hall.

[172] Pöschel, J. (1993) *Nekhoroshev estimates for quasi-convex Hamiltonian systems*. Math. Z. **213**, no. 2, 187–216.

[173] Reed, C. (2022) *A note on Newton's shell-point equivalency theorem*. Am. J. Phys. **90**, 394–396.

[174] Reed M. and Simon, S. (1979) **Scattering Theory**, Methods of Modern Mathematical Physics, vol. 3, Academic Press.

[175] Roberts, G. (2007) *Linear stability analysis of the figure-eight orbit in the three-body problem*. Erg. Th. Dyn. Sys. **27**, 1947–1963.

[176] Robinson, C. (1984) *Homoclinic orbits and oscillations for the planar three-body problem*. J. Diff. Eq. **52**, 356–377.

[177] Rose, D. (2015) *Geometric phase and periodic orbits of the equal-mass, planar three-body problem with vanishing angular momentum*. PhD thesis, University of Sydney.

[178] Royden, H. L. (1968) **Real Analysis**, 2nd ed., Macmillan.

[179] Rutherford, E. (1911) *The scattering of α and β particles by matter and the structure of the atom*. Phil. Mag. Series 6, **21**, 669–688.

[180] Saari, D. (1971) *Improbability of collisions in Newtonian gravitational systems*. Trans. AMS **162**, 267–271.

[181] Saari, D. (1975) *Collisions are of first category*. Proc. AMS **47**, 442–445.

[182] Saari, D. (1977) *A global existence theorem for the four body problem of Newtonian mechanics*. J. Diff. Eq. **26**, 80–111.

[183] Saari, D. (1984) *From rotation and inclination to zero configurational velocity surfaces, I, a natural rotating coordinate system*. Celest. Mech. **33**, 299–318.

[184] Saari, D. (1988) *Symmetry in n-particle systems*. In **Hamiltonian Dynamical Systems**, Contemp. Math. **81**, pp. 23–42, AMS.

[185] Saari, D. (2005) **Collisions, Rings, and Other Newtonian N-Body Problems**, CBMS conference series, no. 104, AMS.

[186] Saari, D. (2015) *Mathematics and the dark matter puzzle*. Am. Math. Mon. **122**, no. 5, 407–423.

[187] Saari, D. and Hulkower, N. (1981) *On the manifold of total collapse orbits and of complete parabolic orbits for the n-body problem*. J. Diff. Eq. **41**, 27–43.

[188] Schubart, J. (1956) *Nulerische Aufsuchung periodischer Lösungen im Dreikorperproblem*. Astronomische Nachriften **283**, 17–22.

[189] Seifert, H. (1948) *Periodische Bewegungen Mechanischer System*. Math. Z, **51**, 197–216. See also, *Periodic motions of mechanical systems*. Translated by Bill McCain. https://peopleweb.prd.web.aws.ucsc.edu/~rmont/papers/periodicMcCain.pdf.

[190] Shub, M. (1970) *Appendix to Smale's paper: Diagonals and relative equilibria*. In **Manifolds**, Springer Lecture Notes in Mathematics **197**, 199–201.

[191] Simó, C. (1978) *Relative equilibrium solutions in the four-body problem*. Celest. Mech. **18**, 165–184.

[192] Simó, C. (2000) *Dynamical properties of the figure eight solution of the three-body problem*. In Proc. Cele. Mech. Conference dedicated to D. Saari for his 60th birthday, pp 209–228, American Mathematical Society.

[193] Simó, C. (2001) *New families of solutions in N-body problems.* In Proc. 3rd European Congress of Mathematics, Progress in Mathematics series **201**, 101–115, Birkhäuser.

[194] Simó, C. (2001) *Periodic orbits of the planar N-body problem with equal masses and all bodies on the same path.* In **The Restless Universe: Applications of N-Body Gravitational Dynamics to Planetary, Stellar and Galactic Systems**, NATO Advanced Study Institute, IOP Publishing.

[195] Siegel, C. and Moser, J. (1971) **Lectures on Celestial Mechanics**, Springer-Verlag.

[196] Singer, S. (2003) **Symmetry in Mechanics: A Gentle, Modern Introduction**, Birkhäuser.

[197] Smale, S. (1970) *Topology and mechanics, II.* Invent. Math. **11**, 45–64.

[198] Smale, S. (1998) *Mathematical problems for the next century.* Math. Intell. **20**, no. 2, 7–15.

[199] Stiefel, E. L. and Scheifele, G. (1971) **Linear and Regular Celestial Mechanics**, Grundlehren der mathematischen Wissenschaften series, **174**, Springer.

[200] Strömgren, E. (1933) *Connaissance actuelle des orbites dans le problème des trois corps.* Bull. Astronomique **9**, 87–130.

[201] Sundman, K. (1906) *Recherches sur le probléme de trois corps.* Acta Soc. Scientiarum Fennicae **34**, 1–43.

[202] Sundman, K. (1913) *Mémoire sur le probléme de trois corps.* Acta Math. **36**, 105–179.

[203] Sussman, H. (1973) *Orbits of families of vector fields and integrability of distributions.* Trans. AMS **180**, 171–188.

[204] Szebehely, V. and Peters, C. (1967) *Complete solution of a general problem of three bodies.* AJ **72**, 876–882. https://articles.adsabs.harvard.edu/pdf/1967AJ.....72..876S.

[205] Tanikawa, K. and Mikkola, S. (2000) *One-dimensional three-body problem via symbolic dynamics.* Chaos 10, no. 3, 649–657. https://ui.adsabs.harvard.edu/abs/2000Chaos..10..649T/abstract.

[206] Tanikawa, K. and Mikkola, S. (2000) *Triple collisions in the one-dimensional three-body problem.* Celest. Mech. Dyn. Astr. **76**, 23–34.

[207] Terracini, S. (2006) *On the variational approach to the periodic n-body problem.* Celest. Mech. Dyn. Astr. **95**, 3–25.

[208] Todhunter, M. A. (1871) **Researches in the Calculus of Variations, Principally on the Theory of Discontinuous Solutions**, Cambridge, Macmillan and Co.

[209] Todhunter, M. A. (1871) **History of the Calculus of Variations**, Cambridge, Macmillan and Co.

[210] Venturelli, A. (2002) *Application de la minimisation de l'action au Probléme des N corps dans le plan et dans l'espace.* PhD thesis, University of Paris.

[211] Waldvogel, J. (1972) *A new regularization of the problem of three bodies.* Celest. Mech. **6**, 221–231.

[212] Weinstein, A. (1981) *Symplectic Geometry.* Bull. Amer. Math. Soc. **5**, no. 1, 1–14.

[213] Weinstein, A. (1984) *The local structure of Poisson manifolds.* J. Diff. Geom. **18**, 523–557.

[214] Wikipedia. (2023) *Stability of the solar system.* https://en.wikipedia.org/wiki/ Stability_of_the_Solar_System.

[215] Wintner, A. (1941) **The Analytical Foundations of Celestial Mechanics**, Princeton University Press. Reprinted by Dover Press, 2014.

[216] Xia, Z. (1991) *Central configurations with many small masses.* J. Diff. Eq. **91**, no.1, 168–179.

[217] Xia, Z. (1992) *The existence of noncollision singularities in Newtonian systems.* Ann. Math. **135**, no. 3, 411–468.

[218] Young, L. C. (1980) **Lectures on the Calculus of Variations and Optimal Control Theory**, Chelsea Publishing Company.

[219] Yu, X. (2022) *On the stability of Lagrange relative equilibrium in the planar three-body problem.* arXiv: 1911.12269.

[220] Zehnder, E. (2010) **Lectures on Dynamical Systems**, European Mathematical Society.

Index

nited States
or Publisher Services